PREPARATIVE
INORGANIC
REACTIONS

Volume 5

PREPARATIVE INORGANIC REACTIONS

Volume 5

Editor

WILLIAM L. JOLLY

Department of Chemistry
University of California
and
Inorganic Materials Research Division
Lawrence Radiation Laboratory
Berkeley, California

1968
INTERSCIENCE PUBLISHERS
a division of John Wiley & Sons
New York · London · Sydney · Toronto

Copyright © 1968 by John Wiley & Sons, Inc.

All Rights Reserved. No part of this book
may be reproduced by any means, nor transmitted,
nor translated into a machine language without
the written permission of the publisher.

Library of Congress Catalog Card Number 64-17052
        SBN 470   446926

PRINTED IN THE UNITED STATES OF AMERICA

CONTENTS

Synthesis of Coordination Compounds   1
    John L. Burmeister, *University of Delaware, Newark, Delaware*
    and Fredrick Basalo, *Northwestern University, Evanston, Illinois*

The Boron Hydrides   45
    R. W. Parry and M. K. Walter, *Department of Chemistry, University of Michigan, Ann Arbor, Michigan*

Compounds Containing P—P Bonds   103
    Ekkehard Fluck, *Institut für Anorganische Chemie der Universität, Stuttgart, Germany*

Condensed Phosphoric Acids and Condensed Phosphates   157
    C. Y. Shen and D. R. Dyroff, *Inorganic Chemicals Division, Monsanto Company, St. Louis, Missouri*

Author Index   223

Subject Index   241

Cumulative Index, Volumes 1-5   247

The Application of Reaction Mechanisms to the Synthesis of Coordination Compounds

JOHN L. BURMEISTER

University of Delaware, Newark, Delaware

and

FRED BASOLO

Northwestern University, Evanston, Illinois

I. Introduction	1
II. Oxidation–Reduction Reactions	4
A. Inner Sphere Mechanism	4
B. Outer Sphere Mechanism	13
III. Substitution Reactions	17
A. Hydrolysis and Anation Reactions of Octahedral Complexes	18
B. Substitutions without Metal–Ligand Cleavage	20
C. The *trans* Effect	25
D. Reactions of Metal Carbonyl and Nitrosyl Complexes	29
E. Reactions of Square Planar Complexes	31
IV. Miscellaneous Applications	36
A. Stabilization of Metal Complexes by Large Counterions	36
B. Resolution of Optically Active Amines	37
References	38

I. INTRODUCTION

This chapter represents a departure from the others published in this series. Rather than a critical discussion of the synthesis of a specific class of compounds (to do so for coordination compounds would require at least an entire volume), it is an attempt to document a synthetic concept.

The development of synthetic coordination chemistry has taken place in two broadly overlapping stages. For the greater part of almost two centuries, beginning with Tassaert's discovery of $CoCl_3 \cdot 6NH_3$ in 1798, the syntheses have largely been accomplished on an empirical basis, utilizing variations in the amounts and order of addition of reactants, temperature, solvents, etc. Many coordination complexes are still being prepared according to the original procedures (sometimes slightly modified) developed by

TABLE I
Known Linkage Isomers of Metal Complexes

Stable isomer	Refs.	Stable isomer	Refs.
M—ONO and M—NO$_2$		M—SCN and M—NCS	
[Co(NH$_3$)$_5$NO$_2$]$^{2+}$	1,2[a]	[Cr(H$_2$O)$_5$NCS]$^{2+}$	19,20[b]
[Co(NH$_3$)$_2$(py)$_2$(NO$_2$)$_2$]$^+$	3[a]	[Pd(Et$_4$dien)NCS]$^+$	21,22[a]
[Co(en)$_2$(NO$_2$)$_2$]$^+$	3–5[a]	[Pd(4,7-diphenylphen)(SCN)$_2$]	23[a]
[Rh(NH$_3$)$_5$NO$_2$]$^{2+}$	6,7[a]	[Cu(tripyam)(NCS)$_2$]	24[a]
[Ir(NH$_3$)$_5$NO$_2$]$^{2+}$	6,7[a]	[(C$_5$H$_5$)Fe(CO)$_2$NCS]	25[a]
[Pt(NH$_3$)$_5$NO$_2$]$^{3+}$	7[a]	[(C$_5$H$_5$)Mo(CO)$_3$NCS]	25[a]
[Co(CN)$_5$NO$_2$]$^{3-}$	8[b]	{Pd[P(OCH$_3$)$_3$]$_2$(NCS)$_2$}	26[e]
[Ni(Me$_2$en)$_2$(ONO)$_2$]	9[b]		
[Ni(EtenEt)$_2$(ONO)$_2$]	9[b]	M—SSO$_3$ and M—OS$_2$O$_2$	
		[Co(NH$_3$)$_5$OS$_2$O$_2$]$^+$	27[e]
M—SCN and M—NCS		M—NC and M—CN	
{Pd[As(C$_6$H$_5$)$_3$]$_2$(NCS)$_2$}	10,11[a]	[Co(CN)$_5$CN]$^{3-}$	8[b]
[Pd(bipy)(NCS)$_2$]	10,11[a]	[Cr(H$_2$O)$_5$CN]$^{2+}$	28[b]
[Cd(CNS)$_4$]$^{2-}$	12,13[c]	M—SeCN and M—NCSe	
[Mn(CO)$_5$SCN]	14[b]	[Pd(Et$_4$dien)NCSe]$^+$	29[a]
{Pd[As(n-C$_4$H$_9$)$_3$]$_2$(SCN)$_2$}	15[d]	M—OSO$_2$ and M—SO$_3$	
[Rh(NH$_3$)$_5$NCS]$^{2+}$	16,17[a]	[Co(NH$_3$)$_5$SO$_3$]$^+$	30[a]
[Ir(NH$_3$)$_5$NCS]$^{2+}$	17,18[a]		

SYNTHESIS OF COORDINATION COMPOUNDS

Individual complexes containing both bonding modes

[Co(en)$_2$(NO$_2$)(ONO)]$^+$	4,5	
[Pd(Me$_2$bipy)(SCN)(NCS)]	23	[Cu(tripyam)(SCN)(NCS)]	24
[Ni(NO$_2$)$_4$(ONO)$_2$]$^{4-}$	31	{Pd[(C$_6$H$_5$)$_2$As(o-C$_6$H$_4$)P(C$_6$H$_5$)$_2$](SCN)(NCS)}	32

Linkage isomers involving bridging ligands

(n-C$_3$H$_7$)$_3$P–Pt(NCS)–Cl–Pt(Cl)(SCN)–P(n-C$_3$H$_7$)$_3$ 33[a]

K[Fe^{II}(CN)$_6$Cr^{III}]   34[a]

Linkage isomers involving chelating ligand

H–C(=O)–C(H$_3$C)=C(H$_3$C)–Co(acac)$_2$ → H$_3$C–C(=O)–C(H)=C(H$_3$C)–Co(acac)$_2$ 35[a]

---

[a] Both isomers isolated.
[b] Unstable isomer detected in solution, but not isolated.
[c] S- and N-bonded forms are in equilibrium in solution, but solid (as K$_2$[Cd(SCN)$_4$]·2H$_2$O) contains two bridging and two S-bonded thiocyanates [Z. V. Zvonkova, *Zh. Fiz. Khim.*, **26**, 1798 (1952)].
[d] Partial isomerization to N-bonded isomer in molten state.
[e] Mixture of isomers initially isolated in solid state.

Abbreviations: py = pyridine, en = ethylenediamine, Me$_2$en = N,N-dimethylethylenediamine, EtenEt = N,N'-diethylethylenediamine, bipy = 2,2'-bipyridine, Et$_4$dien = N,N,N',N'-tetraethyldiethylenetriamine, 4,7-diphenylphen = 4,7-diphenyl-1,10-phenanthroline, tripyam = tri(2-pyridyl)amine (bidentate), Me$_2$bipy = 4,4'-dimethylbipyridine, acac = acetylacetonate.

such pioneers as S. M. Jørgensen and Alfred Werner. However, the development of modern instrumentation and techniques has permitted detailed mechanistic studies of the reactions involved in these syntheses, the great majority of the work having been accomplished within the past twenty years. The results of such studies are being used with increasing frequency in the synthesis of new coordination complexes and in the development of better procedures for synthesizing known complexes although, compared to organic chemistry, this stage is still in relative infancy.

The phenomenon of inorganic linkage isomerism (the existence of two complexes differing only in the mode of attachment of an ambidentate ligand to the metal atom) serves as a graphic illustration, in terms of the number of known examples, of the burgeoning importance of this approach. Although the first linkage isomeric pair of metal complexes was isolated more than one hundred years ago[1] [the isoxantho and xantho complexes of cobalt(III)], their true nature was not ascertained until almost forty years later,[2,3] and a second example of this phenomenon was not discovered until some sixty years had elapsed.[6,7] However, as shown in Table I, at least thirty such isomeric pairs have now been reported, the majority having been synthesized as a result of the application of known reaction mechanisms. Specific references to many of these examples will be made in the following discussion.

Metal complexes undergo both oxidation–reduction and substitution reactions. The kinetics and mechanisms of both types of reactions have been studied in great detail and several books and reviews have been written on the subject.[36–44]

II. OXIDATION–REDUCTION REACTIONS

Two fundamentally different paths appear to be available in oxidation–reduction reactions. One is usually accompanied by ligand migration between the oxidant and reductant, although this is not a prerequisite, and is called the inner sphere mechanism. The other takes place via an outer sphere activated complex, and involves only electron transfer. Each of these approaches can be used in synthesizing metal complexes.

A. Inner Sphere Mechanism

The inner sphere, or bridged activated complex, mechanism has been found to be operative in a large number of oxidation–reduction reactions involving coordination complexes. As exemplified by reaction 1, the

$$[Co(NH_3)_5X]^{2+} + [Co(CN)_5]^{3-} \longrightarrow [(H_3N)_5Co^{III}-X-Co^{II}(CN)_5]^-$$
$$\xrightarrow{H_2O} [Co(H_2O)_6]^{2+} + 5NH_3 + [Co(CN)_5X]^{3-} \quad (1)$$

mechanism requires the presence of a ligand, X, which is capable of forming a bridge between the oxidant and reductant, thereby facilitating the redox reaction. Ideally, to permit identification of the mechanism simply by product identification, the oxidant and reductant should be, respectively, substitution inert and substitution labile and the substitution characteristics of the products should be reversed. Thus, the bridging ligand is retained by the oxidized form of the reductant.

The synthetic possibilities inherent in this mechanism are readily apparent. Within the limits specified, various combinations of bridging ligands and metal complex reductants can be employed with a given metal oxidant. Several examples (by no means a complete tabulation) are shown in Table II. In addition to those shown in Table II, monoacido complexes of pentaamminecobalt(III) have also been reduced by metal ions such as iron(II),[59,60] vanadium(II),[61,62] and europium(II),[62] for which, because of the substitution lability of their oxidized forms, the product criterion of mechanism is not applicable.

TABLE II
Examples of Inner Sphere Redox Reactions

Oxidant	Reductant	Bridging groups, X, and refs.
$[Co^{III}(NH_3)_5X]^{n+}$	$[Cr(H_2O)_6]^{2+}$	F^-, Cl^-, Br^-, I^-, SO_4^{2-} [45]; N_3^-, PO_4^{3-}, $P_2O_7^{4-}$, acetate, butyrate, crotonate, oxalate, succinate, maleate[46]; $-NCS^-$ [46,47]; H_2O[48]; urethane, methyl glycinate, benzocaine, ethyl nicotinate, ethyl isonicotinate, ethyl-4-aminobutyrate[49]; $-SO_3^{2-}$, $-OS_2O_2^{2-}$, $-SSO_3^{2-}$ [27]; $-ONO^-$, $-NO_2^-$ [50]; $-CN^-$ [28]
$[Co^{III}(NH_3)_5X]^{n+}$	$[Co(CN)_5]^{3-}$	$S_2O_3^{2-}$ [51]; Br^-, I^- [52]; Cl^-, N_3^-, OH^- [53]; $-NCS^-$ [53,54]; $-ONO^-$, $-NO_2^-$ [8]; $-CN^-$ [8,55]
$[Co^{III}(en)_2(NCS)X]^{n+}$	$[Cr(H_2O)_6]^{2+}$	H_2O, NH_3, Cl^-, $-NCS^-$ [20]
$[Fe(H_2O)_5NCS]^{2+}$	$[Cr(H_2O)_6]^{2+}$	19
$[Fe(CN)_6]^{3-}$	$[Cr(H_2O)_6]^{2+}$	45
$[Cr(H_2O)_5X]^{2+}$	$[Cr(H_2O)_6]^{2+}$	F^-, Cl^-, Br^-, $-NCS^-$, N_3^- [56]
X_2	$[Co(CN)_5]^{3-}$	Br_2, I_2 [52]; O_2 [57]
$[Fe(CN)_6]^{3-}$	$[Co(CN)_5]^{3-}$	57
$[Hg(SeCN)_4]^{2-}$	$[Co(CN)_5]^{3-}$	58

Ambidentate ligands, when used as the bridging group, X, offer the intriguing possibility of synthesizing coordination complexes wherein the bonding mode of the ambidentate ligand has been changed from that existing in the oxidant. Such is the case when NCS^- is used as the bridging ligand in reaction 1, the Co^{III}—NCS—Co^{II} bridged activated complex leading to the formation of $[Co(CN)_5SCN]^{3-}$. This isomer has been found to be stable with respect to isomerization.[54] Such behavior is to be contrasted with that observed when $—NO_2^-$ or $—CN^-$ is used as the bridging group, the initially formed $[Co(CN)_5ONO]^{3-}$ and $[Co(CN)_5NC]^{3-}$ products isomerizing rapidly to N-bonded (nitro) and C-bonded complexes, respectively.[8]

Although the effects of nonbridging ligands, originally in the coordination sphere of the oxidant, on the rate of electron transfer via bridging ligands has received considerable study,[46,61,63] comparatively little work has been done on the effects of nonbridging ligands originally in the coordination sphere of the reductant. There is some evidence that, in the case of ambidentate bridging groups, these nonbridging ligands play a very important role in determining the nature of the metal–ambidentate ligand bond in the final product. For example, whereas the pentaammines $[Co(NH_3)_5NCS]^{2+}$ [64] and $[Co(NH_3)_5NCSe]^{2+}$ [58] both contain N-bonded pseudohalides, the pentacyanides $[Co(CN)_5SCN]^{3-}$ [54] and $[Co(CN)_5NCSe]^{3-}$ [58] (the latter obtained by the $[Co(CN)_5]^{3-}$ reduction of $[Hg(SeCN)_4]^{2-}$) are, respectively, S- and N-bonded. Consequently, when oxidants {$[Hg(SCN)_4]^{2-}$ and $[Co(NH_3)_5NCSe]^{2+}$} were chosen which would result in the initial formation of the unstable $[Co(CN)_5NCS]^{3-}$ and $[Co(CN)_5SeCN]^{3-}$ isomers, the compounds

$$K_4[(NC)_4Co\underset{SCN}{\overset{NCS}{\diamond}}Co(CN)_4]\cdot 5H_2O \text{ and } K_2[Co(CN)_5H_2O]$$

were isolated instead.[58]

In an interesting reversal of emphasis, the discovery of two different reaction rates for the chromium(II) reduction of thiosulfatopentaamminecobalt(III) has been cited[27] as evidence for the existence of both O- (90%) and S-bonded isomers in the complex as it is usually prepared.

Although most of the examples cited in Table II involve net transfer of the bridging group to the oxidized reductant, there are a few cases where the bridged activated complex has been isolated. Thus, when $[Fe(CN)_6]^{3-}$ is reduced by $[Cr(H_2O)_6]^{2+}$ or $[Co(CN)_5]^{3-}$, the bridged species $K[(NC)_5Fe^{II}—CN—Cr^{III}(H_2O)_5]$[45] and $Ba_3[(NC)_5Fe^{II}—CN—Co^{III}(CN)_5]\cdot 16H_2O$[57] result, due primarily to the fact that iron(II), chromium(III), and cobalt(III) are all substitution inert. Similar considerations explain the stability of the bridged complexes $[(NC)_5Co-O_2-Co(CN)_5]^{6-}$, $[(NC)_5Co-SO_2-Co(CN)_5]^{6-}$, and $[(NC)_5Co-SnCl_2-Co(CN)_5]^{6-}$, resulting from the reaction of $[Co(CN)_5]^{3-}$ with, respectively, oxygen,[57] sulfur dioxide,[65] and tin(II) chloride.[65] The reaction between $[IrCl_6]^{2-}$ and $[Cr(H_2O)_6]^{2+}$ lies at the opposite extreme with no net transfer of chloride taking place.[45]

Although the reductants shown in Table II are all of the one-electron type, this is not a limitation of the mechanism. Two systems have been studied which involve, presumably, two-electron reductants and give promise of more extensive synthetic applications. Both involve six-coordinated low-spin d^6 complexes of platinum metals which, generally, are very substitution inert. However, the substitution reactivity of these systems can be markedly increased by the presence of catalytic amounts of the metal ion in a different (lower) oxidation state, suggesting that a redox process provides a low-energy path for ligand replacement reactions.

In the first case, the exchange of radioactive chloride with *trans*-$[Pt(en)_2Cl_2]^{2+}$ has been found to take place very slowly. However, in the presence of $[Pt(en)_2]^{2+}$, the rate is greatly enhanced. The rate law, rate $= k$ [Pt(IV)][Pt(II)][Cl$^-$], determined from the results of kinetic studies,[66] suggests the mechanistic scheme shown in reactions 2 and 3.

$$[Pt(en)_2]^{2+} + {}^*Cl^- \underset{}{\overset{fast}{\rightleftharpoons}} [Pt(en)_2{}^*Cl]^+ \quad (2)$$

$$[Pt(en)_2{}^*Cl]^+ + trans\text{-}[Pt(en)_2Cl_2]^{2+} \underset{}{\overset{slow}{\rightleftharpoons}} \begin{array}{c} en \quad en \\ [{}^*Cl-Pt-Cl-Pt-Cl]^{3+} \\ en \quad en \end{array} \quad (3)$$

$$\Big\updownarrow slow$$

$$trans\text{-}[Pt(en)_2({}^*Cl)Cl]^{2+} + [Pt(en)_2Cl]^+$$

The mechanism involves a bridged activated complex two-electron redox reaction which labilizes the chloro groups along the z axis but preserves the xy plane of $[Pt(en)_2]^{2+}$. Repetition of the sequence of events with $[Pt(en)_2{}^*Cl]^+$ and *trans*-$[Pt(en)_2({}^*Cl)Cl]^{2+}$ as the reactants completes the exchange.

The elucidation of this mechanism had immediate synthetic applications. Normally, *trans*-$[Pt(en)_2X_2]^{2+}$ complexes had been prepared by the oxidation of $[Pt(en)_2]^{2+}$ with X_2, the products being limited by the oxidizing strengths of X_2, e.g., *trans*-$[Pt(en)_2(SCN)_2]^{2+}$ cannot be prepared in this manner. However, the application[67] of reaction 4 has led to the

$$trans\text{-}[Pt(en)_2Cl_2]^{2+} + 2X^- \xrightarrow{[Pt(en)_2]^{2+}} trans\text{-}[Pt(en)_2X_2]^{2+} + 2Cl^- \quad (4)$$

synthesis of several complexes of this type (X = Br$^-$, —SCN$^-$). The method is not general for any X, no reaction taking place for X = SO_4^{2-}, ClO_4^-, $C_2H_3O_2^-$, F$^-$, and NO_3^-, i.e., ligands which are poorly coordinating and have little tendency to form bridges between two metal atoms. However, the method is not specific for bis(ethylenediamine)platinum(IV) complexes, since, for example, the same mechanism accounts for the fact that, although $[Pt(NH_3)_5Cl]^{3+}$ reacts only very slowly with aqueous HCl, the reaction occurs readily to give good yields of trans-$[Pt(NH_3)_4Cl_2]^{2+}$ in the presence of catalytic amounts of $[Pt(NH_3)_4]^{2+}$.[68] The chloro-bridged intermediate has, along its z axis, Cl—Pt^{II}—Cl—Pt^{IV}—NH$_3$, so that cleavage of the Cl—Pt^{IV} bond results in the formation of Cl—Pt^{IV}—Cl, the trans-dichloro product. The reverse reaction between trans-$[Pt(NH_3)_4Cl_2]^{2+}$ and ammonia and the substitution reactions of trans-$[Pt(NH_3)_4X_2]^{2+}$ complexes in general have been found to occur via Pt^{II}—X—Pt^{IV} bridged activated complex mechanisms.[69-71]

Kinetic studies of the second system chosen for discussion were prompted by earlier observations published by Delepine,[72,73] wherein he reported that substitution reactions of rhodium(III) complexes appear to be catalyzed by alcohols. He found that the reaction of $Na_3[RhCl_6]$ with pyridine led to the formation of several chloropyridine rhodium(III) complexes.

However, because of the insolubility of $[Rh(py)_3Cl_3]$, it was difficult to prepare complexes containing more than three molecules of pyridine. For example, the preparation of a small yield of $[Rh(py)_4Cl_2]Cl$ required the heating of a slurry of the complexes in water containing excess pyridine for ten hours. In order to provide a better solvent for the nonionic complex and thus more readily generate the desired cationic complexes, Delepine added alcohol to an aqueous solution of $Na_3[RhCl_6]$ prior to introducing pyridine. Quite unexpectedly, an immediate reaction took place at room temperature, giving $[Rh(py)_4Cl_2]Cl$ in high yield. Further study showed that both primary and secondary alcohols were effective in promoting the reaction, but tertiary alcohols, ether, dioxane, and acetone were ineffective.

Rund et al. have investigated this system in detail,[74] and have found that no detectable amount of alcohol is consumed in the reaction. In fact, reducing agents such as tin(II), BH_4^-, and hydrazine are more effective catalysts than is alcohol. It appears that the function of the additives is to generate catalytic amounts of a lower valent rhodium species which then provides a path for the reaction of the rhodium(III) complexes. Kinetic studies of reaction 5 showed that, in excess pyridine, the rate of reaction is

$$[Rh(H_2O)Cl_5]^{2-} + 4py \xrightarrow{\text{catalyst}} [Rh(py)_4Cl_2]^+ + 3Cl^- \quad (5)$$

first order in both the rhodium(III) and catalyst concentrations. Although the lower valent rhodium species was not identified, the assumption that it is rhodium(I) permitted the postulation of a mechanism similar to that proposed for the platinum(II) catalysis of platinum(IV) reactions. Support for this assumption is found in the observation[75,76] that $RhCl_3$ can be reduced by alcohol to the $+1$ state in the presence of potassium hydroxide and ligands which stabilize that oxidation state. The mechanism proposed is shown in reactions 6–10. These results suggest that the use of lower

$$[Rh(H_2O)Cl_5]^{2-} \xrightarrow{reduce} Rh(I) \quad (6)$$

$$Rh(I) + 4py \xrightarrow{fast} [Rh(py)_4]^+ \quad (7)$$

$$[Rh(py)_4]^+ + [Rh(H_2O)Cl_5]^{2-} + H_2O \xrightarrow{slow} [H_2O-Rh(py)_4-Cl-RhCl_4-OH_2]^- \quad (8)$$

$$[H_2O-Rh(py)_4-Cl-RhCl_4-OH_2]^- \xrightarrow{fast} trans\text{-}[Rh(py)_4(H_2O)Cl]^{2+} + Rh(I) \quad (9)$$

$$trans\text{-}[Rh(py)_4(H_2O)Cl]^{2+} + Cl^- \xrightarrow{fast} trans\text{-}[Rh(py)_4Cl_2]^+ + H_2O \quad (10)$$

valent rhodium species as catalysts for reactions of rhodium(III) can aid in the otherwise difficult synthesis of certain rhodium(III) complexes, and that this general approach can be helpful in the synthetic chemistry of the complexes of the platinum metals. A recent example of the utility of this approach is seen in the synthesis of several trans-$[Rh(en)_2XY]^{n+}$ complexes by Baker and Gillard.[77]

It is seldom found, in modern inorganic chemistry, that a compound's synthetic utility is of such importance that its discoverer's name becomes synonymous with its formula. A striking exception to this generalization is Vaska's compound, trans-$[Ir(P(C_6H_5)_3)_2(CO)Cl]$, discovered in 1961.[78] The remarkable reactivity of this compound with a variety of oxidizing agents has generated a substantial amount of research interest in what is now recognized to be a very general phenomenon—the oxidative addition reactions of metal complexes having a d^8 configuration. Although they are usually not thought of as being inner sphere redox reactions in the sense of those examples just discussed, they do involve the inner coordination sphere of the reductant and can be appropriately considered in this section. The consequences of the reaction are twofold: An increase in the coordination number of the reductant (usually to six) resulting from the coordination of the constituent parts of the oxidant molecule and an increase of two units in the formal oxidation state of the metal, i.e., conversion to a d^6 configuration. The oxidant molecule is usually split into two parts, although, as will be seen shortly, several examples are known wherein the molecule adds to the metal without molecular dissociation and the assignment of a formal oxidation state to the metal becomes difficult.

An appreciation of the scope of the reaction may be gained by considering the examples shown in Table III. Although the majority of the references cited pertain to results published since the discovery of Vaska's compound, the prototype of these reactions, the oxidation of platinum(II) ammine complexes by halogens, was first reported many years ago.[79] A trend, first pointed out by Nyholm and Vrieze,[95] has been found to exist whereby the tendency to form stable adducts having the d^6 configuration increases going from the first- to the third-row transition elements and from right to left within group VIII.

TABLE III
Examples of Oxidative Addition Reactions in d^8 Systems

Reductant	Oxidants and refs.
$[Pt(NH_3)_4]^{2+}$	Cl_2 [79]; Br_2 [80]; H_2O_2 [81]
$[Pt(en)_2]^{2+}$	Cl_2 [82]; Br_2 [83]; H_2O_2 [82]
trans-$[Pt(P(C_2H_5)_3)_2(CH_3)I]$	CH_3I [84]
cis-$[Pt(P(C_2H_5)_3)_2(CH_3)_2]$	Cl_2 [84]
cis or trans-$[Pt(P(C_2H_5)_3)_2(C_6H_5)_2]$	Cl_2, I_2 [85]
trans-$[Pt(P(C_2H_5)_3)_2(H)Cl]$	HCl [86]
$[Pt(P(C_2H_5)_3)_2(Ge(C_6H_5)_3)_2]$	HCl [87]
trans-$[Rh(P(n-C_4H_9)_3)_2(CO)Cl]$	$C_6H_5CH_2Br$, CH_3I, $CH_3OC(O)CH_2I$ [88]
$[Rh(C_5H_5)(CO)_2]$	CF_3I, C_2F_5I, C_3F_7I [89]
trans-$[Ir(P(C_6H_5)_3)_2(CO)Cl]$	HX [78,90]; H_2 [91–93]; Cl_2 [91]; Cl_3SiH, $RSiCl_2H$, $(C_2H_5O)_3SiH$ [94]; HgX_2 [95]; D_2 [91,93]; CH_3I, $CH_3OC(O)CH_2I$, CH_2=$CHCH_2Cl$ [88]; F_2C=CF_2, F_3CC≡CCF_3 [96]; CH_2=CH_2, HC≡CH [92]; RSO_2Cl [97]; O_2, NO(excess in air), SO_2(excess)[93]; $Hg(C$≡$CR)_2$, HC≡$CCO_2C_2H_5$ [98]
$[Fe(CO)_5]$	X_2 [99]; CF_3I, C_2F_5I, C_3F_7I [100]
$[Fe(PR_3)_2(CO)_3]$	Br_2 [101]
$[Fe[(CH_3)_2PCH_2CH_2P(CH_3)_2]_2]$	I_2 [102]
$[Ru(P(C_6H_5)_3)_2(CO)_3]$	HCl, HBr, CF_3COOH, I_2, HgX_2, CH_3I [103]
$[Os(P(C_6H_5)_3)_2(CO)_3]$	HCl, HBr, HI, Br_2, I_2 [104]

The stereochemistry involved in the addition reactions has proved to be a subject of some interest. Unlike the stereospecific trans additions generally observed[82] when platinum(II) ammine complexes are oxidized by halogens, cis additions [as exemplified by reactions 11[90] and 12[104]] have been observed to result from some oxidations of Vaska's compound and bipyramidal, five-coordinate d^8 complexes. Whether or not this is a

$$\begin{array}{c}\text{P}\text{Cl}\\\diagdown\diagup\\\text{Ir}\\\diagup\diagdown\\_\text{O}\text{C}\text{P (s)}\end{array} + \text{HBr(g)} \longrightarrow \begin{array}{c}\text{Cl}\text{P}\text{Br}\\\diagdown|\diagup\\\text{Ir}\\\diagup|\diagdown\\_\text{O}\text{C}\text{P}\text{H (s)}\end{array} \qquad (11)$$

$$\begin{array}{c}_\text{O}\text{C}\text{P}\\\diagdown|\\\text{Os—CO}\\\diagup|\\_\text{O}\text{C}\text{P}\end{array} + \text{Br}_2 \longrightarrow \begin{array}{c}_\text{O}\text{C}\text{P}\text{Br}\\\diagdown|\diagup\\\text{Os}\\\diagup|\diagdown\\_\text{O}\text{C}\text{P}\text{Br}\end{array} + \text{CO} \qquad (12)$$

(P = triphenylphosphine)

general phenomenon is not yet known. A number of molecules, e.g., acetylenes,[92,96,98] olefins,[92,96] carbon monoxide,[93] trifluorophosphine,[93] nitric oxide,[93] sulfur dioxide,[93,105,106] and oxygen,[93,107] add to Vaska's compound without molecular dissociation taking place, and the assignment of a coordination number for the metal becomes, in some cases, a problem of semantics. The structure of these complexes has been found[108] to be quite sensitive to both the nature of the added ligand and to the nature of the halide ion.

Encouraged by the versatility of Vaska's complex, Collman and Kang[109] attempted to oxidatively add a series of organic acid azides, but were surprised to obtain the same golden yellow compound whenever the reactions were carried out at 0° in $CHCl_3$ containing water or an alcohol, regardless of the azide employed. The compound proved to be the second known coordination complex containing molecular nitrogen, the first being $[Ru(NH_3)_5N_2]^{2+}$.[110] Further work[111] by Collman and co-workers has led to the proposed mechanism for the formation of trans-$[Ir(P(C_6H_5)_3)_2(N_2)Cl]$, shown in reaction 13. If an alcohol is present, the acyl isocyanate formed from the collapse of the kinetically undetectable intermediate is irreversibly intercepted; otherwise, it reacts with the nitrogen complex to form the π-bonded isocyanate. Collman et al. have pointed out[111] several important ramifications of this mechanism. They suggest that nitrogen complexes can be formed from the reaction of other metal carbonyls (preferably kinetically labile, but thermodynamically stable) with organic azides. Alternatively, under more vigorous conditions, the coordinated nitrogen could act as a reactive intermediate to trap other π-bonding ligands, thereby serving as a method for preparing new catalyst systems. Other systems which are electronically similar to azides, e.g., nitrous oxide, diazoalkanes, and carbodiimides, may react by analogous pathways. Finally, they speculate that other ligands, such as tertiary phosphines, which react with organic azides can be replaced by nitrogen in this manner. Thus, the cycle is completed: A known, fruitful reaction type yields an unexpected

(13)

result, the elucidation of which promises, perhaps, to yield even greater benefits.

For example, Collman et al. have shown[112] that the addition of CO in reaction 13 regenerates the original carbonyl complex which can then react with more organic azide so that the overall process is catalytic, as shown in reaction 14. In the absence of a metal complex, CO and organic

$$ArN_3 + CO \xrightarrow{[Ir(P(C_6H_5)_3)_2(CO)Cl]} ArNCO + N_2 \qquad (14)$$

azides do not react at a measurable rate. They suggest[112] that this is an example of a more general class of transition metal reactions: Metal ion promoted atom-transfer oxidation–reduction reactions. These authors hypothesize that incorporation of both reactants into *cis* positions in the coordination sphere of a metal complex might provide a mechanism to facilitate such reactions and cite several other examples, drawn from their own work and from the literature, in support of their hypothesis. Nitrobenzene reacts with iron pentacarbonyl (thermally or photochemically activated to produce a coordinatively unsaturated complex) to form carbon dioxide and azo and nitroso compounds.[113,114] Tertiary phosphines can be catalytically oxidized by O_2 in the presence of d^{10} complexes of the type $[M(P(C_6H_5)_3)_4]$ [M = Ni(O), Pd(O), Pt(O)].[115] Similarly, the oxidation of cyclohexene to form cyclohexen-3-one is catalyzed by the d^8 complexes $[Ir(P(C_6H_5)_3)_2(CO)I]$, $[Ir(P(C_6H_5)_3)_2(N_2)Cl]$, and $[Rh(P(C_6H_5)_3)_3Cl]$.[112] Other examples involve substitution in the coordination sphere of the metal without metal–ligand bond cleavage and will be discussed later in the appropriate section.

B. Outer Sphere Mechanism

As the name implies, redox reactions proceeding via this mechanism do not involve substitution into the coordination spheres of either the oxidant or reductant complexes. This mechanism, sometimes called electron transfer, is found to predominate when both the oxidant and reductant are substitution inert and when both are completely coordinated or when no bridging ligands are available for substitution into the coordination sphere of a labile oxidant or reductant, as exemplified by reactions 15 and 16.

$$[*Fe(bipy)_3]^{3+} + [Fe(bipy)_3]^{2+} \rightleftharpoons [*Fe(bipy)_3]^{2+} + [Fe(bipy)_3]^{3+} \quad (15)$$

$$[Co(NH_3)_6]^{3+} + [Cr(H_2O)_6]^{2+} \longrightarrow \{[Co(NH_3)_6][Cr(H_2O)_6]\}^{5+} \quad (16)$$

$$\xrightarrow[H^+]{H_2O} [Co(H_2O)_6]^{2+} + 6NH_4^+ + [Cr(H_2O)_6]^{3+}$$

This type of reaction has not been used as extensively for synthetic purposes as that discussed in the preceding section, but it is of value when it is desired to oxidize or reduce a coordination complex without changing its first coordination sphere. A particularly intriguing application of this mechanism has been reported by Busch.[116] Both Werner[117] and Dwyer and Gyarfas[118] had shown that when one of the diastereoisomers of a salt of an optically labile complex ion is much less soluble than the other, virtually all of the complex may be obtained as the less soluble form. This occurs because the more soluble isomer racemizes, as the less soluble isomer crystallizes, thus producing more of the less soluble form, which then crystallizes. The process is termed a second-order asymmetric induction because a second substance, the precipitating ion, is necessary for its promotion. Asymmetric induction such as this would normally not be possible with a substance such as $[Co(en)_3]^{3+}$, which, being substitution inert, does not tend to racemize or otherwise undergo changes in configuration. However, Busch was able to isolate in high yield the diastereoisomeric salt $(+)[Co^{III}(en)_3]Cl(d\text{-tartrate})$ by adding a catalytic amount of $[Co(en)_3]^{2+}$ to a solution of *racemic* $[Co(en)_3]Cl_3$ containing the resolving agent d-tartrate ion, as shown in reaction 17. Because the cobalt(II) complex

$$(+,-)[Co(en)_3]^{3+} + Cl^- + d\text{-tartrate}^{2-} \xrightleftharpoons[]{[Co(en)_3]^{2+}} (-)[Co(en)_3]^{3+} \text{ (solution)}$$

$$\searrow (+)[Co(en)_3]Cl(d\text{-tartrate}) \text{ (solid)} \quad (17)$$

is labile and because it undergoes electron transfer with the cobalt(III) complex, the latter is, in effect, labilized. The result is a path for the conversion of $(-)[Co(en)_3]^{3+}$ into its mirror image form which then crystallizes from solution as the less soluble diastereoisomeric salt. This method yielded 150% of the *dextro* complex, based on the amount of the isomer originally present in $(+,-)[Co(en)_3]^{3+}$. Theoretically, all of the racemic mixture can be converted to the *dextro* isomer. Utilization of *l*-tartrate in the same process gives a high yield of $(-)[Co^{III}(en)_3]Cl(l\text{-tartrate})$.

Recently, Lee et al. succeeded[119] in directly resolving, for the first time, optically active tervalent complex ions of the type $[M(phen)_3]^{3+}$ [M = Cr(III), Co(III)]. Resolution was achieved by utilizing potassium antimonyl-*d*-tartrate as the resolving agent in an aqueous ethanol medium, and the cobalt(III) complex was not obtained optically pure. The results of the work of Busch, coupled with the rapid electron transfer observed[120,121] in $[M(phen)_3]^{2+,3+}$ systems, suggest that the addition of catalytic amounts of $[M(phen)_3]^{2+}$ should also allow resolution of the $[M(phen)_3]^{3+}$ ions in high yields. This may obviate the necessity of employing fractional crystallization to effect the separation of the diastereoisomers and may yield the optically pure cobalt(III) complex.

Adamson[122] has pointed out that the product obtained from the oxidation of some reductants, e.g., $[Co^{II}(EDTA)]^{2-}$, depends on the oxidant used. As shown by Schwarzenbach,[123] oxidation with Br_2 yields $[Co^{III}(HEDTA)Br]^-$, presumably as the result of reaction 18, which in-

$$[Co(EDTA)]^{2-} + Br_2 + H^+ \longrightarrow [(HEDTA)Co-Br-Br]^- \longrightarrow \\ [Co(HEDTA)Br]^- + Br \quad (18)$$

volves atom transfer. The use of $[Fe(CN)_6]^{3-}$ as the oxidizing agent results, as expected, in the formation of $[Co(EDTA)]^-$. This was expected to generate the sexadentated product because it was assumed that it would react by an outer sphere process. However, more recent studies[124] suggest that the reaction also proceeds via a bridged activated complex, as shown in reaction 19.

$$[Co(EDTA)]^{2-} + [Fe(CN)_6]^{3-} \longrightarrow [(EDTA)Co-NC-Fe(CN)_5]^{5-} \\ \downarrow \\ [Co(EDTA)]^- + [Fe(CN)_6]^{4-} \quad (19)$$

The product of a redox reaction may, at times, depend on whether a one- or a two-electron oxidizing or reducing agent is used. Adamson et al. have shown[125] that, when a solution of cobalt(II) in excess oxalate is oxidized by the one-electron oxidant cerium(IV), $[Co(C_2O_4)_3]^{3-}$ results whereas the action of two-electron, oxygen carrying oxidants, such as hydrogen peroxide and hypochlorite, yields, instead, Durant's salt.

$$K_4[(C_2O_4)_2Co\underset{\underset{H}{O}}{\overset{\overset{H}{O}}{\diamond}}Co(C_2O_4)_2]\cdot 2H_2O$$

Saffir and Taube[126] have studied the oxidation of coordinated oxalate ion in $[Co(NH_3)_5C_2O_4]^+$ with one- and two-electron oxidizing agents. The oxidation of $C_2O_4^{2-}$ to CO_2 requires the removal of two electrons. For a one-electron oxidant, it and the cobalt(III) in the complex cooperate in accepting the two electrons. Thus, the reaction between $[Co(NH_3)_5C_2O_4]^+$ and cerium(IV) exhibits the stoichiometry shown in eq. 20. However,

$$[Co(NH_3)_5(HC_2O_4)]^{2+} + Ce^{4+} + 4H^+ \longrightarrow$$
$$Co^{2+} + Ce^{3+} + 2CO_2 + 5NH_4^+ \quad (20)$$

as shown in reaction 21, the participation of the cobalt(III) is not required when a two-electron oxidant is used.

$$[Co(NH_3)_5(HC_2O_4)]^{2+} + H_2O_2 + H^+ \xrightarrow{Mo(VI)}$$
$$[Co(NH_3)_5H_2O]^{3+} + 2CO_2 + 2H_2O \quad (21)$$

This information was used by Fraser and Taube[127] in their preparation of

$$[Co(NH_3)_5(OOC-\!\!\!\bigcirc\!\!\!-COOH)]^{2+}$$

The usual method for the preparation of monocarboxylato complexes of the type $[Co(NH_3)_5OOCR]^{2+}$ is to allow $[Co(NH_3)_5H_2O]^{3+}$ to react with a buffered solution of the acid. For terephthalic acid, this method yields, instead, primarily the dimeric complex

$$[(H_3N)_5Co-OOC-\!\!\!\bigcirc\!\!\!-COO-Co(NH_3)_5]^{4+}$$

The desired monomeric complex was synthesized by oxidizing the aldehyde group of the *p*-aldehydobenzoatopentaamminecobalt(III) complex with a two-electron oxidant, according to reaction 22. One-electron oxidizing agents could not be used because, as in reaction 20, these involve the

$$[Co(NH_3)_5(OOC\text{-}C_6H_4\text{-}CHO)]^{2+} + Cl_2 + H_2O \longrightarrow$$

$$[Co(NH_3)_5(OOC\text{-}C_6H_4\text{-}COOH)]^{2+} + 2Cl^- + 2H^+ \quad (22)$$

cobalt(III), reducing it to cobalt(II), thus causing the decomposition of the complex.

Haim and Taube have extensively studied the oxidation of $[Co(NH_3)_5I]^{2+}$ with one-electron[128] and two-electron[129] oxidizing agents. They have found intriguing differences in behavior between the two categories and, depending upon the oxidant used, within each category. The generalizations derived from these studies give promise of even broader synthetic applications.

In the reaction with the one-electron oxidant cerium(IV), the oxidation state of the cobalt(III) center is preserved, as shown in reaction 23.

$$[Co(NH_3)_5I]^{2+} + Ce^{4+} + H_2O \longrightarrow [Co(NH_3)_5OH_2]^{3+} + I + Ce^{3+} \quad (23)$$

This behavior is to be contrasted with that shown in reaction 20. Assuming that in each case the external oxidizing agent brings about a one-electron oxidation of the ligand and that intermediates with the corresponding radicals in the coordination sphere of cobalt(III) are formed, the question naturally arises as to why the oxidation of the $C_2O_4^-$ radical ion by the cobalt(III) center is favored over its expulsion from the coordination sphere while the opposite situation arises in the case of the iodine atoms. Haim and Taube suggest that the difference in behavior arises from the fact that $C_2O_4^-$ may be oxidized by simple electron abstraction, whereas the oxidation of I to HOI is (at least) kinetically less favorable for it requires that an I—O bond be formed.

When iodine atoms and hydroxyl or methyl radicals are employed as the one-electron oxidants, the cobalt(III) center is reduced in the process, as exemplified by reaction 24. (It should be pointed out that the iodine

$$[Co(NH_3)_5I]^{2+} + I + 5H^+ \longrightarrow Co^{2+} + I_2 + 5NH_4^+ \quad (24)$$

atoms generated in reaction 23 react with the substrate as shown in reaction 24, accounting for some 50% of the overall reaction.) It would appear that the behavior exhibited by the one-electron oxidizing agents depends upon the ability of the latter to form a bond with the iodine in the coordination sphere of the cobalt(III).

The two-electron oxidants employed, as a group, had no effect on the oxidation state of the cobalt(III) center. However, whereas the oxidants Cl_2

and Br_2 gave quantitative yields of, respectively, $[Co(NH_3)_5Cl]^{2+}$ and $[Co(NH_3)_5Br]^{2+}$, the oxidants HOBr, ICl, O_3, CH_3CO_3H, $S_2O_8^{2-}$, HSO_5^-, and H_2O_2 all gave quantitative yields of $[Co(NH_3)_5OH_2]^{3+}$. The suggestion is made that, in the first case, the retention of the halide ion of the oxidant results from a rapid rearrangement of the $[Co(NH_3)_5IX]^{3+}$ (X = Cl, Br) intermediate to form $[Co(NH_3)_5XI]^{3+}$, followed by the loss of I(I), as, for example, HOI. In the case of the ICl oxidant, the $[Co(NH_3)_5I_2]^{3+}$ intermediate presumably hydrolyzes. The reactions of the remaining oxidants of the second type are believed to involve a $[Co(NH_3)_5IOH]^{3+}$ intermediate which, upon rearrangement to $[Co(NH_3)_5OH]^{3+}$, followed by hydrolysis, yields the common product
I
$[Co(NH_3)_5H_2O]^{3+}$. Reactions of the former type offer the possibility, depending upon the proper choice of substrate and oxidant, of synthesizing $[ML_mX]^{n+}$ complexes from $[ML_mY]^{n+}$ substrates far more rapidly and efficiently than can be accomplished by nonredox, nucleophilic substitution reactions involving substitution inert metal ions. It would be particularly interesting to examine the corresponding reactions of analogous chromium(III) and rhodium(III) complexes. Since the divalent oxidation state is less readily available to these metals than it is to cobalt, total oxidation may be provided by the oxidant and even a one-electron oxidant would not cause the destruction of the chromium(III) and rhodium(III) complexes.

III. SUBSTITUTION REACTIONS

Substitution reactions of metal complexes consist of generalized Lewis acid–base reactions where one ligand replaces another, or one metal ion is replaced by another. The former case involves nucleophilic substitution (S_N), wherein the electrons on the entering ligand seek the nucleus of the coordinated metal, and may be either uni- (S_N1) or bimolecular (S_N2), depending upon whether the rupture of the metal–outgoing ligand bond is more important than the formation of the metal–incoming ligand bond or both are of comparable importance in the transition state of the rate-determining step. In the S_N1 process, the formation of a lower coordinated species is rate-determining, whereas the formation of a higher coordinated species is rate-determining in the S_N2 process. Nucleophilic substitution reactions of metal complexes have been found to be more common and have been the subject of more extensive studies than those involving electrophilic substitution (S_E), wherein the nucleus of a metal seeks the electrons of a coordinated ligand, such as the reaction between $[Co(NH_3)_5Cl]^{2+}$ and Ag^+.

A. Hydrolysis and Anation Reactions of Octahedral Complexes

The hydrolysis reactions of metal complexes, particularly those of cobalt(III) [a vivid example of which is given in reaction 25], simple though

$$\text{\textit{trans}-[Co(en)}_2\text{Cl}_2]^+ + \text{H}_2\text{O} \longrightarrow [\text{Co(en)}_2(\text{H}_2\text{O})\text{Cl}]^{2+} + \text{Cl}^- \qquad (25)$$
$$\text{(green)} \qquad\qquad\qquad \text{(red)}$$

they may appear at first glance, have proved to be deceptively complicated. The fact that the rate of hydrolysis of some metal ammines depends greatly on pH, e.g., those of cobalt(III) ammine complexes are more than a million times faster in alkaline than in acid solutions, has been the subject of a long, unfinished, mechanistic debate.[130] The observed first-order dependence on the concentration of both the complex and the hydroxide ion was, quite naturally, first interpreted as involving an S_N2 mechanism. However, considerable evidence has also been presented in support of the conjugate base (S_N1CB) mechanism proposed by Garrick[131] and shown in reactions 26–28.

$$[\text{Co(NH}_3)_5\text{Cl}]^{2+} + \text{OH}^- \underset{}{\overset{\text{fast}}{\rightleftharpoons}} [\text{Co(NH}_3)_4(\text{NH}_2)\text{Cl}]^+ + \text{H}_2\text{O} \qquad (26)$$

$$[\text{Co(NH}_3)_4(\text{NH}_2)\text{Cl}]^+ \xrightarrow{\text{slow}} [\text{Co(NH}_3)_4\text{NH}_2]^{2+} + \text{Cl}^- \qquad (27)$$

$$[\text{Co(NH}_3)_4\text{NH}_2]^{2+} + \text{H}_2\text{O} \xrightarrow{\text{fast}} [\text{Co(NH}_3)_5\text{OH}]^{2+} \qquad (28)$$

Although the rate-determining step involves a first-order dissociation of the conjugate base to form a five-coordinate intermediate in reaction 27, the concentration of the conjugate base is dependent upon both the complex and hydroxide concentrations existing in the acid–base equilibrium, giving overall second-order kinetics. Very recently, Gillard[132] has proposed a third mechanism for the base hydrolysis of cobalt(III) complexes involving electron transfer from the hydroxide ion to cobalt(III). This explanation does not seem too plausible, considering that other much better reducing agents than hydroxide ion have little or no effect on the rates of reaction of the cobalt(III) ammines.

Since it is outside the scope of this chapter, we will not attempt to justify our choice of the S_N1CB mechanism and refer the reader to the arguments already advanced in the literature.[130] Instead, we will attempt to briefly discuss some of its synthetic ramifications. The mechanism suggests that base catalysts might be useful in synthesizing certain metal complexes. For example, it explains why the reaction between $CrCl_3$ and liquid ammonia yields chiefly $[Cr(NH_3)_5Cl]Cl_2$ but, if catalytic amounts of KNH_2 are added, a large yield of $[Cr(NH_3)_6]Cl_3$ is obtained. It implies that some group other than the solvent may rapidly enter the complex at the final step [reaction 28]. This occurs only to a slight extent when the solvent is water[133] because water is such a good coordinating solvent and because, being a hydroxylic solvent, proton transfer is very fast. However, the mechanism does suggest that, in a poorly coordinating, nonhydroxylic solvent, the five-coordinated amido complex might react with other ligands, as generalized in reaction 29. It has been shown[134] that dimethylsulfoxide

$$[Co(NH_3)_5Cl]^{2+} + X^- \xrightarrow{OH^-} [Co(NH_3)_5X]^{2+} + Cl^- \qquad (29)$$

is a suitable solvent for reactions of this type. Specifically, the substrate chosen was *trans*-$[Co(en)_2(NO_2)Cl]^+$. The reaction of this complex with various nucleophiles, X^-, in DMSO to yield $[Co(en)_2(NO_2)X]^+$ is very slow (for $X = NO_2^-$, first-order kinetics, nearly independent of nitrite concentration, with a $t_{1/2}$ of 5–6 hr at room temperature was observed). However, in the presence of catalytic quantities of hydroxide ion, the same reaction is complete in less than two minutes. Other bases such as piperidine were also found to act as catalysts. For a given concentration of catalyst, the rate of reaction was found to be independent of the concentration of the nucleophilic reagent and, furthermore, was the same for three different nucleophilic reagents (NO_2^-, NCS^-, N_3^-). These experimental facts are consistent with the S_N1CB mechanism shown in reactions 30–33.

$$[Co(en)_2(NO_2)Cl]^+ + B \underset{}{\overset{fast}{\rightleftharpoons}} [Co(en)(en-H)(NO_2)Cl] + BH^+ \qquad (30)$$

$$[Co(en)(en-H)(NO_2)Cl] \xrightarrow{slow} [Co(en)(en-H)NO_2]^+ + Cl^- \qquad (31)$$

$$[Co(en)(en-H)NO_2]^+ + NO_2^- \xrightarrow{fast} [Co(en)(en-H)(NO_2)_2] \qquad (32)$$

$$[Co(en)(en-H)(NO_2)_2] + BH^+ \xrightarrow{fast} [Co(en)_2(NO_2)_2]^+ + B \qquad (33)$$

Implicit in these observations is the possibility of making use of base catalysis for the synthesis of metal complexes in nonaqueous systems. However, it appears that this approach to synthesis has not yet been utilized.

Except for the S_N1 (limiting) reactions of $[Co(CN)_5X]^{n-}$,[135,136] $[Co(CN)_4(SO_3)OH_2]^{3-}$,[137] and $[Co(NH_3)_5SO_3]^+$,[138] reactions of cobalt(III) of the type shown in reaction 34 appear to first involve formation

$$[Co(NH_3)_5X]^{2+} + Y^- \xrightarrow{H_2O} [Co(NH_3)_5Y]^{2+} + X^- \qquad (34)$$

of $[Co(NH_3)_5H_2O]^{3+}$ and then replacement of water by Y^-.[139] However, Haim and Taube have suggested[140] that the induced aquation of $[Co(NH_3)_5N_3]^{2+}$ by NO^+, shown in reactions 35 and 36, involves the

$$[Co(NH_3)_5N_3]^{2+} + NO^+ \longrightarrow [Co(NH_3)_5N_3NO]^{3+} \qquad (35)$$

$$[Co(NH_3)_5N_3NO]^{3+} + H_2O \longrightarrow [Co(NH_3)_5OH_2]^{3+} + N_2 + N_2O \qquad (36)$$

pentacoordinated species $[Co(NH_3)_5]^{3+}$, which, theoretically, could add either a solvent molecule or some other ligand, Y. In water, the aquo complex mainly forms, but the addition of Y becomes more probable in poorly coordinating solvents. Utilizing these mechanistic considerations, Jordan et al. were able to prepare[141] a series of novel $[Co(NH_3)_5Y]^{n+}$ complexes as shown in reactions 37 and 38. The $[Co(NH_3)_5(O\text{—}\underset{\underset{O}{\|}}{C}\text{—}NH_2)]^{2+}$ ion has

$$[Co(NH_3)_5N_3]^{2+} + NO^+ + Y \xrightarrow{Y} [Co(NH_3)_5Y]^{3+} + N_2 + N_2O \quad (37)$$

$$[Y = OP(OC_2H_5)_3, CH_3CN]$$

$$[Co(NH_3)_5(OP(OC_2H_5)_3)]^{3+} + Y \xrightarrow[\Delta]{Y} [Co(NH_3)_5Y]^{n+} + OP(OC_2H_5)_3 \quad (38)$$

$$[Y = OCH_3^- \text{ (in methanol), NCR, pyridine}]$$

also been used[142,143] in a similar manner as a source of $[Co(NH_3)_5]^{3+}$, the overall reaction in water being that shown in reaction 39.

$$[Co(NH_3)_5(O\text{—}\underset{\underset{O}{\|}}{C}\text{—}NH_2)]^{2+} + NO^+ \longrightarrow$$
$$[Co(NH_3)_5OH_2]^{3+} + CO_2 + N_2 \quad (39)$$

B. Substitutions without Metal–Ligand Cleavage

The majority of substitution reactions, such as the hydrolysis reactions already described, require that the bond between the metal and the ligand initially in the coordination sphere be ruptured. An obvious exception would involve the reaction of organic ligands while they are coordinated, such as the Friedel-Crafts acylation of a cyclopentadienide ring in ferrocene. More germane to the discussion, however, is an example such as reaction 40, the stoichiometry of which implies the simple substitution

$$[(H_3N)_5Co\text{—}OCO_2]^+ + 2H_3{}^*O^+ \longrightarrow$$
$$[(H_3N)_5Co\text{—}OH_2]^{3+} + 2H_2{}^*O + CO_2 \quad (40)$$

of a water molecule for a carbonate ion, with resultant decomposition of the latter to carbon dioxide. However, Hunt et al. have shown,[144] using oxygen-18 labeled water, that the Co—O bond remains intact during the reaction.

In a similar manner, both kinetic studies[5] and oxygen-18 tracer experiments[145] have shown that the formation of $[Co(NH_3)_5ONO]^{2+}$ from $[Co(NH_3)_5H_2O]^{3+}$ in a buffered HNO_2/NO_2^- solution proceeds without Co—O bond cleavage. The proposed mechanism is shown in reactions

41–43. The nitrito complex, once formed, rearranges by an intramolecular process to the stable nitro complex.

$$[Co(NH_3)_5H_2O]^{3+} + OH^- \underset{}{\overset{fast}{\rightleftharpoons}} [Co(NH_3)_5OH]^{2+} + H_2O \quad (41)$$

$$2HNO_2 \xrightarrow{fast} N_2O_3 + H_2O \quad (42)$$

$$[Co(NH_3)_5OH]^{2+} + N_2O_3 \xrightarrow{slow} \begin{bmatrix} (H_3N)_5Co-O\cdots H \\ \vdots \\ \vdots \\ O-N\cdots ONO \end{bmatrix}^{2+} \quad (43)$$

$$\downarrow fast$$

$$[(H_3N)_5Co-ONO]^{2+} + HNO_2$$

The significance of this mechanism in terms of the synthesis of linkage isomers of two different oxy anions has clearly been demonstrated. Previous attempts to prepare nitrito complexes of metals other than cobalt(III) had been unsuccessful, due either to the greater lability of the metal ions employed, or, as in the case of the platinum metals, to a tendency on the part of research workers to employ strenuous conditions (e.g., long reflux periods), since the platinum metals are known to react more slowly than cobalt(III). In either situation, only the thermodynamically more stable nitro isomer was observed. Since, in the case of cobalt(III), the formation of the nitrito complex does not require M—O bond cleavage, it was theorized[6,7] that the metal might be expected to play a less important role and reaction conditions satisfactory for cobalt(III) might also work for the platinum metals. Using this approach, it was possible to synthesize three new nitrito isomers of the type $[M(NH_3)_5ONO]^{n+}$ [M = Rh(III), Ir(III), Pt(IV)]. These kinetic products can readily be separated from reaction mixtures of $[M(NH_3)_5H_2O]^{n+}$ and HNO_2/NO_2^- buffers at 0°. Although all three eventually isomerize to the nitro form, the chromium(III) derivative, prepared in the same manner, showed no tendency to isomerize. Only the nitro form could be isolated for platinum(II), due to its greater lability. In a similar way, Stranks[30] has found that adding sulfur dioxide or sodium sulfite to solutions of $[Co(NH_3)_5H_2O]^{3+}$ at low temperatures and low pH yields the unstable pink oxygen-bonded sulfito complex rather than the stable yellow-brown sulfur-bonded isomer.

The fact that the nitrito complex forms from the aquo complex without Co—O cleavage has also been used to advantage by Dwyer et al. to trap[146] optically active cobalt(III) complexes containing Co—OH$_2$ bonds in a stable, fixed stereochemistry. Of equal importance is the observation that the reaction of mono- and diaquo species such as $[Co(en)_2(NO_2)H_2O]^{2+}$

and $[Co(en)_2(H_2O)_2]^{3+}$ react with nitrite ion with retention of configuration.[145,147] Using these considerations, they were able to show that reaction 44 proceeded via a *trans* displacement process involving both

$$(+)cis\text{-}[Co(en)_2Cl_2]^+ + Ag_2CO_3 \longrightarrow (-)[Co(en)_2CO_3]^+ + 2AgCl \quad (44)$$

hydroxide and silver ions followed by the addition of bicarbonate ion to the dihydroxo complex. By trying each intermediate in the reaction sequence separately, the predominant inversion was found to occur with the release of the first chloride ion which gave 34% inversion, 14% retention, and 52% isomerization to the *trans* product. The inverted chloro hydroxo complex proceeded via the dihydroxo to the carbonato complex with retention of configuration. The intermediate hydroxo complexes were converted to the corresponding aquo complexes by the addition of acid and trapped and identified as the nitrite complexes. Since the addition of carbon dioxide to an aquo complex to form the carbonato complex also must proceed without Co—O bond cleavage, due to the principle of microscopic reversibility, this, too, was used to trap and fix stereochemistry in the Dwyer study.

Cook and Jauhal[148,149] have reported the synthesis of **1** and **2** by the

$$\begin{array}{c} P \diagdown \diagup O \\ Pt \\ P \diagup \diagdown O \end{array} SO_2 \quad \text{and} \quad \begin{array}{c} P \diagdown \diagup ONO_2 \\ Pt \\ P \diagup \diagdown ONO_2 \end{array}$$

(1) (2)

(P = triphenylphosphine)

reaction of $[Pt(P(C_6H_5)_3)_2O_2]$ with sulfur dioxide and nitrogen dioxide. One is tempted to speculate that these reactions take place without Pt—O bond cleavage, although no confirmatory evidence has yet been presented. However, the reaction of the same substrate with nitric oxide has been reported[112] to yield the dinitro (N-bonded) rather than the dinitrito (O-bonded) product.

Although all the examples chosen thus far have involved the retention of a metal–oxygen bond, reactions of this type involving other donor atoms are well known. Werner[150,151] utilized the oxidation of coordinated thiocyanate, as shown in reaction 45, as evidence for the fact that the

$$[Co(en)_2(NCS)_2]Cl + 8Cl_2 + 12H_2O \longrightarrow$$
$$[Co(en)_2(NH_3)_2]Cl_3 + 2CO_2 + 2H_2SO_4 + 14HCl \quad (45)$$

thiocyanates are bonded through the nitrogen atom. In a similar fashion, Chernyaev[152] employed the reduction of coordinated nitrite ion in platinum(II) complexes to show that the nitrite is N-bonded, as exemplified by reaction 46.

$$[Pt(en)(NH_3)NO_2]Cl + 3H_2 + HCl \longrightarrow [Pt(en)(NH_3)_2]Cl_2 + 2H_2O \qquad (46)$$

Kruck and Noack[153] were able to prepare a series of carboalkoxycarbonyl complexes by reaction 47. This suggests that the oxygen atoms of

$$[M(CO)_4L_2]^+ + RO^- \xrightarrow{KOH} [M(CO)_3L_2CO_2R] \qquad (47)$$
$$[M = Mn, Re; L_2 = 2P(C_6H_5)_3, phen; R = CH_3, C_2H_5, C_5H_{11}, CH_2C_6H_5]$$

the coordinated carbon monoxide groups, normally thought to be kinetically inert nuclei, might be susceptible to exchange in water via intermediates of the type $[(OC)_5MCOOH]$ or $[(OC)_5MC(OH)_2]^+$, produced by the attack of OH^- or OH_2 on the carbonyl carbon. This possibility has been verified for $[Re(CO)_6]^+$, where the exchange with the oxygen atoms in water has been found to be moderately fast.[154] In contrast, no detectable exchange takes place between $[Mo(CO)_6]$ and water within a 75 hr period. Muetterties[154] ascribes the difference in behavior to the activation of the ligand in the former complex by the formal positive charge, and suggests the possibility of analogous activation in other metal complexes in which there are unsaturated sites within the ligand and electron delocalization in the metal–ligand system, e.g., nitrogen atoms in cationic metal cyanides with the nitrogen atoms in ammonia solutions and oxygen atoms in metal nitrosyls with oxygen atoms in aqueous systems. Activation of ligand sites toward electrophiles may be achieved in certain anionic complexes, e.g., deuteration of anionic metal derivatives of cyclopentadiene. With $[Re(CO)_6]^+$ itself, the possibility of synthesizing the thiocarbonyl $[Re(CO)_5CS]^+$ by exchange with H_2S also arises. Thiocarbonyl complexes of the type $[Rh(P(C_6H_5)_3)_2(CS)X]$ and $[Rh(P(C_6H_5)_3)_2(CS)X_3]$, formed by the oxidative addition of X_2 ($X = Cl, Br$) to the former, have already been prepared[155] by refluxing $[Rh(P(C_6H_5)_3)_3X]$ with carbon disulfide.

The reaction of a coordinated carbon monoxide group has also been utilized by Fischer and Maasböl[156] in their synthesis of the first transition metal complex of a carbene, as shown in reaction 48.

$$[W(CO)_6] + CH_3^- \longrightarrow [(OC)_5W-\underset{\underset{O}{\|}}{C}-CH_3]^- \xrightarrow[H^+]{CH_2N_2} (OC)_5W\!=\!C\!\begin{array}{c} O-CH_3 \\ \diagdown \\ CH_3 \end{array} \qquad (48)$$

As has been pointed out by King,[157] metal complexes containing both carbon monoxide and cyanide ion, though known, have proved to be difficult to prepare for a variety of reasons, by the direct reaction of the cyanide ion with metal carbonyl derivatives. However, Wannagat and Seyffert[158] have demonstrated that coordinated carbon monoxide can be directly converted to cyanide ion by reaction with sodium bis(trimethylsilyl)amide, as shown in reaction 49. Although the mechanism of this reac-

$$[M(CO)_x] + Na[N(Si(CH_3)_3)_2] \longrightarrow Na[M(CO)_{x-1}CN] + [(CH_3)_3Si]_2O \quad (49)$$

tion is not known, it appears that the driving force for the reaction may be in part the conversion of the Si—N bond to the more stable Si—O bond. Whereas the salts they prepared using this method, $Na[Fe(CO)_4CN]$ and $Na[Ni(CO)_3CN]$, were found to be very sensitive to air oxidation, King[157] utilized the same approach to synthesize the more stable water-soluble salts $Na[M(CO)_5CN]$ (M = Cr, Mo, W). These, in turn, react without ligand cleavage with trimethyltin chloride to yield $[M(CO)_5(CNSn(CH_3)_3)]$ and can be protonated to yield $[M(CO)_5CNH]$, the first metal complexes of hydrogen isocyanide. The related silicon analog $[Mo(CO)_5(CN—Si(CH_3)_3)]$ was also prepared. These results suggest the possibility of preparing a molecular nitrogen complex using the amide as shown in reaction 50. This

$$[L_nM—NO] + Na[N(Si(CH_3)_3)_2] \longrightarrow [L_nM—N_2] + [(CH_3)_3Si]_2O \quad (50)$$

has been attempted[159] with the substrate $[Co(CO)_3NO]$, but, instead of attack on the nitrosyl nitrogen to yield $[Co(CO)_3N_2]^-$, attack occurred on the CO group yielding $[Co(CO)_2(CN)NO]^-$ as the product. Nitrite ion also reacts with coordinated CO to form CO_2 (as complexed carbonate) and coordinated NO.[160]

Lastly, the versatile coordinated CO group has been converted[161] into a cyanate ion, as shown in reaction 51. Beck and Smedal[161] suggest that the reaction occurs via nucleophilic attack of the azide on a CO ligand, loss of nitrogen, and rearrangement to the stable isocyanate.

$$[W(CO)_6] + N_3^- \longrightarrow [W(CO)_5NCO]^- + N_2 \quad (51)$$

The transformation of coordinated azide into cyanate by the attack of free carbon monoxide and expulsion of nitrogen has also been found to occur[162] as shown in reaction 52. Coordination of the azide is a necessary

$$[M(P(C_6H_5)_3)_2(N_3)_2] + 2CO \longrightarrow [M(P(C_6H_5)_3)_2(NCO)_2] + 2N_2 \quad (52)$$
$$[M = Pd(II), Pt(II)]$$

prerequisite for the reaction, since no reaction was observed when carbon monoxide was passed under high pressure into an aqueous solution of sodium azide. Encouraged by these results, Beck and Fehlhammer[162] attempted to add other reagents to the coordinated azide, and succeeded in preparing the unusual complexes **3** and **4** by the addition of carbon disulfide and trifluoroacetonitrile to [Pd(P(C₆H₅)₃)₂(N₃)₂]. Another tetrazolato complex (**5**) possibly containing a metal–carbon bond, was formed

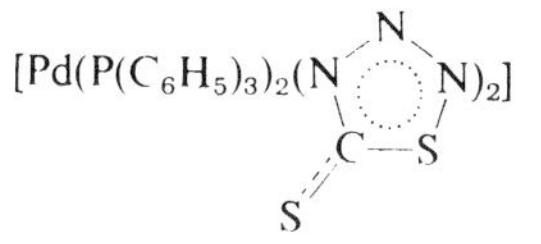

$$[Pd(P(C_6H_5)_3)_2(N\overset{N}{\underset{C-S}{\diagdown\diagup}}N)_2] \qquad [Pd(P(C_6H_5)_3)_2(N\overset{N}{\underset{C-N}{\diagdown\diagup}}N)_2]$$

$$\text{S} \qquad\qquad\qquad\qquad F_3C$$

(3) 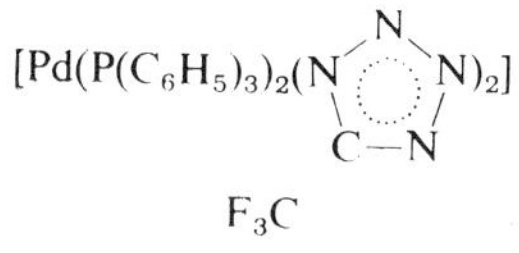 (4)

$$[As(C_6H_5)_4]\,[Au(C\overset{N}{\underset{N\ \ N}{\diagdown\diagup}}N)_4]$$

$$H_{11}C_6$$

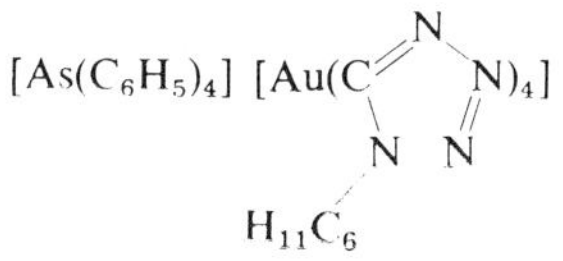

(5)

on treatment of [As(C₆H₅)₄][Au(N₃)₄] with cyclohexyl isocyanide.

C. The *trans* Effect

The phenomenon of the *trans* effect, with regard to both theory and practice, has been the subject of several extensive reviews.[36,163,164] Since an exhaustive coverage is beyond the scope of this chapter, our comments regarding it will be limited to a few illustrative examples.

The *trans* effect is perhaps best defined in terms of the effect of a coordinated group upon the rate of substitution reactions of ligands opposite to it. Metal complexes in which the rate influence of opposite, or *trans* groups, is definitely greater than the influence of adjacent, or *cis* groups, are considered to show a *trans* effect.[164] Thus, for complex **6**, the ligand L

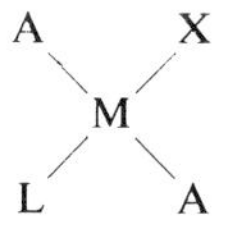

$$\begin{array}{c} A \qquad X \\ \diagdown\,\diagup \\ M \\ \diagup\,\diagdown \\ L \qquad A \end{array}$$

(6)

will have a certain influence on the rate of replacement of X by another group Y. By measuring rates, a series of ligands can be put into an order of decreasing *trans* effects. Such an order would not necessarily be invariant, but might depend on the nature of the metal complex and on the reagent Y.

The great majority of the work on the *trans* effect has been concerned with the square planar complexes of platinum(II). Since the observed reaction products are often determined by kinetic rather than thermodynamic factors, the *trans* effect is of considerable synthetic importance in these systems.

The observed[164] qualitative *trans* effect order for a series of ligands in platinum(II) complexes is as follows: CN^-, CO, C_2H_4, $NO > SC(NH_2)_2$, PR_3, $SR_2 > NO_2^- > I^-$, $SCN^- > Br^- > Cl^- > NH_3$, py, $RNH_2 > OH^- > H_2O$. Utilizing this series, it has been possible to prepare many different isomeric platinum(II) complexes. A simple example[165] of this approach is found in the synthesis of *cis*- and *trans*-$[Pt(NH_3)(NO_2)Cl_2]^-$, shown in reactions 53 and 54. The results obtained are those expected on

$$\begin{array}{c}\text{Cl}\diagdown\diagup\text{Cl}^{2-}\\ \text{Pt}\\ \text{Cl}\diagup\diagdown\text{Cl}\end{array} \xrightarrow{NH_3} \begin{array}{c}\text{Cl}\diagdown\diagup\text{NH}_3^-\\ \text{Pt}\\ \text{Cl}\diagup\diagdown\text{Cl}\end{array} \xrightarrow{NO_2^-} \begin{array}{c}\text{Cl}\diagdown\diagup\text{NH}_3^-\\ \text{Pt}\\ \text{Cl}\diagup\diagdown\text{NO}_2\end{array} \quad (53)$$

$$\begin{array}{c}\text{Cl}\diagdown\diagup\text{Cl}^{2-}\\ \text{Pt}\\ \text{Cl}\diagup\diagdown\text{Cl}\end{array} \xrightarrow{NO_2^-} \begin{array}{c}\text{Cl}\diagdown\diagup\text{NO}_2^{2-}\\ \text{Pt}\\ \text{Cl}\diagup\diagdown\text{Cl}\end{array} \xrightarrow{NH_3} \begin{array}{c}\text{Cl}\diagdown\diagup\text{NO}_2^-\\ \text{Pt}\\ \text{H}_3\text{N}\diagup\diagdown\text{Cl}\end{array} \quad (54)$$

the basis of the *trans* effect order being $NO_2^- > Cl^- > NH_3$. In Kurnakov's test,[166] use is made of the greater *trans* effect of thiourea to assign structures to the geometric isomers of $[PtA_2X_2]$ complexes, as shown in reactions 55 and 56. Because of the large *trans* effect of thiourea, the ammonia molecules are also readily replaced in reaction 55 but not in reaction 56.

$$\begin{array}{c}\text{Cl}\diagdown\diagup\text{NH}_3\\ \text{Pt}\\ \text{Cl}\diagup\diagdown\text{NH}_3\end{array} \xrightarrow{tu} \begin{array}{c}\text{tu}\diagdown\diagup\text{NH}_3^{2+}\\ \text{Pt}\\ \text{tu}\diagup\diagdown\text{NH}_3\end{array} \xrightarrow{tu} \begin{array}{c}\text{tu}\diagdown\diagup\text{tu}^{2+}\\ \text{Pt}\\ \text{tu}\diagup\diagdown\text{tu}\end{array} \quad (55)$$

$$\begin{array}{c}\text{Cl}\diagdown\diagup\text{NH}_3\\ \text{Pt}\\ \text{H}_3\text{N}\diagup\diagdown\text{Cl}\end{array} \xrightarrow{tu} \begin{array}{c}\text{Cl}\diagdown\diagup\text{NH}_3^+\\ \text{Pt}\\ \text{H}_3\text{N}\diagup\diagdown\text{tu}\end{array} \xrightarrow{tu} \begin{array}{c}\text{tu}\diagdown\diagup\text{NH}_3^{2+}\\ \text{Pt}\\ \text{H}_3\text{N}\diagup\diagdown\text{tu}\end{array} \quad (56)$$

Unlike the well-known *cis–trans* isomerism observed for complexes of the type $[PtA_2B_2]$, square planar complexes having four different ligands of the type [PtABCD] are rather uncommon, and still less frequent are examples wherein all three possible geometric isomers have been isolated

SYNTHESIS OF COORDINATION COMPOUNDS

for a given complex. The *trans* effect has proved to be invaluable in the synthesis of such isomers, the most recent example[167] of which is shown in

$$\begin{array}{c}\text{py}\diagdown\diagup\text{Br}\\ \text{Pt}\\ \text{Br}\diagup\diagdown\text{NH}_3\end{array} \xrightarrow{\text{NO}_2^-} \begin{array}{c}\text{py}\diagdown\diagup\text{Br}\\ \text{Pt}\\ \text{O}_2\text{N}\diagup\diagdown\text{NH}_3\end{array} \tag{57}$$

$$\begin{array}{c}\text{Cl}\diagdown\diagup\text{Cl}^-\\ \text{Pt}\\ \text{H}_3\text{N}\diagup\diagdown\text{Cl}\end{array} \xrightarrow{\text{Br}^-} \begin{array}{c}\text{Br}\diagdown\diagup\text{Br}^-\\ \text{Pt}\\ \text{H}_3\text{N}\diagup\diagdown\text{Br}\end{array} \xrightarrow{\text{NO}_2^-} \begin{array}{c}\text{Br}\diagdown\diagup\text{Br}^-\\ \text{Pt}\\ \text{H}_3\text{N}\diagup\diagdown\text{NO}_2\end{array} \xrightarrow{\text{py}} \begin{array}{c}\text{py}\diagdown\diagup\text{Br}\\ \text{Pt}\\ \text{H}_3\text{N}\diagup\diagdown\text{NO}_2\end{array} \tag{58}$$

$$\begin{array}{c}\text{py}\diagdown\diagup\text{Cl}^-\\ \text{Pt}\\ \text{Cl}\diagup\diagdown\text{Cl}\end{array} \xrightarrow{\text{Br}^-} \begin{array}{c}\text{py}\diagdown\diagup\text{Br}^-\\ \text{Pt}\\ \text{Br}\diagup\diagdown\text{Br}\end{array} \xrightarrow{\text{NO}_2^-} \begin{array}{c}\text{py}\diagdown\diagup\text{NO}_2^-\\ \text{Pt}\\ \text{Br}\diagup\diagdown\text{Br}\end{array} \xrightarrow{\text{NH}_3} \begin{array}{c}\text{py}\diagdown\diagup\text{NO}_2\\ \text{Pt}\\ \text{H}_3\text{N}\diagup\diagdown\text{Br}\end{array} \tag{59}$$

reactions 57–59. The products of reactions 58 and 59 were then used to synthesize two more geometric isomers of the complex

$$[\text{Pt(py)(NH}_3)(\text{NO}_2)(\text{Cl})(\text{Br})\text{I}]$$

by means of reactions 60 and 61. The two new isomers represent the sixth and seventh thus far prepared of the fifteen which are theoretically possible.

$$\begin{array}{c}\text{py}\diagdown\diagup\text{Br}\\ \text{Pt}\\ \text{H}_3\text{N}\diagup\diagdown\text{NO}_2\end{array} \xrightarrow{\text{Cl}_2} \begin{array}{c}\text{Cl}\\ \text{py}\diagdown|\diagup\text{Br}\\ \text{Pt}\\ \text{H}_3\text{N}\diagup|\diagdown\text{NO}_2\\ \text{Cl}\end{array} \xrightarrow{\text{I}^-} \begin{array}{c}\text{Cl}\\ \text{py}\diagdown|\diagup\text{Br}\\ \text{Pt}\\ \text{H}_3\text{N}\diagup|\diagdown\text{NO}_2\\ \text{I}\end{array} \tag{60}$$

$$\begin{array}{c}\text{py}\diagdown\diagup\text{NO}_2\\ \text{Pt}\\ \text{H}_3\text{N}\diagup\diagdown\text{Br}\end{array} \xrightarrow{\text{Cl}_2} \begin{array}{c}\text{Cl}\\ \text{py}\diagdown|\diagup\text{NO}_2\\ \text{Pt}\\ \text{H}_3\text{N}\diagup|\diagdown\text{Br}\\ \text{Cl}\end{array} \xrightarrow{\text{I}^-} \begin{array}{c}\text{Cl}\\ \text{py}\diagdown|\diagup\text{NO}_2\\ \text{Pt}\\ \text{H}_3\text{N}\diagup|\diagdown\text{Br}\\ \text{I}\end{array} \tag{61}$$

Unlike the reactions of platinum(II) complexes, reactions of six-coordinated cobalt(III) complexes often result in extensive rearrangement.

A striking exception to this generalization is found in the synthesis[168] of *cis*-dichloro-*trans*-diammine(ethylenediamine)cobalt(III) chloride, shown in reaction 62. The stereospecificity evidently arises because of the very

$$\begin{array}{c}\text{NH}_3\\\text{H}_3\text{N}\overset{|}{\underset{|}{\underset{\text{NH}_3}{\text{Co}}}}\text{SO}_3\\\text{H}_3\text{N}\phantom{\overset{|}{\text{Co}}}\text{SO}_3\end{array}\xrightarrow{\text{en}}\left(\begin{array}{c}\text{NH}_3\\\text{N}\overset{|}{\underset{|}{\underset{\text{NH}_3}{\text{Co}}}}\text{SO}_3\\\text{N}\phantom{\overset{|}{\text{Co}}}\text{SO}_3\end{array}\right)^{-}\xrightarrow[\text{alcohol}]{\text{HCl}}\left(\begin{array}{c}\text{NH}_3\\\text{N}\overset{|}{\underset{|}{\underset{\text{NH}_3}{\text{Co}}}}\text{Cl}\\\text{N}\phantom{\overset{|}{\text{Co}}}\text{Cl}\end{array}\right)^{+} \quad (62)$$

large *trans* effect of the SO_3^{2-} groups, which labelizes the ammonia molecules. This observation, coupled with the results of the work of Halpern et al. on the $S_N1(\text{lim})$ reactions[138] of $[Co(NH_3)_5SO_3]^+$ has enabled Buckingham et al. to synthesize[169] *trans*-$[Co(NH_3)_4(^{15}NH_3)Cl]^{2+}$ via reaction 63,

$$\begin{array}{c}\text{SO}_3\\\text{H}_3\text{N}\overset{|}{\underset{|}{\underset{\text{NH}_3}{\text{Co}}}}\text{NH}_3\\\text{H}_3\text{N}\phantom{\text{Co}}\text{NH}_3\end{array}^+\xrightarrow{-\text{NH}_3}\left[\begin{array}{c}\text{SO}_3\\\text{H}_3\text{N}\overset{|}{\underset{|}{\text{Co}}}\text{NH}_3\\\text{H}_3\text{N}\phantom{\text{Co}}\text{NH}_3\end{array}\right]^+\xrightarrow{^{15}\text{NH}_3}$$

$$\begin{array}{c}\text{SO}_3\\\text{H}_3\text{N}\overset{|}{\underset{|}{\underset{^{15}\text{NH}_3}{\text{Co}}}}\text{NH}_3\\\text{H}_3\text{N}\phantom{\text{Co}}\text{NH}_3\end{array}^+\xrightarrow{\text{HCl}}\begin{array}{c}\text{Cl}\\\text{H}_3\text{N}\overset{|}{\underset{|}{\underset{^{15}\text{NH}_3}{\text{Co}}}}\text{NH}_3\\\text{H}_3\text{N}\phantom{\text{Co}}\text{NH}_3\end{array}^{2+} \quad (63)$$

which must take place without rearrangement. The *trans* structure of the reaction product was readily established by its proton NMR spectrum.

The large *trans* effect of carbon monoxide (relative to that of an incoming ligand, L) has also been used to synthetic advantage in octahedral systems. Thus, the reaction of $[Mn(CO)_5Br]$ with a variety of neutral[170] and negatively charged[171,172] ligands yields products of the form **7** and **8**.

$$\begin{array}{c}\text{Br}\\\text{OC}\overset{|}{\underset{|}{\underset{\text{CO}}{\text{Mn}}}}\text{L}\\\text{OC}\phantom{\text{Mn}}\text{CO}\end{array}\qquad\begin{array}{c}\text{Br}\\\text{OC}\overset{|}{\underset{|}{\underset{\text{CO}}{\text{Mn}}}}\text{L}\\\text{OC}\phantom{\text{Mn}}\text{L}\end{array}$$

(7) (8)

Similarly, [Cr(CO)$_4$bipy] and [Mn(CO)$_4$hfac] (hfac = hexafluoroacetylacetonate) yield,[173,174] upon reaction with L, complexes having the structure in **9**.

$$\begin{array}{c} L \\ OC \diagdown | \diagup \\ M \\ OC \diagup | \diagdown \\ C \\ O \end{array}$$

(9)

D. Reactions of Metal Carbonyl and Nitrosyl Complexes

The synthesis of metal carbonyls has already been discussed in this series[175] and the exposition which follows consists of several examples which are, in addition to those already presented, illustrative of the application of the central theme of this chapter to metal carbonyls and nitrosyls, rather than a comprehensive survey.

Metals in such complexes are in low oxidation states and therefore are classified as class b[176] or soft,[177] responding to nucleophiles in the order N < P > As > Sb. The complex [Co(CO)$_3$NO] behaves[178] in this manner. This system reacts at about the same rate in toluene as in tetrahydrofuran and exhibits no reaction with halide ions.

However, Morris has found[159] that the similarly constituted iron derivative [Fe(NO)$_2$(CO)$_2$] reacts via a first-order process with ^{14}CO and triphenylarsine (both very poor reagents) in tetrahydrofuran at rates which are 10^3 times faster than in toluene. It also reacts with halide ions, but in the order Cl$^-$ > Br$^-$ > I$^-$. It therefore appears that the hard oxygen atom of tetrahydrofuran and the hard chloride ion do not attack the soft iron(-II), but rather the carbon atom of the carbonyl group. This suggests the use of a hard base (solvent or reagent) to catalyze the reaction of a poor reagent. In point of fact, Morris found this to suffice very nicely as a method for the preparation of [Fe(NO)$_2$(CO)As(C$_6$H$_5$)$_3$]. The proposed reaction scheme is shown in reaction 64. The reason that this effect is operative for [Fe(NO)$_2$(CO)$_2$], but not for [Co(CO)$_3$NO], may lie in the fact that the two NO groups in the former complex can better delocalize charge in resonance structure **9** than can the lone NO in the cobalt system.

The rate-determining step in the reaction of noncarbonyl complexes with chelating ligands, e.g., [M(H$_2$O)$_6$]$^{n+}$ + en, involves the removal of

$(ON)_2(OC)Fe=C=O \xrightarrow{B:}$

$$(ON)_2(OC)\bar{F}e-C\begin{matrix}O\\ \\B^+\end{matrix} \longleftrightarrow (ON)_2(OC)Fe=C\begin{matrix}O^-\\ \\B^+\end{matrix} \quad (64)$$

$\downarrow Y[As(C_6H_5)_3]$

$[Fe(NO)_2(CO)As(C_6H_5)_3]$

the first water molecule, followed by rapid chelation[179] preventing the formation of bridged species. By way of contrast, the rate of reaction 65 was found[180] to be greater than that of reaction 66, presumably due to the

$[Fe(NO)_2(CO)_2] + \text{diphos} \xrightarrow{\text{fast}} [Fe(NO)_2(CO)\text{diphos}] + CO \quad (65)$
monodentate complex

$[Fe(NO)_2(CO)\text{diphos}] \xrightarrow{\text{slow}} [Fe(NO)_2\text{diphos}] + CO \quad (66)$
chelate complex

diphos = 1,2-bis(diphenylphosphino)ethane

increased π bonding possible between the metal and the remaining carbon monoxide group in the monodentate complex. The slow rate of conversion of the monodentate complex to the chelate complex suggests that, if given the opportunity, the former will react preferentially with more of the dicarbonyl complex to form a bridged complex. This suggestion was realized[180] in the synthesis of the bridged complexes $[Fe(NO)_2CO]_2(\text{diphos})$, $[Co(NO)(CO)_2]_2(\text{diphos})$, $[CH_3COMn(CO)_4]_2(\text{diphos})$, $[Ni(CO)_3]_2(\text{diphos})$, and $[Co(NO)(CO)_2(\text{diphos})Fe(NO)_2(CO)]$. Note that the last compound is an example of a heterometal bridged complex.

Efforts to prepare $[Mn(CO)_4P(C_6H_5)_3]$ by the reaction of triphenylphosphine with $[Mn_2(CO)_{10}]$ have been thwarted by the stability of the Mn—Mn bond. The sequence of reactions 67–69 has been observed[181]

$(OC)_5Mn-Mn(CO)_5 + P(C_6H_5)_3 \longrightarrow (OC)_5Mn-Mn(CO)_4P(C_6H_5)_3 + CO \quad (67)$

$(OC)_5Mn-Mn(CO)_4P(C_6H_5)_3 + P(C_6H_5)_3 \longrightarrow$
$(C_6H_5)_3P(OC)_4Mn-Mn(CO)_4P(C_6H_5)_3 + CO \quad (68)$

$Mn_2(CO)_8(P(C_6H_5)_3)_2 \xrightarrow{\text{heat}} Mn_2(CO)_9P(C_6H_5)_3 + \text{paramagnetic residue} \quad (69)$

instead. It was also observed that $Mn_2(CO_8[P(C_6H_5)_3]_2$ reacts with CO to yield $Mn_2(CO)_9P(C_6H_5)_3$ by a first-order process. Therefore, it appeared that NO might be expected to undergo a similar reaction to yield initially $Mn_2(CO)_8(NO)P(C_6H_5)_3$. However, since NO is a three-electron donor, this system would contain one electron in excess of the "magic" number of 18 for each manganese and might be unstable with respect to Mn—Mn bond cleavage. With this approach in mind, it was possible[182] to prepare $Mn(CO)_4NO$ in good yield by reaction 70. Presumably, the sequence

$$Mn_2(CO)_8[P(C_6H_5)_3]_2 + 2NO \longrightarrow$$
$$Mn(CO)_4NO + Mn(CO)_3(NO)P(C_6H_5)_3 + CO + P(C_6H_5)_3 \quad (70)$$

of reactions involved are those represented by reactions 71–74. The

$$Mn_2(CO)_8[P(C_6H_5)_3]_2 \rightleftharpoons Mn_2(CO)_8P(C_6H_5)_3 + P(C_6H_5)_3 \quad (71)$$

$$Mn_2(CO)_8P(C_6H_5)_3 + NO \rightleftharpoons Mn_2(CO)_8(NO)P(C_6H_5)_3 \quad (72)$$

$$Mn_2(CO)_8(NO)P(C_6H_5)_3 \longrightarrow Mn(CO)_4NO + Mn(CO)_4P(C_6H_5)_3 \quad (73)$$

$$Mn(CO)_4P(C_6H_5)_3 + NO \longrightarrow Mn(CO)_3(NO)P(C_6H_5)_3 + CO \quad (74)$$

starting material used in this reaction gives better results than $Mn_2(CO)_{10}$ or some of its other derivatives because of its ease of dissociation [reaction 71], which permits the addition of NO to initiate the overall reaction.

E. Reactions of Square Planar Complexes

Square planar geometry is most often found for complexes of metal ions having a low-spin d^8 electronic configuration, e.g., palladium(II), platinum(II), gold(III), rhodium(I), iridium(I), and some complexes of nickel(II). Of these, the substitution reactions of platinum(II) have received the most extensive study.[36,163,164] These reactions generally follow a two-term rate law. Thus, for a reaction such as 75, the rate law is given by

$$[MA_3X]^+ + Y^- \longrightarrow [MA_3Y]^+ + X^- \quad (75)$$

reaction 76. The first term is also bimolecular in character, since it reflects a

$$\text{Rate} = k_1[MA_3X^+] + k_2[MA_3X^+][Y^-] \quad (76)$$

solvent displacement path for the substitution. That such reactions proceed via an S_N2 mechanism is to be expected because stable five-coordinated complexes are known.[183]

The low-spin d^8 complex $[Pd(Et_4dien)Cl]^+$ provides[184] a particularly striking exception to the generalizations derived regarding the mechanism of square planar substitutions. The intrusion of the ethyl groups into the space above and below the square plane causes the complex to look like and react like an octahedral complex. The reaction of its nonsubstituted analog, $[Pd(dien)Cl]^+$, with hydroxide ion or bromide ion is too fast to measure at room temperature by the stopped flow method, indicating that it has a $t_{1/2} < 10^{-3}$ sec. However, $[Pd(Et_4dien)Cl]^+$ reacts with various reagents at 25° with $t_{1/2} = 6$ min. Of the reagents examined, except for the hydroxide and thiosulfate[185] ions, the rate of reaction 77 is independent of

$$[Pd(Et_4dien)Cl]^+ + L^- \longrightarrow [Pd(Et_4dien)L]^+ + Cl^- \qquad (77)$$

both the nature and concentration of L^-. The unique behavior of hydroxide ion is thought to be due to a rapid acid–base preequilibrium, forming the more reactive amido species $[Pd(Et_4dien-H)Cl]$, an S_N1CB mechanism. Evidence in support of this mechanism is found in the observation that hydroxide ion has no effect on the rate of reaction of $[Pd(MeEt_4dien)Cl]^+$, which contains no N—H hydrogen. In contrast to this, the thiosulfate ion continues to react at a rate which is dependent upon its concentration. Apparently, this is due to its double negative charge and its strong nucleophilic tendency. Thus, substitution processes in such sterically hindered systems seem to be intermediate between the processes for octahedral and square planar systems.

The sterically crowded environment occasioned by the presence of the Et_4dien ligand, coupled with the low temperature isolation of kinetic products, has permitted the recent syntheses of two linkage isomeric $[Pd(Et_4dien)X]^+$ pairs $[X = -SCN, -NCS^{21,22}; -SeCN, -NCSe^{29}]$. The rationale employed was based on the fact that M—S linkages

$$\diagdown C \diagdown N$$

have, in general, been found to be nonlinear, whereas M—NCS linkages are either linear or exhibit very large bond angles.[186] It was assumed, by analogy, that the same difference is exhibited by Se- versus N-coordinated selenocyanate. In each case, the nonlinear M—X linkage would

$$\diagdown C \diagdown N$$

be expected to be sterically unfavorable in a $[Pd(Et_4dien)XCN]^+$ complex and, hence, would rearrange to the N-bonded isomer, as shown in reaction 78. It should be noted that the sterically promoted existence of the stable N-bonded isomers is in opposition to the normal M(class a, or hard)—NCX, M(class b, or soft)—XCN bonding pattern exhibited by these ligands.[176,177,187]

The first preparation[10,11] of linkage isomers of the thiocyanate ion utilized the same kinetic considerations (rapid formation and isolation of the unstable isomer), but was predicated on the basis of electronic, rather than steric, considerations, although, in retrospect, the latter may also be

$$\begin{array}{c}\text{NCX}\diagdown\phantom{\text{Pd}}\diagup\text{XCN}^{2-}\\ \text{Pd}\\ \text{NCX}\diagup\phantom{\text{Pd}}\diagdown\text{XCN}\end{array} + \text{Et}_4\text{dien} \xrightarrow[\text{temperature}]{\text{low}} \begin{array}{c}\text{H}\\ \diagdown\\ \text{N}\phantom{\text{Pd}}\text{N(C}_2\text{H}_5)_2{}^+\\ \big(\phantom{\text{x}}\text{Pd}\phantom{\text{xx}}\big)\\ (\text{H}_5\text{C}_2)_2\text{N}\phantom{\text{Pd}}\text{XCN}\end{array}\quad\text{(isolate)}$$

$$\downarrow \text{(isomerize in solution)} \quad (78)$$

$$\begin{array}{c}\text{H}\\ \diagdown\\ \text{N}\phantom{\text{Pd}}\text{N(C}_2\text{H}_5)_2{}^+\\ \big(\phantom{\text{x}}\text{Pd}\phantom{\text{xx}}\big)\\ (\text{H}_5\text{C}_2)_2\text{N}\phantom{\text{Pd}}\text{N}\\ \diagdown\\ \text{C}\\ \diagdown\\ \text{X}\end{array}\quad\text{(isolate)}$$

involved. Earlier, Turco and Pecile had shown[188] that the nature of the metal–thiocyanate bond is sensitive to the presence of certain other ligands in the coordination sphere of palladium(II) and platinum(II) complexes, e.g., $[M(SCN)_4]^{2-}$ versus $[M(NH_3)_2(SCN)_2]$ versus $[M(PR_3)_2(NCS)_2]$. They ascribed the reversal in bonding in the last case to the π-electron withdrawal by the d orbitals of the phosphorus atoms, which decreases the d_π–d_π stabilization of the M—S bond to the point that the more ionic M—N bond is favored. With the exception of the Et$_4$dien case just discussed, all of the ligands which were subsequently found[189] to cause the same reversal in palladium(II) and platinum(II) systems have at least a moderate π-acceptor capacity. This suggested that there should be some borderline ligands for which the energy difference between the M—SCN and M—NCS isomers is small, permitting the isolation of both. The observation[11] that the triphenylphosphine derivatives are N-bonded, whereas those of triphenylstibine are S-bonded suggested the use of triphenylarsine. The requirement that the kinetic product be isolated prior to its rearrangement dictated the choice of palladium(II) rather than platinum(II), because the reactions of the former are at least 10^4 times faster for analogously constituted complexes. Secondly, a system wherein the reaction product would be insoluble and immediately separate from the reaction mixture was desirable. Both of these requirements were met by the [PdL$_2$X$_2$] system, and the linkage isomers were subsequently prepared according to reaction 79. Much the same results were obtained using the bidentate ligand 2,2'-bipyridine, although a lower temperature ($-78°$) was required to permit isolation of the unstable S-bonded isomer. The consequences of varying the π-acceptor properties of the L ligand are strikingly

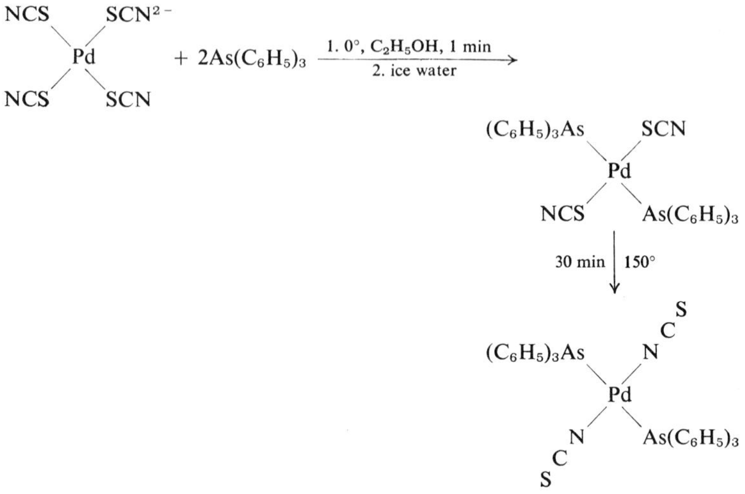

$$\begin{array}{c}\text{NCS}\\\text{NCS}\end{array}\!\!\diagup\!\!\text{Pd}\!\!\diagdown\!\!\begin{array}{c}\text{SCN}\\\text{SCN}\end{array}^{2-} + 2\text{As}(C_6H_5)_3 \xrightarrow[\text{2. ice water}]{1.\ 0°,\ C_2H_5OH,\ 1\ \text{min}}$$

$$(C_6H_5)_3As\diagdown\!\!\text{Pd}\!\!\diagup\!\!\begin{array}{c}SCN\\As(C_6H_5)_3\end{array}$$
$$NCS\diagup$$

$\Big\downarrow$ 30 min | 150°   (79)

$$(C_6H_5)_3As\diagdown\!\!\text{Pd}\!\!\diagup\!\!\begin{array}{c}S\\C\\N\\As(C_6H_5)_3\end{array}$$
$$\begin{array}{c}N\\C\\S\end{array}\diagup$$

evident in the complexes shown here, all of which have been found to be stable with respect to isomerization in the solid state:

$[Pd\{(C_6H_5)_2As(o-C_6H_4)P(C_6H_5)_2\}(SCN)(NCS)],^{32}$
$[Pd(As(n-C_4H_9)_3)_2(SCN)_2],^{15}$
$[Pd(Me_2bipy)(SCN)(NCS)],^{23}$
$[Pd(py)_2(SCN)_2],^{15}$
$[Pd(5-NO_2-1,10\text{-phenanthroline})(NCS)_2],^{15}$
$[Pd(1,10\text{-phenanthroline})(SCN)_2].^{11}$

It is reasonable to assume that similar considerations should be operative in other low-spin d^8 systems, and the $[RhL_2(CO)NCS]$, $[RhL_3(NCS)]$ (L = phosphine, arsine, stibine, or phosphite), and $[Rh(CO)_2(NCS)_2]^-$ complexes recently reported by Jennings and Wojcicki[190] would appear to verify this possibility.

It should be pointed out, however, that octahedral systems have not been found to behave in this manner. Thus, the stable isomer of $[Mn(CO)_5SCN]$ is S-bonded,[14] and the substitution of *weaker* π-bonding ligands for two of the CO groups, as in cis-$[Mn(CO)_3(bipy)NCS]$,[191] generally yields N-bonded complexes, the two exceptions being explained on the basis of steric interactions. Similarly, an increase in negative charge on the metal, as in $[Cr(CO)_5NCS]^-$ [192] results in a switch to N-bonding, despite the fact that the soft character of the metal has increased. Obviously, the results in the square planar and octahedral systems need to be reconciled.

A second mechanistic consideration also proved to be feasible in the synthesis of the isomers of $[Pd(bipy)(NCS)_2]$. By analogy to organic substitution reactions, the nucleophilicity of the sulfur end of the thiocyanate would be expected to be greater than that of the nitrogen end, e.g., for benzyl thiocyanate (S-bonded), the rate of exchange[193] exceeds the rate of isomerization by a factor of 10^3. Consequently, the initial product in reaction 80 was expected and found[194] to be S-bonded.

$$[Pd(bipy)(OH_2)_2]^{2+} + 2SCN^- \xrightarrow[C_2H_5OH/H_2O]{-78°} [Pd(bipy)(SCN)_2] \xrightarrow[30 \text{ min}]{150°} [Pd(bipy)(NCS)_2] \quad (80)$$

The rate of reaction of some platinum(II) complexes with nucleophilic reagents has been shown[195] to be subject to electrophilic catalysis. For example, the rate of reaction 81 in methanol strongly increases in the

$$\textit{trans-}[Pt(pip)_2Cl_2] + Y^- \longrightarrow \textit{trans-}[Pt(pip)_2(Y)Cl] + Cl^- \quad (81)$$
$$(pip = piperidine; Y^- = {}^{36}Cl^-, NO_2^-)$$

presence of nitrous acid or boric acid. Unlike the general rate reaction 76 observed for the uncatalyzed reaction of the nonsterically hindered complexes of platinum(II), a three-term rate law has been found to be operative [eq. 82].

$$\text{rate} = k_1[\text{complex}] + k_2[\text{complex}][Y^-] + k_3[\text{complex}][HA][Y^-] \quad (82)$$

A mechanism to explain the third term in the rate law is shown in reaction 83. The enhancement of rate is ascribed to the increase in the concentration

(83)

of the six-coordinate intermediate containing both HNO_2 and Y^- which, in turn, results from the withdrawal of π electrons from platinum by the HNO_2, making it easier for the metal to accept σ electrons from the nucleophile Y^-. This presupposes that platinum(II) complexes which already contain π-acceptor ligands in their coordination sphere should not be as susceptible to acid catalysis by HNO_2 and, indeed, such was found to be the case for $[Pt(P(C_6H_5)_3)_2Cl_2]$, $[Pt(As(C_6H_5)_3)_2Cl_2]$, $[Pt(pip)_2(NO_2)Cl]$, and $[Pt(pip)_2Br_2]$ (all *trans*). Similarly, $[Pt(Et_4dien)Cl]^+$ does not undergo acid catalysis because the ethyl groups prevent the interaction with HNO_2 along the z axis.

The fact that *trans*-$[Pt(pip)_2Cl_2]$ is subject to acid catalysis whereas *trans*-$[Pt(pip)_2(NO_2)Cl]$ is not, permitted the first synthesis of the latter complex. In the absence of HNO_2, the reaction of *trans*-$[Pt(pip)_2Cl_2]$ with NO_2^- yields *trans*-$[Pt(pip)_2(NO_2)_2]$, because of the strong *trans* effect of the nitrite ion.

The effect just described may be similar to the H^+ catalysis of carbon monoxide exchange with $[Fe(CO)_5]$.[196] This has not yet been used for synthetic purposes, and may not be operative in the case of σ donors, since they, being stronger bases than carbon monoxide, may consume the H^+. However, for π donors such as olefins, it may prove useful.

IV. MISCELLANEOUS APPLICATIONS

A. Stabilization of Metal Complexes by Large Counterions

The isolation[22,29] of the unstable $[Pd(Et_4dien)XCN]^+$ (X = S, Se) complexes, discussed in the preceding section, as, respectively, the PF_6^- and $B(C_6H_5)_4^-$ salts is a good example of a valuable empirical generalization: *metal complex ions which are difficult to isolate can often be isolated as salts of large ions having an equal, but opposite, charge*. Not only does this facilitate the isolation of complexes, but it can also stabilize certain complexes in the solid state.

This has been demonstrated by Dwyer and his students[197] in their resolutions of optically active metal complexes and, more recently, in the isolation, by Raymond and Basolo, of the elusive $[Ni(CN)_5]^{3-}$ ion.[198] This species has been shown[199] to form upon the addition of excess cyanide ion to solutions of $[Ni(CN)_4]^{2-}$. However, evaporation of the solution yields only $K_2[Ni(CN)_4] \cdot H_2O$, even from solutions saturated with potassium cyanide.[200] The use[198] of cationic complexes of chromium(III) as the counterions, as opposed to those of cobalt(III), was dictated by mechanistic considerations. The appropriate cobalt(III) complexes, e.g., $[Co(NH_3)_6]^{3+}$ and $[Co(en)_3]^{3+}$ undergo substitution in the $4M$ potassium cyanide solutions which are required for the formation of $[Ni(CN)_5]^{3-}$. This may be due to catalysis by cobalt(II) or hydroxide ion. Chromium(III) systems, however, are less sensitive to base hydrolysis and the possibility of contamination by the easily oxidized chromium(II) would be remote. Accordingly, the compounds $[Cr(NH_3)_6][Ni(CN)_5]\cdot 2H_2O$ and $[Cr(en)_3][Ni(CN)_5]\cdot 1.5H_2O$ were prepared, both being stable at room temperature. The stabilizing influence of the chromium(III) cations became readily apparent when, subsequently, the compound $K_3[Ni(CN)_5]\cdot 2H_2O$ was isolated[201] by a low-temperature equilibration technique. Although stable when stored in a stoppered container at $-15°$, it rapidly decomposes to a mixture of $K_2[Ni(CN)_4]$ and KCN at room temperature.

Additional examples of the utility of this approach, taken from the recent literature, are shown in Table IV. Each compound cited has a feature of special interest, and these are also noted in the table.

TABLE IV
Examples of Counterion Stabilization

Compound	Point of interest and ref.
$[Os(CO)_3(P(C_6H_5)_3)_2H][PF_6]$	Evidence of intermediate hydride formation in oxidative addition of HX to d^8 osmium(0) complex[104]
$[As(C_6H_5)_4][Au(C(N)(N-N)C_6H_{11})_4]$	Possible Au—C bond[162]
$[B(\gamma\text{-pic})(N(CH_3)_3)(Cl)H]PF_6$	First boron cation resolved into optically active isomers[202]
$[As(C_6H_5)_4][MF_5]$	M = Si(IV), Ge(IV); abnormal coordination numbers[203]
$[Co(tetren)Cl][ZnCl_4]$	Optically active cation containing quinquidentate amine[204]
$[Co(pn)_3][MCl_6]$	M = Cr(III), Mn(III), Fe(III); hexachloro complexes stabilized in solid state[205]
$[Cr(NH_3)_6][CuCl_5]$	Trigonal bipyramidal structure with axial Cu—Cl bond distances shorter than equatorial[206]

Abbreviations: γ-pic = 4-methylpyridine, tetren = tetraethylenepentamine, pn = 1,2-propanediamine.

B. Resolution of Optically Active Amines

In contrast to the vast number of known optically active carbon compounds, attempts to resolve tertiary amines of the type $RR'R''N$ and salts of quaternary ammonium ions of the form $RR'R''NH^+$ have consistently met with failure. This has been attributed to the rapid inversion experienced by amines and the rapid proton exchange in aqueous media (thereby making inversion possible) experienced by quaternary ammonium ions. However, if one of the R groups is a metal complex, the rate of exchange is lowered substantially.[207,208]

This suggested the possibility of resolving an optically active metal complex wherein the sole source of activity arises from a coordinated asymmetric nitrogen atom. Sargeson and co-workers have succeeded in resolving three such complexes:

$$[Co(NH_3)_4(CH_3NHCH_2COO)]^{2+} \ [209]$$
$$[Co(NH_3)_4(CH_3NH(CH_2)_2NH_2)]^{3+} \ [210]$$
$$trans,trans\text{-}[Co(CH_3NH(CH_2)_2NH_2)_2(NO_2)_2]^{+} \ [211]$$

Both the racemization and hydrogen-exchange processes follow the rate law shown in reaction 84,

$$R = k[\text{complex}][\text{OH}^- \text{ or OD}^-] \quad (84)$$

which is consistent with the proposal that the rate-determining step is the abstraction of the N—H proton by OH^- or OD^-. The large deuteration rate constant/racemization rate constant ratios observed ($k_D/k_R = 4 \times 10^3$, 10^5, 9×10^4 for the three complexes in the order cited) indicates that the coordinated amines have considerable optical stability, i.e., the intermediate preserves its original configuration most of the time before and after it is reprotonated by a solvent molecule.

References

1. W. Gibbs and F. A. Genth, *Am. J. Sci.*, **24**, 86 (1857).
2. S. M. Jørgensen, *Z. Anorg. Allgem. Chem.*, **5**, 168 (1894).
3. A. Werner, *Ber.*, **40**, 765 (1907).
4. F. Basolo, B. D. Stone, J. G. Bergmann, and R. G. Pearson, *J. Am. Chem. Soc.*, **76**, 3079 (1954).
5. R. G. Pearson, P. M. Henry, J. G. Bergmann, and F. Basolo, *J. Am. Chem. Soc.*, **76**, 5920 (1954).
6. F. Basolo and G. S. Hammaker, *J. Am. Chem. Soc.*, **82**, 1001 (1960).
7. F. Basolo and G. S. Hammaker, *Inorg. Chem.*, **1**, 1 (1962).
8. J. Halpern and S. Nakamura, *J. Am. Chem. Soc.*, **87**, 3002 (1965).
9. D. M. L. Goodgame and M. A. Hitchman, *Inorg. Chem.*, **5**, 1303 (1966).
10. F. Basolo, J. L. Burmeister, and A. J. Poë, *J. Am. Chem. Soc.*, **85**, 1700 (1963).
11. J. L. Burmeister and F. Basolo, *Inorg. Chem.*, **3**, 1587 (1964).
12. A. Tramer, in *Theory and Structure of Complex Compounds*, B. Jezowska-Trzebiatowska, Ed., Pergamon, New York, 1964, p. 225.
13. O. W. Howarth, R. E. Richards, and L. M. Venanzi, *J. Chem. Soc.*, **1964**, 3335.
14. M. F. Farona and A. Wojcicki, *Inorg. Chem.*, **4**, 857 (1965).

15. A. Sabatini and I. Bertini, *Inorg. Chem.*, **4**, 1665 (1965).
16. H.-H. Schmidtke, *Z. Physik. Chem. (Frankfurt)*, **45**, 305 (1965).
17. H.-H. Schmidtke, *J. Am. Chem. Soc.*, **87**, 2522 (1965).
18. H.-H. Schmidtke, *Inorg. Chem.*, **5**, 1682 (1966).
19. A. Haim and N. Sutin, *J. Am. Chem. Soc.*, **87**, 4210 (1965).
20. A. Haim and N. Sutin, *J. Am. Chem. Soc.*, **88**, 434 (1966).
21. F. Basolo, W. H. Baddley, and J. L. Burmeister, *Inorg. Chem.*, **3**, 1202 (1964).
22. F. Basolo, W. H. Baddley, and K. J. Weidenbaum, *J. Am. Chem. Soc.*, **88**, 1576 (1966).
23. I. Bertini and A. Sabatini, *Inorg. Chem.*, **5**, 1025 (1966).
24. G. C. Kulasingam and W. R. McWhinnie, *Chem. Ind. (London)*, **1966**, 2200.
25. T. E. Sloan and A. Wojcicki, Abstracts of Papers, 153rd American Chemical Society Meeting, Miami Beach, Florida, 1967, p. L-83.
26. J. L. Burmeister and H. J. Gysling, *Inorg. Chim. Acta*, **1**, 100 (1967).
27. D. E. Peters and R. T. M. Fraser, *J. Am. Chem. Soc.*, **87**, 2758 (1965).
28. J. H. Espenson and J. P. Birk, *J. Am. Chem. Soc.*, **87**, 3280 (1965).
29. J. L. Burmeister and H. J. Gysling, *Chem. Commun. (London)*, **1967**, 543.
30. D. R. Stranks, quoted by R. T. M. Fraser, "Linkage Isomerism," in *Werner Centennial*, Advances in Chemistry Series No. 62, American Chemical Society, Washington, D.C., 1967, pp. 296, 298.
31. D. M. L. Goodgame and M. A. Hitchman, *Inorg. Chem.*, **6**, 813 (1967).
32. P. Nicpon and D. W. Meek, *Inorg. Chem.*, **6**, 145 (1967).
33. J. Chatt and F. A. Hart, *J. Chem. Soc.*, **1961**, 1416.
34. D. F. Shriver, S. A. Shriver, and S. E. Anderson, *Inorg. Chem.*, **4**, 725 (1965).
35. J. P. Collman and J. Sun, *Inorg. Chem.*, **4**, 1273 (1965).
36. F. Basolo and R. G. Pearson, *Mechanisms of Inorganic Reactions*, 2nd ed., Wiley, New York, 1967.
37. H. Taube, "Mechanisms of Redox Reactions of Simple Chemistry," in *Advances in Inorganic Chemistry and Radiochemistry*, Vol. 1, H. J. Emeléus and A. G. Sharpe, Eds., Academic Press, New York, 1959, pp. 1–50.
38. D. R. Stranks, in *Modern Coordination Chemistry*, J. Lewis and R. G. Wilkins, Eds., Interscience, New York, 1960, pp. 78–166.
39. F. Basolo and R. G. Pearson, "Mechanisms of Substitution Reactions of Metal Complexes," in *Advances in Inorganic Chemistry and Radiochemistry*, Vol. 3, H. J. Emeléus and A. G. Sharpe, Eds., Academic Press, New York, 1961, pp. 1–85.
40. J. Halpern, *Quart. Rev. (London)*, **15**, 207 (1961).
41. R. G. Wilkins, *Quart. Rev. (London)*, **16**, 316 (1962).
42. J. O. Edwards, *Inorganic Reaction Mechanisms: An Introduction*, Benjamin, New York, 1964.
43. C. H. Langford and H. B. Gray, *Ligand Substitution Processes*, Benjamin, New York, 1966.
44. W. L. Reynolds and R. W. Lumry, *Mechanisms of Electron Transfer*, Ronald, New York, 1966.
45. H. Taube and H. Meyers, *J. Am. Chem. Soc.*, **76**, 2103 (1954).
46. H. Taube, *J. Am. Chem. Soc.*, **77**, 4481 (1955).
47. R. L. Carlin and J. O. Edwards, *J. Inorg. Nucl. Chem.*, **6**, 217 (1958).
48. W. Kruse and H. Taube, *J. Am. Chem. Soc.*, **82**, 526 (1960).
49. R. T. M. Fraser, *Inorg. Chem.*, **3**, 1561 (1964).
50. R. T. M. Fraser, *J. Chem. Soc.*, **1965**, 3641.

51. P. R. Ray and N. K. Dutt, *Z. Anorg. Allgem. Chem.*, **234**, 65 (1937).
52. A. W. Adamson, *J. Am. Chem. Soc.*, **78**, 4260 (1956).
53. J. P. Candlin, J. Halpern, and S. Nakamura, *J. Am. Chem. Soc.*, **85**, 2517 (1963).
54. J. L. Burmeister, *Inorg. Chem.*, **3**, 919 (1964).
55. J. L. Burmeister and D. Sutherland, *Chem. Commun. (London)*, **1965**, 175.
56. D. L. Ball and E. L. King, *J. Am. Chem. Soc.*, **80**, 1091 (1958).
57. A. Haim and W. K. Wilmarth, *J. Am. Chem. Soc.*, **83**, 509 (1961).
58. J. L. Burmeister and M. Y. Al-Janabi, *Inorg. Chem.*, **4**, 962 (1965).
59. A. Haim, *J. Am. Chem. Soc.*, **85**, 1016 (1963).
60. J. H. Espenson, *Inorg. Chem.*, **4**, 121 (1965).
61. R. T. M. Fraser, *Inorg. Chem.*, **2**, 954 (1963).
62. J. P. Candlin, J. Halpern, and D. L. Trimm, *J. Am. Chem. Soc.*, **86**, 1019 (1964).
63. C. Bifano and R. G. Linck, *J. Am. Chem. Soc.*, **89**, 3945 (1967), and references contained therein.
64. M. M. Chamberlain and J. C. Bailar, Jr., *J. Am. Chem. Soc.*, **81**, 6412 (1959).
65. A. A. Vlcek and F. Basolo, *Inorg. Chem.*, **5**, 156 (1966).
66. F. Basolo, M. L. Morris, and R. G. Pearson, *Discussions Faraday Soc.*, **29**, 80 (1960).
67. R. C. Johnson and F. Basolo, *J. Inorg. Nucl. Chem.*, **13**, 36 (1960).
68. W. R. Mason, III, and R. C. Johnson, *Inorg. Chem.*, **4**, 1258 (1965).
69. R. C. Johnson and E. R. Berger, *Inorg. Chem.*, **4**, 1262 (1965).
70. R. R. Rettew and R. C. Johnson, *Inorg. Chem.*, **4**, 1565 (1965).
71. W. R. Mason, III, E. R. Berger, and R. C. Johnson, *Inorg. Chem.*, **6**, 248 (1967).
72. M. Delepine, *Bull. Soc. Chim.*, **45**, 235 (1929).
73. M. Delepine, *Compt. Rend.*, **236**, 559 (1953).
74. J. V. Rund, F. Basolo, and R. G. Pearson, *Inorg. Chem.*, **3**, 658 (1964).
75. J. Chatt and B. L. Shaw, *Chem. Ind. (London)*, **1960**, 931.
76. J. Chatt and B. L. Shaw, *Chem. Ind. (London)*, **1961**, 290.
77. D. J. Baker and R. D. Gillard, *Chem. Commun. (London)*, **1967**, 520.
78. L. Vaska and J. W. DiLuzio, *J. Am. Chem. Soc.*, **83**, 2784 (1961).
79. J. Reiset, *Compt. Rend.*, **10**, 870 (1840).
80. A. Werner, *Ber.*, **40**, 4093 (1907).
81. O. Carlgren and P. T. Cleve, *Z. Anorg. Allgem. Chem.*, **1**, 65 (1892).
82. F. Basolo, J. C. Bailar, Jr., and B. R. Tarr, *J. Am. Chem. Soc.*, **72**, 2433 (1950).
83. A. Schleicher, H. Henkel, and L. Spies, *J. Prakt. Chem.*, **105**, 31 (1922).
84. J. Chatt and B. L. Shaw, *J. Chem. Soc.*, **1959**, 705.
85. J. Chatt and B. L. Shaw, *J. Chem. Soc.*, **1959**, 4020.
86. J. Chatt and B. L. Shaw, *J. Chem. Soc.*, **1962**, 5075.
87. R. J. Cross and F. Glockling, *Proc. Chem. Soc.*, **1964**, 143.
88. R. F. Heck, *J. Am. Chem. Soc.*, **86**, 2796 (1964).
89. J. A. McCleverty and G. Wilkinson, *J. Chem. Soc.*, **1964**, 4200.
90. L. Vaska, *J. Am. Chem. Soc.*, **88**, 5325 (1966).
91. L. Vaska and J. W. DiLuzio, *J. Am. Chem. Soc.*, **84**, 679 (1962).
92. L. Vaska and R. E. Rhodes, *J. Am. Chem. Soc.*, **87**, 4970 (1965).
93. L. Vaska and D. L. Catone, *J. Am. Chem. Soc.*, **88**, 5324 (1966).
94. A. J. Chalk and J. F. Harrod, *J. Am. Chem. Soc.*, **87**, 16 (1965).
95. R. S. Nyholm and K. Vrieze, *J. Chem. Soc.*, **1965**, 5337.
96. R. Cramer and G. W. Parshall, *J. Am. Chem. Soc.*, **87**, 1392 (1965).
97. J. P. Collman and W. R. Roper, *J. Am. Chem. Soc.*, **88**, 180 (1966).
98. J. P. Collman and J. W. Kang, *J. Am. Chem. Soc.*, **89**, 844 (1967).

99. W. Hieber and G. Bader, *Ber.*, **61**, 1717 (1928).
100. R. B. King, S. L. Stafford, P. M. Treichel, and F. G. A. Stone, *J. Am. Chem. Soc.*, **83**, 3604 (1961).
101. W. Hieber and J. Muschi, *Ber.*, **98**, 3931 (1965).
102. J. Chatt and H. R. Watson, *J. Chem. Soc.*, **1962**, 2545.
103. J. P. Collman and W. R. Roper, *J. Am. Chem. Soc.*, **87**, 4008 (1965).
104. J. P. Collman and W. R. Roper, *J. Am. Chem. Soc.*, **88**, 3504 (1966).
105. L. Vaska and S. S. Bath, *J. Am. Chem. Soc.*, **88**, 1333 (1966).
106. S. J. LaPlaca and J. A. Ibers, *Inorg. Chem.*, **5**, 405 (1966).
107. S. J. LaPlaca and J. A. Ibers, *J. Am. Chem. Soc.*, **87**, 2581 (1965).
108. J. A. Ibers, J. McGinnety, and N. Kime, *Proceedings of the 10th International Conference on Coordination Chemistry*, K. Yamasaki, Ed., The Chemical Society of Japan, Tokyo, 1967, p. 93.
109. J. P. Collman and J. W. Kang, *J. Am. Chem. Soc.*, **88**, 3459 (1966).
110. A. D. Allen and C. V. Senoff, *Chem. Commun. (London)*, **1965**, 621.
111. J. P. Collman, M. Kubota, J.-Y. Sun, and F. Vastine, *J. Am. Chem. Soc.*, **89**, 169 (1967).
112. J. P. Collman, M. Kubota, and J. W. Hosking, *J. Am. Chem. Soc.*, **89**, 4809 (1967).
113. E. A. Koerner von Gustorf, Abstracts of Papers, 153rd American Chemical Society Meeting, Miami Beach, Florida, 1967, p. L–43.
114. J. E. Kimlick, *J. Org. Chem.*, **30**, 2014 (1965).
115. G. Wilke, H. Schott, and P. Heimbach, *Angew. Chem. Intern. Ed.*, **6**, 92 (1967).
116. D. H. Busch, *J. Am. Chem. Soc.*, **77**, 2747 (1955).
117. A. Werner, *Ber.*, **45**, 3061 (1912).
118. F. P. Dwyer and E. C. Gyarfas, *J. Proc. Roy. Soc. N.S. Wales*, **83**, 263 (1950).
119. C. S. Lee, E. M. Gorton, H. M. Neumann, and H. R. Hunt, Jr., *Inorg. Chem.*, **5**, 1397 (1966).
120. L. Eimer and A. I. Medalia, *J. Am. Chem. Soc.*, **74**, 1592 (1952).
121. P. Ellis, R. G. Wilkins, and M. J. G. Williams, *J. Chem. Soc.*, **1957**, 4456.
122. A. W. Adamson, *Rec. Trav. Chim.*, **75**, 809 (1956).
123. G. Schwarzenbach, *Helv. Chim. Acta*, **32**, 839 (1949).
124. A. W. Adamson and E. Gonick, *Inorg. Chem.*, **2**, 129 (1963).
125. A. W. Adamson, H. Ogata, J. Grossman, and R. Newbury, *J. Inorg. Nucl. Chem.*, **6**, 319 (1958).
126. P. Saffir and H. Taube, *J. Am. Chem. Soc.*, **82**, 13 (1960).
127. R. T. M. Fraser and H. Taube, *J. Am. Chem. Soc.*, **82**, 4152 (1960).
128. A. Haim and H. Taube, *J. Am. Chem. Soc.*, **85**, 495 (1963).
129. A. Haim and H. Taube, *J. Am. Chem. Soc.*, **85**, 3108 (1963).
130. See, for example, references 36 and 43 and references contained therein.
131. F. J. Garrick, *Nature*, **139**, 507 (1937).
132. R. D. Gillard, *J. Chem. Soc. (A)*, **1967**, 917.
133. D. A. Buckingham, I. I. Olsen, and A. M. Sargeson, *J. Am. Chem. Soc.*, **88**, 5443 (1966).
134. R. G. Pearson, H.-H. Schmidtke, and F. Basolo, *J. Am. Chem. Soc.*, **82**, 4434 (1960).
135. R. Grassi, A. Haim, and W. K. Wilmarth, *Inorg. Chem.*, **6**, 237 (1967).
136. R. Barca, J. Ellis, M.-S. Tsao, and W. K. Wilmarth, *Inorg. Chem.*, **6**, 243 (1967).
137. P. H. Tewari, R. W. Gaver, H. K. Wilcox, and W. K. Wilmarth, *Inorg. Chem.*, **6**, 611 (1967).

138. J. Halpern, R. A. Palmer, and L. M. Blakley, *J. Am. Chem. Soc.*, **88**, 2877 (1966).
139. R. G. Pearson and J. W. Moore, *Inorg. Chem.*, **3**, 1334 (1964).
140. A. Haim and H. Taube, *Inorg. Chem.*, **2**, 1199 (1963).
141. R. B. Jordan, A. M. Sargeson, and H. Taube, *Inorg. Chem.*, **5**, 1091 (1966).
142. A. M. Sargeson and H. Taube, *Inorg. Chem.*, **5**, 1094 (1966).
143. D. A. Buckingham, I. I. Olsen, A. M. Sargeson, and H. Satrapa, *Inorg. Chem.*, **6**, 1027 (1967).
144. J. P. Hunt, A. C. Rutenberg, and H. Taube, *J. Am. Chem. Soc.*, **74**, 268 (1952).
145. R. K. Murmann and H. Taube, *J. Am. Chem. Soc.*, **78**, 4886 (1956).
146. F. P. Dwyer, A. M. Sargeson, and I. K. Reid, *J. Am. Chem. Soc.*, **85**, 1215 (1963).
147. R. K. Murmann, *J. Am. Chem. Soc.*, **77**, 5190 (1955).
148. C. D. Cook and G. S. Jauhal, *Inorg. Nucl. Chem. Letters*, **3**, 31 (1967).
149. C. D. Cook and G. S. Jauhal, *J. Am. Chem. Soc.*, **89**, 3066 (1967).
150. A. Werner, *Z. Anorg. Allgem. Chem.*, **22**, 91 (1900).
151. A. Werner, *Ann. Chem.*, **351**, 65 (1907).
152. I. I. Chernyaev, *Ann. Inst. Platine*, **1929**, 52; through *Chem. Abstr.*, **24**, 2684 (1930).
153. T. Kruck and M. Noack, *Chem. Ber.*, **97**, 1693 (1964).
154. E. L. Muetterties, *Inorg. Chem.*, **4**, 1841 (1965).
155. M. C. Baird and G. Wilkinson, *Chem. Commun. (London)*, **1966**, 267.
156. E. O. Fischer and A. Maasböl, *Angew. Chem. Intern. Ed.*, **3**, 580 (1964).
157. R. B. King, *Inorg. Chem.*, **6**, 25 (1967), and references therein.
158. U. Wannagat and H. Seyffert, *Angew. Chem.*, **77**, 457 (1965).
159. D. E. Morris, private communication.
160. W. Hieber and K. Beutner, *Z. Naturforsch.*, **15b**, 323 (1960).
161. W. Beck and H. S. Smedal, *Angew. Chem. Intern. Ed.*, **5**, 253 (1966).
162. W. Beck and W. P. Fehlhammer, *Angew. Chem. Intern. Ed.*, **6**, 169 (1967).
163. J. V. Quagliano and L. Schubert, *Chem. Rev.*, **40**, 201 (1952).
164. F. Basolo and R. G. Pearson, "The *Trans* Effect in Metal Complexes," in *Progress in Inorganic Chemistry*, Vol. 4, F. A. Cotton, Ed., Interscience, New York, 1962, pp. 381–454.
165. I. I. Chernyaev, *Ann. Inst. Platine*, **5**, 102, 118 (1927).
166. N. S. Kurnakov, *J. Prakt. Chem.*, **50**, 483 (1894).
167. A. D. Gellman and L. N. Essen, *Proceedings of the 10th International Conference on Coordination Chemistry*, K. Yamasaki, Ed., The Chemical Society of Japan, Tokyo, 1967, p. 385.
168. J. C. Bailar, Jr., and D. F. Peppard, *J. Am. Chem. Soc.*, **62**, 105 (1940).
169. D. A. Buckingham, I. I. Olsen, and A. M. Sargeson, *J. Am. Chem. Soc.*, **89**, 5129 (1967).
170. R. J. Angelici, F. Basolo, and A. J. Poë, *J. Am. Chem. Soc.*, **85**, 2215 (1963), and references contained therein.
171. E. W. Abel and I. S. Butler, *J. Chem. Soc.*, **1964**, 434.
172. R. J. Angelici, *Inorg. Chem.*, **3**, 1099 (1964).
173. R. J. Angelici and J. R. Graham, *J. Am. Chem. Soc.*, **87**, 5586 (1965).
174. F. A. Hartman, M. Kilner, and A. Wojcicki, *Inorg. Chem.*, **6**, 34 (1967).
175. J. C. Hileman, "Metal Carbonyls," in *Preparative Inorganic Reactions*, Vol. 1, W. L. Jolly, Ed., Interscience, New York, 1964, pp. 77–120.
176. S. Ahrland, J. Chatt, and N. R. Davies, *Quart. Rev. (London)*, **12**, 265 (1958).
177. R. G. Pearson, *J. Am. Chem. Soc.*, **85**, 3533 (1963).
178. E. M. Thorsteinson and F. Basolo, *J. Am. Chem. Soc.*, **88**, 3929 (1966).

179. R. H. Holyer, C. D. Hubbard, S. F. A. Kettle, and R. G. Wilkins, *Inorg. Chem.*, **5**, 622 (1966).
180. R. J. Mawby, D. Morris, E. M. Thorsteinson, and F. Basolo, *Inorg. Chem.*, **5**, 27 (1966).
181. H. Wawersik and F. Basolo, *Chem. Commun. (London)*, **1966**, 366.
182. H. Wawersik and F. Basolo, *Inorg. Chem.*, **6**, 1066 (1967).
183. E. L. Muetterties and R. A. Schunn, *Quart. Rev. (London)*, **20**, 245 (1966).
184. W. H. Baddley and F. Basolo, *J. Am. Chem. Soc.*, **86**, 2075 (1964).
185. J. B. Goddard, private communication.
186. J. Lewis, R. S. Nyholm, and P. W. Smith, *J. Chem. Soc.*, **1961**, 4590.
187. J. L. Burmeister, *Coordin. Chem. Rev.*, **1**, 205 (1966), and references contained therein.
188. A. Turco and C. Pecile, *Nature*, **191**, 66 (1961).
189. See reference 26, and references contained therein.
190. M. A. Jennings and A. Wojcicki, *Inorg. Chem.*, **6**, 1854 (1967).
191. M. F. Farona and A. Wojcicki, *Inorg. Chem.*, **4**, 1402 (1965).
192. A. Wojcicki and M. F. Farona, *J. Inorg. Nucl. Chem.*, **26**, 2289 (1964).
193. G. Caprioli and A. Iliceto, *Ric. Sci.*, **26**, 2714 (1956).
194. J. L. Burmeister, Ph.D. thesis, Northwestern University, Evanston, Ill., 1964, pp. 127, 129.
195. U. Belluco, L. Cattalini, F. Basolo, R. G. Pearson, and A. Turco, *Inorg. Chem.*, **4**, 925 (1965); and references contained therein.
196. F. Basolo, A. T. Brault, and A. J. Poë, *J. Chem. Soc.*, **1964**, 676.
197. F. P. Dwyer and F. L. Garvan, *Inorg. Syn.*, **6**, 192 (1960).
198. K. N. Raymond and F. Basolo, *Inorg. Chem.*, **5**, 949 (1966).
199. R. L. McCullough, L. H. Jones, and R. A. Penneman, *J. Inorg. Nucl. Chem.*, **13**, 286 (1960).
200. J. S. Coleman, H. Peterson, Jr., and R. A. Penneman, *Inorg. Chem.*, **4**, 135 (1965).
201. W. C. Andersen and R. H. Harris, *Inorg. Nucl. Chem. Letters*, **2**, 315 (1966).
202. G. E. Ryschkewitsch and J. M. Garret, *J. Am. Chem. Soc.*, **89**, 4241 (1967).
203. H. C. Clark and K. R. Dixon, *Chem. Commun. (London)*, **1967**, 717.
204. D. A. House and C. S. Garner, *Inorg. Chem.*, **5**, 2097 (1966).
205. W. E. Hatfield, R. C. Fay, C. E. Pfluger, and T. S. Piper, *J. Am. Chem. Soc.*, **85**, 265 (1963).
206. M. Mori, Y. Saito, and T. Watanabe, *Bull. Chem. Soc. (Japan)*, **34**, 295 (1961).
207. J. S. Anderson, H. V. A. Briscoe, and N. L. Spoor, *J. Chem. Soc.*, **1943**, 361.
208. J. W. Palmer and F. Basolo, *J. Phys. Chem.*, **64**, 778 (1960).
209. B. Halpern, A. M. Sargeson, and K. R. Turnbull, *J. Am. Chem. Soc.*, **88**, 4630 (1966).
210. D. A. Buckingham, L. G. Marzilli, and A. M. Sargeson, *J. Am. Chem. Soc.*, **89**, 825 (1967).
211. D. A. Buckingham, L. G. Marzilli, and A. M. Sargeson, *J. Am. Chem. Soc.*, **89**, 3428 (1967).

The Boron Hydrides

R. W. PARRY and M. K. WALTER

Department of Chemistry, University of Michigan
Ann Arbor, Michigan

CONTENTS

I. Introduction	46
II. The Preparation of Diborane	48
A. Methods Based on Binary Metal Hydrides or Solubilized Forms of Metal Hydrides—$LiAlH_4$, $LiHB(OCH_3)_3$, etc.	48
1. Reactions of MH and BX_3 without Solvents	49
2. Reactions of MH and BX_3 Compounds in Solution	50
B. Methods Based on Borohydride Decomposition	54
1. Methods for the Synthesis of Alkali Metal Borohydrides	54
2. The Reaction of Metal Borohydrides with Protonic Acids	55
3. The Reaction of Metal Borohydrides with Lewis Acids	56
4. The Electrolysis of Borohydrides	57
C. Miscellaneous Methods of Diborane Synthesis	58
1. By Displacement of BH_3 from a Borane Complex	58
2. Hydrogenation of Boron Halides in an Electrolysis Cell	58
3. The Direct Hydrogenation of Chloroboranes and Alkyl Chloroboranes	59
4. Other Reducing Agents	59
D. Isotopically Substituted Diborane	60
E. Alkylated Diboranes	60
F. Special Notes on the Purification of Diborane	60
III. The Preparation of the Higher Boron Hydrides	60
A. A Summary of Known Hydrides	60
B. The Synthesis of B_4H_{10}	61
1. The Conversion of B_2H_6 to B_4H_{10} at High Pressures	64
2. Pyrolysis of B_2H_6 in a Hot–Cold Reactor	64
3. Preparation of B_4H_{10} from MB_3H_8 Salts and from MB_9H_{14} Salts	65
4. Reactions for B_4H_{10} Preparation Which Have Little Value in Synthesis	65
5. The Preparation of Isotopically Labeled B_4H_{10} Molecules	66
6. The Preparation of Alkyl and Halo Derivatives of B_4H_{10}	66
C. The Pentaboranes (B_5H_9 and B_5H_{11}) and Their Alkyl and Halo Derivatives	67
1. The Synthesis of Pentaborane-11	67
2. The Synthesis of Pentaborane-9	69

D. The Hexaboranes (B_6H_{10}, Isomeric B_6H_{10}, and B_6H_{12}). 73
   1. Hexaborane-10 (B_6H_{10}) 73
   2. Isomeric Hexaborane-10 74
   3. Hexaborane-12 (B_6H_{12}) 75
E. Heptaboranes and Octaboranes (B_8H_{12}, B_8H_{14}, B_8H_{18}) 76
   1. Heptaboranes—Not Identified 76
   2. Octaborane-12 (B_8H_{12}) 77
   3. Octaborane-14 (B_8H_{14}) 79
   4. Octaborane-18 (B_8H_{18}) 79
F. The Nonaboranes (B_9H_{15} and i-B_9H_{15}) 79
   1. Normal B_9H_{15} 79
   2. Isomeric B_9H_{15} 81
G. The Decaboranes 82
   1. Decaborane-14 ($B_{10}H_{14}$). 82
   2. Decaborane-16 ($B_{10}H_{16}$) 86
   3. Decaborane-9 Free Radical ($B_{10}H_9\cdot$) 87
H. Undecaboranes ($B_{11}H_x$) 87
I. Unidentified Hydrides (Possible $B_{12}H_x$, $B_{13}H_x$, $B_{14}H_x$) 88
J. Pentadecaborane ($B_{15}H_{22}$) 88
K. The Octadecaboranes (n-$B_{18}H_{22}$ and i-$B_{18}H_{22}$) 88
   1. Normal $B_{18}H_{22}$ 88
   2. Isomeric-$B_{18}H_{22}$ 89
L. The Icosaboranes ($B_{20}H_x$) 90
   1. Icosaborane-16 ($B_{20}H_{16}$) 90
   2. Icosaborane-26 ($B_{20}H_{26}$) 90
M. Higher Hydrides 91
IV. The Mechanism of Boron Hydride Conversions 91
  A. General Considerations 91
  B. The Pyrolysis of Diborane 92
  C. Transitions and Exchanges Involving B_4H_{10} and B_5H_{11} 95
  D. Polyhedral Rearrangements and Exchange Reactions of Higher Hydrides. 96
References 97

I. INTRODUCTION

The higher boron hydrides were first prepared by Stock[1] using the reaction of magnesium boride with aqueous hydrochloric acid. It is perhaps ironic that in Stock's original procedure, diborane, the simplest of the boranes,* was obtained only as a low-yield decomposition product of the higher hydrides. In contrast, diborane is made directly today and serves as the starting material for the syntheses of all of the higher hydrides of boron.

Since Stock's system utilized aqueous acid, those hydrides with greatest sensitivity to moisture were decomposed while those which hy-

---

* While BH, BH_2, and BH_3 are well-known species in spectroscopy,[142] they are not stable hydrides for purposes of this discussion. Diborane is considered to be the simplest stable boron hydride.

drolyzed more slowly were recovered. Stock's crude hydride mixture was predominantly B_4H_{10} with smaller amounts of B_5H_9 plus B_6H_{10} and traces of $B_{10}H_{14}$. The total yield of all hydrides recovered from the reaction of acids and borides ranged from 4 to 11% based on the metal boride used. At best, the magnesium boride process was *very inefficient*.

A spectacular advance in the synthesis of boron hydrides resulted when Schlesinger, Brown, Burg, Sanderson, and their co-workers[2] abandoned the use of acidic hydrogens and negative boride clusters and substituted in their stead the hydridic hydrogens of LiH, $LiAlH_4$, NaH, etc., and the more positive boron units of the boron halides or boron alkoxides. Under proper conditions, the boron of BCl_3 can be converted to B_2H_6 in almost quantitative yields. The fact that yields increased when compounds with hydridic hydrogens were the starting point rather than compounds with acidic hydrogens illustrates a generalization from the periodic table. *Hydridic compounds* are not easily prepared from acidic reagents.* For example, the synthesis of hydridic LiH, NaH, or CaH_2 is *not* achieved by treatment of lithium, sodium, or calcium salts with acids. On the other hand, acidic materials such as HCl of HF are obtained in good yield by treatment of NaCl or NaF with H_2SO_4. It is well to remember that boron is closer to Li and Na in the period table than it is to F or Cl. The acid route to B_2H_6 is quite inappropriate.

From B_2H_6, one can obtain higher hydrides, the polyhedral borane anions, and the carboranes. In no case is the mechanism of a given conversion clearly established, but fruitful speculation is usually abundant. In this paper, methods for the synthesis of each of the boron hydrides will be critically reviewed. Interest will center on chemical relationships. Mechanistic, stereochemical, and bonding arguments will be considered in a rationalization of the chemical arguments presented; but an extensive development of "electron deficient" bonding theory does not seem appropriate here. Excellent reviews[3-6] appearing in recent years have given an ample development of structure and bonding. For those interested in the details of the reactions of the boron hydrides, the comprehensive monograph by Adams[4] is recommended for a review of work prior to

---

* Obtaining an unambiguous quantitative criterion for acidic or hydridic character of hydrogens in a compound is not always easy, particularly in borderline cases. For example, it is frequently noted that rapid exchange of protons with acid or water implies acidic character of the hydrogens. Recently, Davis[50] found that $(CH_3)_3NBH_3$ exchanges hydrogens on the boron rapidly in acidic heavy water:

$$R_3NBH_3(\text{soln}) \xrightarrow[\text{0.5-5.0}M \text{ DCl}]{\text{large excess } D_2O} R_3NBD_3$$

Clearly, more detailed mechanistic analysis must be used. A process other than ionization of H^+ from R_3NBH_3 is almost certainly significant here.

1964. Other very useful reviews on boron compounds and boranes have appeared more recently.[5,8,9]

II. THE PREPARATION OF DIBORANE

Diborane, B_2H_6, the simplest of the free boron hydrides, melts at $-165°C$, boils at $-90°C$, and has the bridge structure shown in Figure 1. The literature on diborane preparation is extensive. Furthermore, plants to make tons of B_2H_6 per day have been designed and built. (See, for example, ref. 89.) Despite this fact, it is frequently more convenient to generate B_2H_6 in the laboratory than it is to obtain and handle the commercial material. In this review, useful methods of synthesis will be considered under four general headings: (*1*) methods based on binary metal hydrides, (*2*) methods based on borohydrides, (*3*) methods based on base displacement processes, and (*4*) other procedures.

A. Methods Based on Binary Metal Hydrides or Solubilized Forms of Metal Hydrides—$LiAlH_4$, $LiHB(OCH_3)_3$, etc.

The first really effective laboratory procedure for the preparation of B_2H_6 appeared in the open literature in 1947. The synthesis, reported by Finholt, Bond, and Schlesinger, was based on the reaction of BCl_3 with the then new reagent, $LiAlH_4$:[7]

$$3LiAlH_4(\text{soln}) + 4BCl_3(\text{soln}) \xrightarrow{(C_2H_5)_2O} 2B_2H_6(g) + 3LiCl(s) + 3AlCl_3(\text{soln})$$

Yields of 99% were reported. Since $LiAlH_4$ can be synthesized in a relatively straightforward fashion from LiH and $AlCl_3$, the overall process can be considered as a synthesis from LiH and BCl_3 with $AlCl_3$ present as a solubilizing agent for LiH. The appropriate equations are shown in Sequence 1.

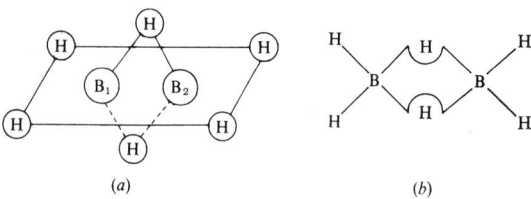

Fig. 1. (*a*) Diborane, B_2H_6, in perspective; (*b*) B_2H_6, planar projection.

$$12\text{LiH} + 3\text{AlCl}_3 \longrightarrow 3\text{LiAlH}_4 + 9\text{LiCl}$$
$$3\text{LiAlH}_4 + 4\text{BCl}_3 \xrightarrow{\text{diethyl ether}} 2\text{B}_2\text{H}_6 + 3\text{AlCl}_3 + 3\text{LiCl}$$

Overall:
$$12\text{LiH} + 4\text{BCl}_3 \xrightarrow{\frac{(\text{C}_2\text{H}_5)_2\text{O}}{\text{AlCl}_3}} 2\text{B}_2\text{H}_6 + 12\text{LiCl}$$

Sequence 1

The overall equation given above has broad significance in that it suggests the synthesis of diborane by an anion exchange process of the general form:

$$6\text{MH} + 2\text{BX}_3 \longrightarrow \text{B}_2\text{H}_6 + 6\text{MX}$$

All of the most widely used processes for diborane synthesis have this general form. If the synthesis is run in a solvent such as an ether, the reaction is accelerated by the presence of a reagent which will solubilize the normally insoluble metal hydrides. A variety of solubilizing agents can be used which form complex metal hydrides. If the process is carried out in the absence of a solvent, high temperatures are usually required. Both approaches have been studied extensively.

1. Reactions of MH and BX_3 without Solvents

In most reactions of this type, an active metal is allowed to react directly with H_2 in the presence of a boron acid. Presumably, MH forms first and then reacts immediately to give B_2H_6 and the metal halide:

$$2\text{M} + \text{H}_2 \longrightarrow 2\text{MH}$$
$$6\text{MH} + 2\text{BX}_3 \longrightarrow \text{B}_2\text{H}_6 + 6\text{MX}$$

Hurd[24] describes a process in which a mixture of H_2 and boron halide was passed through a bed of an active metal at 200–400°C. Metals used included the alkali metals, as well as magnesium, zinc, and aluminum; the products were diborane and monohaloborane. Total yields of diborane plus monohaloborane amounted to about 50%. Many variations in this basic procedure have been recorded. An alloy of the composition Al_2Cu was used[25]; boroxin and alkyl borates replaced boron halides[27]; boric oxide and metal borates were reduced to B_2H_6 by H_2 and an active metal–metal oxide combination at 1000–1300°C.[26] Reactions of the above type have been most effectively used in the preparation of borohydrides (see p. 54).

2. Reactions of MH and BX_3 Compounds in Solution

Metal hydrides are insoluble in all solvents other than those protonic materials which decompose them. This problem is the central one surrounding the reaction of MH and BX_3 in solution:

$$6MH(s) + 2BX_3 \xrightarrow[\text{solubilizing agent}]{\text{solvent}} 6MX + B_2H_6$$

The problem has been attacked by varying: (a) the metal M, (b) the solvent used, (c) the solubilizing agent, and (d) the temperature. The nature of the compound BX_3 is also important for many reasons including hydride solubilization.

Ethers are frequently the solvent of choice. Polyethers appear to be better solvents than simple ethers for many reactions. The activity (solubility) of simple and complex alkali metal hydrides in ethers appears to decrease in the order LiH, NaH, KH; hence, the utility of the alkali metal hydrides as reducing agents for BX_3 decreases in the order LiH > NaH > KH. Schlesinger, Brown, Gilbreath, and Katz[8] found that the reduction of BF_3 in diethylether goes more rapidly and proceeds more nearly toward completion when LiH is used rather than NaH. Further, both rate of reaction and yield are increased if the hydride is finely ground and if grinding action is maintained throughout the experiment. Yields of approximately 90% were reported under the most favorable conditions.

Adams and Pearson [9a] patented dimethylethers of polyethylene glycols, $CH_3O(CH_2CH_2O)_nCH_3$, as solvents for the reaction of NaH with boron trifluoride. The improved solvent properties of the polyethers rendered the reaction practical. Strangely enough, the use of BCl_3 instead of BF_3 gave a reaction which was too slow to be useful. Again, solubility problems would appear to be dominant. The NaF formed could react with BF_3 to give $NaBF_4$ which is sufficiently soluble in ether to prevent coating of the NaH by NaF. On the other hand, the lower stability of $NaBCl_4$ would retard removal of NaCl and permit the formation of an inactive coating over the NaH. A detailed, experimentally supported explanation has not been advanced. The addition of mineral oil to dilute the polyethers has been suggested.[9b]

The question of low hydride solubility and reactivity can be handled most effectively by adding a Lewis acid which converts the nonreactive simple binary hydride into a soluble and reactive complex hydride. The process is illustrated by the previously discussed use of $AlCl_3$ in the synthesis of B_2H_6.[7] Appropriate equations are:

4LiH(s) + 4(AlCl$_3$)(soln) $\longrightarrow$ 4Li[HAlCl$_3$](soln)

$\longrightarrow$ Li(AlH$_4$)(soln) + 3LiCl(s) + 3AlCl$_3$(soln)

The LiAlH$_4$ can be isolated and studied as a separate species, but no direct evidence for LiHAlCl$_3$ has yet been presented. Its existence as an intermediate is, however, strongly suggested by analogy to cases involving boron acids.

The reaction of LiAlH$_4$ with boron halides has been studied extensively. Shapiro and colleagues[11] identified two well-defined steps in the reduction of BF$_3$ by LiAlH$_4$ in diethylether. Because of the first step, an induction period is noticed before B$_2$H$_6$ evolution takes place. In some

3LiAlH$_4$(soln) + 3BF$_3$(soln) $\xrightarrow{(C_2H_5)_2O}$ 3LiBH$_4$(soln) + 3AlF$_3$(s)

3LiBH$_4$(soln) + BF$_3$(soln) $\xrightarrow{(C_2H_5)_2O}$ 2B$_2$H$_6$(g) + 3LiF(s)

Overall: 3LiAlH$_4$(soln) + 4BF$_3$(soln) $\xrightarrow{(C_2H_5)_2O}$ 3LiF(s) + 3AlF$_3$(s) + 2B$_2$H$_6$(g)

Sequence 2

cases chemists have not added enough BF$_3$ to the system. The second reaction, generating B$_2$H$_6$, then occurs to only a small extent and yields of B$_2$H$_6$ are low. More BF$_3$ increases the yields to above 90%. The reaction between LiAlH$_4$ and boron halides is one of the best procedures for the laboratory preparation of B$_2$H$_6$.

Aluminum chloride has been used to promote the reaction between NaH and BCl$_3$ in benzene solution as well as in ether solution. Yields above 84% were reported.[10a] Because NaAlCl$_4$ is soluble in benzene, AlCl$_3$ also helped to prevent the coating of NaH with insoluble NaCl. CaH$_2$ and MgH$_2$ were useful as hydride sources and AlBr$_3$ and GaCl$_3$ were useful solubilizing agents for the hydrides.[10b] Alkylated benzenes were also effective solvents.[10b]

A number of Lewis acids other than AlCl$_3$ have been used effectively as solubilizing agents for metal hydrides. Trialkylborates have been very effective[8,12,18]:

MH + B(OCH$_3$)$_3$ $\longrightarrow$ MHB(OCH$_3$)$_3$

MH + B(OC$_2$H$_5$)$_3$ $\longrightarrow$ MHB(OC$_2$H$_5$)$_3$

Disproportionation of M[HB(OR)$_3$] under suitable conditions gives MBH$_4$, MOR, and B(OR)$_3$, but M[HB(OR)$_3$] forms easily and is well characterized as a distinct species.[14] The following reaction gives essentially quantitative yields of B$_2$H$_6$ [15]:

3M[HB(OR)$_3$] + BX$_3$ $\xrightarrow[50-75°C]{solvent}$ $\frac{1}{2}$B$_2$H$_6$ + 3MX + 3B(OR)$_3$

When R was methyl, M could be Li, Na, or K. X was fluorine and the solvent was dimethyl, diethyl, or di-*n*-propyl ethers as well as dioxane. About twice the theoretical quantity of BX_3 was used.

In a closely related process —X is —OR instead of F (i.e., excess methyl borate is used in place of boron trifluoride). The *overall* equation is then:

$$6NaH + 6B(OCH_3)_3 \xrightarrow{\text{mineral oil}} 6Na[B(OCH_3)_4] + 6HB(OCH_3)_2$$
$$\downarrow$$
$$B_2H_6 + 4B(OCH_3)_3$$

$Na[HB(OCH_3)_3]$ has been clearly identified as an intermediate (ref. 89, p. 38). The process was patented by Bush, Carpenter, and Schechter[16] and was the basis for the commercial preparation of diborane by Callery Chemical Co. A 94% yield of $HB(OCH_3)_2$ was reported, based on the NaH used. The disproportionation of $HB(OCH_3)_2$ to give B_2H_6 was essentially quantitative.

The fact that $B(OCH_3)_3$ solubilizes NaH suggests that other boron acids such as BF_3, BCl_3, or even the product BH_3, might be effective in dissolving metal hydrides and promoting diborane formation. In the direct reaction of LiH and BF_3 in ether, excess BF_3 is used as a Lewis acid instead of $AlCl_3$. This reaction was the basis for the first commercial diborane process developed by General Electric.

An intensive study of the LiH–BF_3 process by Elliott, Boldebuck, and Roedel[17] revealed that LiH and BF_3 can react in at least two different ways depending upon conditions. *If no solubilizing agent other than BF_3 is present and B_2H_6 is removed as rapidly as it is formed*, the overall stoichiometry is given by the equation:

$$6LiH + 8BF_3 \xrightarrow{(C_2H_5)_2O} 6LiBF_4 + B_2H_6$$

This stoichiometry can be rationalized by the reactions in Sequence 3. In

$$12LiH(s) + 12BF_3(soln) \xrightarrow[25°C]{(C_2H_5)_2O} 12Li[HBF_3](soln)$$
$$\longrightarrow 3LiBH_4(soln) + 9LiBF_4(soln)$$

$$3LiBH_4(soln) + LiBF_4(soln) \xrightarrow[25°C]{(C_2H_5)_2O} 4LiF(s) + 2B_2H_6$$

$$4LiF(s) + 4BF_3(soln) \xrightarrow[25°C]{(C_2H_5)_2O} 4LiBF_4(soln)$$

Overall: $12LiH + 16BF_3 \longrightarrow 12LiBF_4 + 2B_2H_6$

Sequence 3

THE BORON HYDRIDES

the absence of any other effective solubilizing agent for LiH, the reaction between LiH and BF_3 is expected to give $LiHBF_3$. The related compound, $KHBF_3$, was isolated and characterized by Aftandalian, Miller, and Muetterties.[18] The reaction between $LiBH_4$ and $LiBF_4$ is known to go as indicated.[17,20] The reaction between LiF and BF_3 is also known to go as indicated in ether.[17] It is appropriate to note that the reaction $LiH + LiBF_4$ does *not* go, as would be required by Sequence 3.

On the other hand, if a solubilizing agent more effective than BF_3 is present, a different sequence is anticipated. If $B(OCH_3)_3$ is added or if diborane is held in the system by pressurization, about one-fourth as much BF_3 is required to generate the mole of diborane. The overall stoichiometry is given by the equation:

$$6LiH + 2BF_3 \xrightarrow[\substack{B(OCH_3)_3 \\ 10-25°C}]{(C_2H_5)_2O} 6LiF + B_2H_6$$

and is rationalized by the reactions in Sequence 4.

$$18LiH(s) + 18B(OCH_3)_3(soln) \xrightarrow{(C_5H_5)_2O} 18LiHB(OCH_3)_3(soln)$$

$$18LiHB(OCH_3)_3(soln) + 6BF_3(soln) \xrightarrow{(C_2H_5)_2O}$$
$$3B_2H_6(g) + 18LiF(s) + 18B(OCH_3)_3(soln)$$

Overall: $\quad 18LiH + 6BF_3 \longrightarrow 18LiF + 3B_2H_6$

$$6LiH(s) + 3B_2H_6(soln) \longrightarrow 6LiBH_4(soln)$$
$$2LiF(s) + 2BF_3(soln) \longrightarrow 2LiBF_4(soln)$$
$$6LiBH_4(soln) + 2LiBF_4(soln) \longrightarrow 4B_2H_6(g) + 8LiF$$

Overall: $\quad 24LiH + 8BF_3 \xrightarrow[10-25°C]{(C_2H_5)_2O} 4B_2H_6 + 24LiF$

Sequence 4

Close examination indicates that the prime point of difference between the two processes is the reagent used to dissolve the LiH into an active complex: viz., $LiHBF_3$, $LiHB(OCH_3)_3$, or $LiBH_4$. Boron trifluoride is less effective as a solubilizing agent than either $B(OCH_3)_3$ or BH_3. The reaction in the presence of $B(OR)_3$ or BH_3 is known to proceed in two distinct steps; indeed, a plant was built in which the two steps were separated.[19] The first step with BF_3 gives $LiBH_4$:

$$24LiH + 6BF_3 \longrightarrow 6LiBH_4 + 18LiF$$

The second stage involves the reaction of the $LiBH_4$ with BF_3, either directly or as $LiBF_4$:

$$6LiBH_4(soln) + 2LiBF_4(soln) \longrightarrow 4B_2H_6(g) + 8LiF$$

Overall yields of B_2H_6 up to 88% (based on BF_3 used) were obtained when a promoter was used. If tetrahydrofuran were used as a solvent in place of diethyl ether, no promoter was needed because of the greater solubility of the B_2H_6 in the cyclic ether. Yields of 70% were reported.[21]

If BCl_3 and LiH were allowed to react in $(C_2H_5)_2O$, yields up to 77% were reported even without a promoter. This fact suggests that BCl_3 is more effective in the dissolution of LiH in diethyl ether than is BF_3.

The second reaction between a metal borohydride and a Lewis acid or a fluoroborate complex will be treated in more detail in the next section. A single bridged complex, $NaBH_4 \cdot BH_3$, has been recognized in a study of this reaction.[22]

B. Methods Based on Borohydride Decomposition

Diborane may be easily generated in the laboratory by the reaction between an acid and an alkali metal borohydride:

$$2MBH_4 + \text{acid} \longrightarrow B_2H_6(g) + \text{other products}$$

Metal borohydrides are easily available commercially and either Brönsted-Lowry or Lewis acids may be used. It is convenient to discuss these two cases separately after a brief review of methods for borohydride production.

1. Methods for the Synthesis of Alkali Metal Borohydrides

Although alkali metal borohydrides are available commercially, methods for their synthesis are summarized briefly here. Those procedures which require diborane as a raw material[28] are of some historic importance, but are of little value for purposes of this review.

Several procedures for borohydride synthesis resemble the high-temperature processes described for the preparation of diborane from an active metal and a volatile boron compound. A typical reaction is that of Schlesinger, Brown, and Finholt[30] represented by the equation:

$$4NaH(s) + 4B(OR)_3(l) \xrightarrow[\text{high pressure}]{\text{no solvent}} NaBH_4(s) + 3Na[B(OR)_4]$$

A perturbation of this general procedure uses sodium dispersed in mineral oil plus gaseous hydrogen in place of NaH. Methyl borate, boron trifluoride, sodium tetrafluoroborate, or trimethoxyboroxin can be used as the boron source (ref. 4, p. 392). In other perturbations, SiH_4 or Al and H_2 have been used to reduce $NaB(OCH_3)_4$ in tetrahydrofuran or diglyme.[32b,33a]

High-temperature processes, frequently carried out in a pressurized ball mill,[31] have been described for reduction of boric oxide or alkali metal borates (ref. 4, p. 394).

$$4NaH + 2B_2O_3 \longrightarrow NaBH_4 + 3NaBO_2$$
$$2CaH_2 + NaBO_2 \longrightarrow 2CaO + NaBH_4 + 2800 \text{ cal (ref. 32a)}$$
$$3NaBO_2 + 4Al + 6H_2 \longrightarrow 3NaBH_4 + 2Al_2O_3 \text{ (ref. 33b)}$$

A modification of this type of high-temperature procedure is believed to be the basis for a successful commercial $NaBH_4$ process. The equation is:

$$Na_2B_4O_7 + 7SiO_2 + 16Na + 8H_2 \xrightarrow[\text{temp. and pressure}]{\text{high}} 4NaBH_4 + 7Na_2SiO_3$$

Separation problems for $NaBH_4$ would be minimal in this process, whereas they are serious in most other procedures. Patents on the process were granted to Broja, Schlabacher, Schubert, Lang, and Gorreig.[34]

2. The Reaction of Metal Borohydrides with Protonic Acids

The action of a protonic acid on an alkali metal borohydride can be represented most simply as:

$$BH_4^- + H^+ \longrightarrow H_2 + BH_3$$
$$\longrightarrow \tfrac{1}{2}B_2H_6$$

Since B_2H_6 may react further with protons, it is essential that reaction conditions be selected which minimize this secondary process. Aqueous acid attacks B_2H_6 rapidly at room temperature,* hence aqueous solutions are usually avoided. Gaseous acids are used, or solvents such as ether,[37] polyether[37] or chlorobenzene[38] are employed. A representative equation for the reaction of gaseous HCl, HBr, or HI with $LiBH_4$ is[29c,36]

$$LiBH_4(3) + HX(g) \longrightarrow LiX(s) + H_2(g) + \tfrac{1}{2}B_2H_6(g)$$

A small amount of the secondary reaction product B_2H_5X is frequently produced.[29e,36] With $NaBH_4$, the solid–gas phase reaction is so slow that a ball mill has been used as a reaction vessel.

* Although aqueous acid attacks diborane rapidly at room temperature to give $B(OH)_3$ and H_2, Jolly and Schmitt[35] have found that at $-70°C$ $8M$ HCl reacts slowly with B_2H_6 to give evidence for the cation $H_2B(OH_2)_2^+$ or $BH_2^+(aq)$. Above $-20°C$ decomposition ensues as indicated by the equation:

$$H_2B^+(aq) + 3H_2O \longrightarrow 2H_2 + B(OH)_3 + H^+$$

Unacidified alcohol–water solutions at $-75°C$ give evidence for $HB(OH)_2$ while alkaline solutions at $-30°C$ give evidence for:

$$H_2O + 2KOH + B_2H_6 \longrightarrow H_2 + KBH_4 + KBH(OH)_3$$

If $NaBH_4$ is dropped into concentrated (96–100%) H_2SO_4, B_2H_6 can be generated in yields ranging up to 80%.[39] Dilution of the acid with water (79% H_2SO_4) cut the yields of B_2H_6 to 50%. On the other hand, at high concentrations of acid, the product gas consisted of as much as 20% SO_2. At lower acid concentration, the SO_2 concentration dropped to 1%. Some problems with flaming were reported. When methanesulfonic acid, CH_3SO_3H replaced H_2SO_4, no SO_2 was reported in the product and yields were about 75%. Other acids used include H_2PO_3F (79% yield),[40] and H_3PO_4 (50–70% yields).[41] The H_3PO_4 procedure was reported as superior to the H_2SO_4 reaction.[41]

3. The Reaction of Metal Borohydrides with Lewis Acids

The reaction of a typical Lewis acid, BF_3, with a metal borohydride, $NaBH_4$, is represented by the equation[36,42]:

$$3NaBH_4 + 4BF_3 \xrightarrow{\text{polyether*}} 2B_2H_6 + 3NaBF_4$$

The reaction sequence for this process is not clear, but a likely route would involve acid displacement followed by disproportionation.

$$4NaBH_4 + 4BF_3 \xrightarrow{\text{polyether}} 4NaHBF_3 + 2B_2H_6$$

$$NaBH_4 \longleftrightarrow 3NaBF_4$$

Overall: $3NaBH_4 + 4BF_3 \longrightarrow 2B_2H_6 + 3NaBF_4$

Sequence 5

In a polyether solution, diborane evolution may be retarded by formation of $NaBH_4 \cdot BH_3$. This is prevented in practice by adding the $NaBH_4$ solution to the BF_3 solution. Since all $NaBH_4$ is then used up, no $NaBH_4 \cdot BH_3$ complex forms. Yields of up to 90% are recorded[42] with BF_3 and $NaBH_4$ and 91% with BF_3 and KBH_4.[43] $KBH_4 \cdot BH_3$ does not seem to form under the conditions used.[43] Halides of Zn, Al, or Sn are reported to enhance the reaction of KBH_4 and BF_3 in diethyl ether.[44] When BCl_3 in polyether is dropped into $NaBH_4$ in polyether, yields of 100% have been recorded,[42] but ether cleavage causes product contamination.

Elliott, Boldebuck, and Roedel[17] obtained yields near 100% from the addition of F_3B etherate to a saturated diethyl ether solution of $LiBH_4$ at 25°C. Tetrahydrofuran as a solvent reduced the *recovered yield* of B_2H_6 to 32%. Schlesinger and Brown[15] also found that BF_3 reacts with $LiBH_4$,

* The polyether is the dimethyl ether of diethylene glycol (diglyme).

$NaBH_4$, or KBH_4 in dioxane, diethyl, dimethyl, or di-*n*-propyl ethers at 50–75°C to give quantitative yields of B_2H_6.

Other Lewis acids which have been used (ref. 4, p. 569) and the pertinent equations are summarized below:

$$FeCl_3 + LiBH_4 \xrightarrow[-80°C]{(C_2H_5)_2O} FeCl_2 + \tfrac{1}{2}B_2H_6 + \tfrac{1}{2}H_2 + LiCl$$

$$FeCl_2 + 2LiBH_4 \xrightarrow[0°C]{(C_2H_5)_2O} Fe + B_2H_6 + H_2 + 2LiCl$$

$$CuCl_2 + LiBH_4 \xrightarrow[-45°C]{(C_2H_5)_2O} CuCl + LiCl + \tfrac{1}{2}B_2H_6 + \tfrac{1}{2}H_2$$

$$CuCl + LiBH_4 \xrightarrow[0°C]{(C_2H_5)_2O} Cu + \tfrac{1}{2}H_2 + \tfrac{1}{2}B_2H_6 + LiCl$$

$$NaBH_4 + FeCl_3 \xrightarrow[-5°C]{(C_2H_5)_2O} FeCl_2 + \tfrac{1}{2}B_2H_6 + \tfrac{1}{2}H_2 + NaCl$$

Yields ran as high as 80 to 90%. Other acids of similar type include $ZnCl_2$ in $(C_2H_5)_2O$, $AlCl_3$ in polyether, $AlCl_3$ in $(C_2H_5)_2O$, $SnCl_2$, $BiCl_2$, and rare earth chlorides.[4]

Trimethyl borate is also effective in displacing B_2H_6 from $NaBH_4$ if boric acid is present.[45] Unfortunately, the solid residue is very rich in boron.

$$3NaBH_4 + 6B_2O_3 + 12B(OCH_3)_3 \xrightarrow[100°C]{diglyme} 12HB(OCH_3)_2 + 3NaB_5O_7(OCH_3)_2 + 3(CH_3)_2O$$

$$8B(OCH_3)_3 \longleftrightarrow 2B_2H_6$$

$$3NaBH_4 + 6B_2O_3 + 4B(OCH_3)_3 \xrightarrow[100°C]{diglyme} 2B_2H_6 + 3NaB_5O_7(OCH_3)_2 + 3(CH_3)_2O$$

Sequence 6

Other reagents used to displace B_2H_6 from $NaBH_4$ include: $BR_3 + B_2O_3$ (where R is an alkyl), $BCl_3 + AlCl_3$ in benzene, CH_3Br, $(CH_3)_2SO_4$, and haloborazine $(ClBNH)_3$ (ref. 4, p. 569).

Finally, it has been reported[17] that B_2H_6 may be produced in 90% yield from the addition of $LiBF_4$ to an ether solution of $LiBH_4$. This may well involve the intermediate generation of $F_3BO(C_2H_5)_2$. No B_2H_6 is produced from the reaction of $NaBH_4$ and $NaBF_4$.[49]

4. The Electrolysis of Borohydrides

Yields of diborane approaching 100% have been reported (ref. 36, p. 570) for the process:

$$NaBH_4 \xrightarrow[elec.]{solvent} Na + \tfrac{1}{2}H_2 + \tfrac{1}{2}B_2H_6$$

Effective solvents are triethyleneglycol dimethyl ether, dimethylformamide, and low-temperature salt melts.

C. Miscellaneous Methods of Diborane Synthesis

1. By Displacement of BH_3 from a Borane Complex

A number of processes are available for the direct synthesis of borane complexes *without the use of diborane*. Equations representing several such processes (ref. 4, p. 571) are:

$$3NaH + BCl_3 + R_3N \longrightarrow R_3NBH_3 + 3NaCl$$

$$(RO)_3BNR_3 + \tfrac{3}{2}H_2 + Al \xrightarrow[200 \text{ psi}]{180°C} H_3BNR_3 + Al(OR)_3$$

$$R_3BNR_3 + 3H_2 \longrightarrow 3RH + H_3BNR_3$$

$$Cl_3BNR_3 + 3H_2 + 3NR_3 \longrightarrow H_3BNR_3 + [R_3NH]Cl \quad \text{(ref. 4, p. 560)}$$

All complexes have the general form $L:BH_3$ where $L:$ is an amine. Diborane can be displaced from these amine complexes by BF_3 which is a stronger acid toward amines than is BH_3

$$H_3BNR_3 + BF_3 \longrightarrow \tfrac{1}{2}B_2H_6 + F_3BNR_3 \quad \text{(see ref. 4, p. 571)}$$

The corresponding reaction involving trialkyphosphine boranes will not go because BH_3 is a stronger acid toward phosphines than is BF_3.

$$R_3PBH_3 + BF_3 \longrightarrow \text{N.R.}$$

A rather elaborate scheme for diborane synthesis has been suggested by Koester and Ziegler.[46] It is fundamentally a base displacement process of the type described above. The relevant equations are shown in Sequence 7.

$$2Al + 3H_2 + 6C_2H_4 \longrightarrow 2Al(C_2H_5)_3$$
$$2Al(C_2H_5)_3 + 2BF_3 \longrightarrow 2AlF_3 + 2B(C_2H_5)_3$$
$$2R_3N + 2B(C_2H_5)_3 \longrightarrow 2R_3NB(C_2H_5)_3$$
$$(C_2H_5)_3BNR_3 + H_2 \longrightarrow 2R_3NBH_3 + 6C_2H_6$$
$$2R_3NBH_3 + 2BF_3 \longrightarrow 2R_3NBF_3 + B_2H_6$$

Overall: $2Al + 9H_2 + 6C_2H_4 + 2BF_3 \longrightarrow 2AlF_3 + 6C_2H_6 + B_2H_6$

Sequence 7

2. Hydrogenation of Boron Halides in an Electrolysis Cell

Various schemes to feed H_2 and BCl_3 into an electrolysis cell generating an alkali metal (ref. 4, p. 566) have been suggested, but the fundamental chemistry can be reduced to the reaction of an active metal hydride with a boron halide.

3. The Direct Hydrogenation of Chloroboranes and Alkyl Chloroboranes

The first real improvement on Stock's metal boride process for making diborane was the direct hydrogenation of BCl_3 in an electric discharge[47]:

$$30H_2 + 12BCl_3 \xrightarrow[\text{discharge}]{\text{electric}} 6B_2H_5Cl + 30HCl$$

$$2BCl_3 \longleftrightarrow 5B_2H_6$$

The monochloroborane produced initially disproportionates to give B_2H_6 and BCl_3. The mechanism of the actual chlorination process, according to Myers and Putnam,[48] involves the $BCl_2\cdot$ radical.

$$BCl_3 + \text{energy} \longrightarrow BCl_2\cdot + Cl\cdot$$
$$BCl_2\cdot + H_2 \longrightarrow HBCl_2 + H\cdot$$

$$Cl_3B \longleftrightarrow B_2H_6$$

Two pilot plant processes (ref. 89, p. 7–8) for diborane synthesis were based on the direct hydrogenation of BCl_3 at temperatures of 950°C in the absence of a catalyst and at 620–700°C in the presence of a silver catalyst.

$$12BCl_3 + 12H_2 \longrightarrow 12HBCl_2 + 12HCl$$
$$12HBCl_2 \longrightarrow 2B_2H_6 + 8BCl_3$$

These processes were not developed as far as those based on hydrogenation with a metal hydride such as LiH or NaH.

A procedure which is related to the direct hydrogenation of BCl_3 has been reported[51] recently from the U.S.S.R. By using an alkyl dichloroborane such as $RBCl_2$ (R = CH_3, C_2H_5, or C_3H_7), an active carbon catalyst, and pressures of 150–200 atm, hydrogenation can be effected at 320–340° rather than at 620–900°C required for the direct hydrogenation of BCl_3. Yields of 50% B_2H_6 were reported. Pyrolysis of the diborane formed was a problem which was reportedly minimized by the presence of BCl_3.

4. Other Reducing Agents

Reducing agents used for the reduction of boron halides in addition to metal hydrides include (ref. 4, p. 567): SiH_4, SiH_2R_2, Si_2H_6, CH_2O, and $(CH_3)_2SbH$.

D. Isotopically Substituted Diborane

The use of $^{10}BCl_3$ or metal deuterides in the processes described earlier permits the synthesis of isotopically labeled diborane for special purposes.

E. Alkylated Diboranes

Alkylated diboranes can be prepared by exchange reactions involving BR_3 and B_2H_6[143]:

$$\frac{x}{3} BR_3 + \left(1 - \frac{x}{6}\right) B_2H_6 \longrightarrow B_2R_xH_{6-x}$$

As expected, the product mixture obtained is sensitive to the diborane–boron alkyl ratio. With a B_2H_6/BR_3 ratio of 5:1 the products are mostly mono- and dialkyldiboranes. If the same ratio is 1:8, the tri- and tetralkyl diboranes predominate. Products can be separated by vacuum line fractionation or by chromatography. The kinetic stabilities of the methyl diboranes are listed by Lutz and Ritter[77] as 1,2 di $\gg$ 1,1 di $>$ tri $>$ mono $>$ tetra.

Alkyl diboranes such as ethyl diboranes can be obtained by the interaction of B_2H_6 with C_2H_4.

F. Special Notes on the Purification of Diborane

When diethyl ether is used as the solvent in the preparation of B_2H_6, C_2H_6 may be a contaminant. The C_2H_6 can be eliminated by the formation of pyridine-borane[4]:

$$B_2H_6 + \text{pyridine} \longrightarrow H_5C_5NBH_3$$

$$H_2C_5NBH_3 + (C_2H_5)_2OBF_3 \xrightarrow{(C_2H_5)_2O} H_5C_5NBF_3 + B_2H_6$$

If $(CH_3)_2O$ is used, the ether is difficult to remove by normal distillation. $AlCl_3$ will remove the ether easily.[4] $NaBH_4$ in polyether will remove BX_3 or HX as contaminants, but some B_2H_6 will be retained as the complex, $NaBH_4 \cdot BH_3$.

III. THE PREPARATION OF THE HIGHER BORON HYDRIDES

A. A Summary of Known Hydrides

At the end of Stock's career, five higher boron hydrides were unequivocally characterized. These were: B_4H_{10}, B_5H_9, B_5H_{11}, B_6H_{10}, and

$B_{10}H_{14}$. The compound B_6H_{12} was listed as probable. By the end of 1967, B_6H_{12} had been unequivocally established[53,54] and nine additional hydrides had been clearly identified. The new hydrides in chronological order of their characterization are: B_9H_{15},[52] $B_{10}H_{16}$,[65] n-$B_{18}H_{22}$,[60] i-$B_{18}H_{22}$,[66] $B_{20}H_{16}$,[61,62] B_8H_{12},[55] B_8H_{18},[56] i-B_9H_{15},[57,144] and the radical $B_{10}H_9\cdot\cdot$.[58] In addition, very good evidence for an isomer of B_6H_{10}[67] and evidence for unstable B_8H_{14} have been presented recently and some mass spectral evidence is available for a $B_{20}H_{26}$, a $B_{11}H_{17}$, and a $B_{15}H_{21}$.[65] Evidence for the heptaboranes is seriously questioned at the present time[82] and the recently reported[59] $B_{11}H_{15}$ and $B_{11}H_{13}$ have been obtained only as etherates, hence they may really be borane fragments such as B_4H_8 and B_3H_7 rather than distinct new boron hydrides. A summary of the boron hydrides is contained in Table I.

In addition to these molecular hydrides, a number of boron hydride fragments will bind to Lewis bases to give borane addition compounds such as $R_3NB_3H_7$, etc. The borane anions such as $B_3H_8^-$, BH_4^-, etc. are special cases of this general class in which the coordinated base is the hydride ion H^-. Carboranes are not included in this review; however, several recent reviews on the carboranes are available.[68]

Many of the higher hydrides are synthesized by diborane pyrolysis.[153] This fact makes it more convenient to discuss the methods for the synthesis of each boron hydride systematically and then consider the most recent overall information on the mechanism of hydride interconversion.

B. The Synthesis of B_4H_{10}

Tetraborane was the principal recovered product in Stock's original magnesium boride procedure for borane synthesis. The structure of this hydride as determined by Nordman and Lipscomb[149] is shown in Figure 2.

In Stock's method, tetraborane yields were as high as 11% when

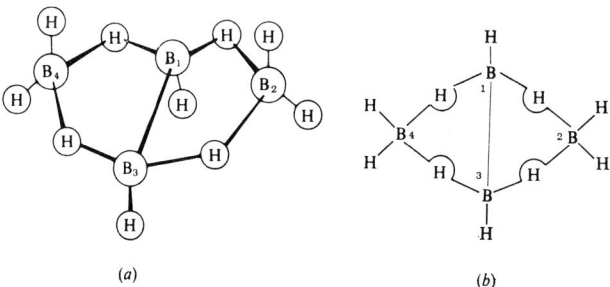

Fig. 2. (a) Tetraborane-10, B_4H_{10}, perspective; (b) B_4H_{10}, planar projection.

TABLE I
The Known Boron Hydrides

Hydride	Structural type	Number H_2B units	mp, °C	bp, °C	VP, mm Hg	Dipole moment	Refs.
A. Characterized							
B_2H_6	Bridged ethylene	2	−165.5	−92	225 at −112°C	0	1
B_4H_{10}	Icosahedral fragment	2	−119.9	16	388 at 0°C	0.56 D	1
B_5H_9	Octahedral fragment	0	−46.1	58	65 at 0°C	2.13 D	1
B_5H_{11}	Icosahedral fragment	3	−123.4	63	52.5 at 0°C	—	1
B_6H_{10}	Icosahedral fragment	0	−62.3	108	7.5 at 0°C	—	1
B_6H_{12}	Icosahedral fragment	(2)?	−82.3	100?	17 at 0°C	—	53,54
B_8H_{12}	Icosahedral fragment	0	−20.0	Very unstable at −20°C		—	55
B_8H_{14}	—	—	Very unstable		—	—	
B_8H_{18}	Two tetraboranes linked	2	2.6	Very low stability	—	—	56
n-B_9H_5	Icosahedral fragment	1	—	—	—	—	52
i-B_9H_{15}	—	0	Very unstable at −30°C		—	—	57,144
$B_{10}H_9\cdot$	Free radical?	0	—	—	—	—	58
$B_{10}H_{14}$	Icosahedral fragment	0	99.7	213	Below 10^{-5} at 0°C	3.5 D	1
$B_{10}H_{16}$	Two B_5H_8 units linked at apex	0	—	—	—	0	65
n-$B_{18}H_{22}$	Two decaborane fragments joined at mouth in line	0	177	—	—	—	60
i-$B_{18}H_{22}$	Two decaborane fragments joined at mouth at an angle	0	—	—	—	—	66
$B_{20}H_{16}$	—	0	195–198	—	—	0	61,62

B. Partially characterized

$i\text{-}B_6H_{10}$	NMR data only	67
B_7H_{13}	Existence doubtful	82
B_7H_{15}	Existence doubtful	82
$B_{11}H_{13}$	May be fragment coordinated to ether	59
$B_{11}H_{15}$	May be fragment coordinated to ether	59
$B_{11}H_{17}$	B_8B_5—B_6H_9 Mass spectral data only	65
$B_{15}H_{21}$	$H_{13}B_{10}$—B_5H_8 Mass spectral data only	65
$B_{20}H_{26}$	$H_{13}B_{10}$—$B_{10}H_{13}$ Mass spectral data only	65

$8N$ H_3PO_4 was used with Mg_3B_2. Separation was achieved by passing gases through traps at -95, -120, and $-196°C$. The condensate in the $-120°C$ trap was almost pure B_4H_{10}, while the condensate in the $-196°C$ trap was about 70% B_4H_{10}. This procedure has now been replaced by procedures based on diborane conversion.

1. The Conversion of B_2H_6 to B_4H_{10} at High Pressures

In Schlesinger's laboratory at the University of Chicago, it was noticed that appreciable concentrations of tetraborane built up when B_2H_6 was stored under high pressure. This observation was developed by Dillard[145] into an effective procedure for tetraborane synthesis. When B_2H_6 was stored for 9 days at 25°C and 40 atm pressure, B_4H_{10} yields of 20% were reported.[145] Other investigators have varied pressure, temperature, and storage time. A pressure of 250 atm at 90°C for 9 hr gave yields of 86% with 28% of the diborane converted.[146] With pressures of 400 to 450 atm at 80°C for 3 hr, yields as high as 95% B_4H_{10} with 17% total diborane conversion were reported.[146] Extensive refinement of the process was done by Wartik and his student at Pennsylvania State[70] and by Hunt[71] at Callery Chemical Co. The Callery process[71] for preparation of 150 g lots of B_4H_{10} is summarized below for the first time. A 2.0–2.5 lb sample of pure B_2H_6 was condensed into a one-gallon stainless steel Hoke cylinder, equipped with a 2200 psi relief valve. This was then stored at 25°C for 10 days in a remote and shielded location. The maximum pressure reached in the cylinder was about 100 atm. To recover the B_4H_{10}, the cylinder was cautiously cooled to $-78°C$ ($p = 400$–600 psi), then the H_2 and some B_2H_6 were allowed to escape until the cylinder pressure was 50 psi at $-78°C$. Using stainless steel tubing, the B_2H_6 was distilled from the reactor at $-50°C$ to a second cylinder at $-78°C$. About 1.8–2.0 lb of B_2H_6 were recovered. The remaining material in the original cylinder was led through a -78 and $-196°C$ trap. The B_4H_{10} stopped in the $-78°C$ trap while B_2H_6 and some B_4H_{10} were stopped at $-196°$. Repeated fractionation produced 150–175 g of B_4H_{10} or about a 15% yield based on the original B_2H_6 charged. The overall equation is:

$$2B_2H_6(g) \longrightarrow B_4H_{10}(g) + H_2(g)$$

2. Pyrolysis of B_2H_6 in a Hot–Cold Reactor

One of the more effective ways for converting B_2H_6 to B_4H_{10} is the hot–cold reactor developed by Klein, Harrison, and Solomon.[72] The reactor consists of a tube about 27 cm long and 8 cm in diameter sealed at the top by a ring seal into an outer tube about 28 cm long and 10 cm in diameter. An entrance tube sealed through the outer unit permits addition and removal of reagents. The volume of the reactor is about 550 cc. The outer tube was immersed in liquid nitrogen; then diborane was admitted and allowed to freeze on the outer wall. The inner tube was filled with ethylene glycol which was maintained at 120°C by an electric heater. The liquid nitrogen bath around the outer tube was replaced with a dry ice slush at $-78°C$, and the reaction started. The operating pressure was about 1700 mm and a blowout manometer set at 2000 mm protected the system. Reaction times ranging from 1–3 hr were used and yields were as high as 80% B_4H_{10}. The B_4H_{10} was condensed in a trap at $-126°C$. Less volatile components were stopped in a trap at $-64°C$ and more volatile components were led into a trap at $-196°C$. The reaction is the same as that for the high-pressure storage of B_2H_6 given above.

3. Preparation of B_4H_{10} from MB_3H_8 Salts and from MB_9H_{13} Salts

The ready conversion[73] of B_2H_6 to $[(CH_3)_4N][B_3H_8]$ gives an easily stored starting material for the preparation of B_4H_{10}. A 40% yield of B_4H_{10} is obtained if $[(CH_3)_4N][B_3H_8]$ is dropped into polyphosphoric acid in a vacuum.[74] Purification procedures similar to those described in Sections 1 and 2 above can be used. This reaction also produces about a 4% yield of B_6H_{12}.

Beall and Lipscomb[107] reported a 37% yield of B_4H_{10} from the action of polyphosphoric acid on $[(C_2H_5)_3NH]B_9H_{14}$ at 25°C in a vacuum. The $B_9H_{14}^-$ salts are prepared from the action of base on $B_{10}H_{14}$[106] (see section on B_6H_{10}).

4. Reactions for B_4H_{10} Preparation Which Have Little Value in Synthesis

Tetraborane can be obtained from several other hydrides in excellent yield, but such reactions have little value in synthesis at present, since the starting hydrides are more difficult to prepare than B_4H_{10}. Typical of these processes are the reactions: (a) B_6H_{12} with a limited amount of water to give approximately one mole of B_4H_{10} for each mole of B_6H_{12} used.[74]

$$B_6H_{12} + 6H_2O \longrightarrow B_4H_{10} + 4H_2 + 2B(OH)_3$$

The slow hydrolysis of B_4H_{10} may account for its appearance in this reaction; (b) B_5H_{11} with alkyl amines or ethers gives yields ranging from 30% in diglyme to 12% in $(CH_3)_3N$; and (c) B_5H_{11} with water gives yields of more than 95% B_4H_{10} based on the equation:

$$B_5H_{11} + 3H_2O \longrightarrow 2H_2 + B(OH)_3 + B_4H_{10}$$

Again, the slow hydrolysis of B_4H_{10} accounts for its appearance in this reaction.

While all of the above processes have little value in tetraborane preparation, they are of significance in any consideration of reaction mechanisms. The topic will be considered later.

Finally, Stock[1] prepared tetraborane in low yields by a Wurz-type coupling reaction:

$$2B_2H_5I + 2Na(amalgam) \longrightarrow B_4H_{10} + 2NaI$$

The process is not of current interest for synthesis processes.

5. The Preparation of Isotopically Labeled B_4H_{10} Molecules

Schaeffer and his students[76,79] found that the reaction of B_5H_{11} with D_2O in a closed tube at 0°C gave B_4H_9D. The deuterium is originally in one of the four bridge positions (NMR), but relatively rapid rearrangement takes place at room temperature.

Using information first obtained by Todd and Koski[80a] on the kinetics of the exchange of deuterium between diborane and tetraborane, Fehlner and Koski[80b] first prepared $B_4H_8D_2$ by the reaction

$$B_2D_6(\text{partial press} = 9mm) + B_4H_{10}(p = 3mm) \xrightarrow{45°} B_4H_8D_2$$

Evidence suggesting deuteration on the 1 and 3 boron atoms was obtained, but the site of deuteration was not unequivocal. Norman and Schaeffer[79] prepared $B_4H_8D_2$ by the action of D_2 on B_4H_8CO. On the basis of NMR data, they concluded that deuteriums were in a bridge position and on the 1-boron atom. Rearrangement of these deuterated species with resultant complete scrambling of deuteriums occurs in the vapor phase over a 30-min period at 25°C.[79,80]

Specific labeling of the borons was achieved by Schaeffer and Tebbe[75] using the reaction between $Na^{10}B_3H_8$, $^{11}B_2H_6$, and HCl:

$$Na^{10}B_3H_8 + HCl + {}^{11}B_2H_6 \longrightarrow NaCl + {}^{11}B^{10}B_3H_{10} + \tfrac{1}{2}{}^{11}B_2H_6 + H_2$$

The ^{11}B went into a 2 position; i.e., it was part of one H_2B group.

6. The Preparation of Alkyl and Halo Derivatives of B_4H_{10}

A few molecules of the general formula B_4H_9X are known where X is an alkyl group or a halogen atom. Lutz and Ritter[77] describe the preparation of $B_4H_9CH_3$ through the equation:

$$B_4H_{10} + B_2H_5CH_3 \longrightarrow B_4H_9CH_3 + B_2H_6$$

It is believed[77] that the CH_3 group is on a 2-boron, though its exact location is not known.

An ethylene tetraborane, $C_2H_4B_4H_8$, was prepared by Harrison, Solomon, Hites, and Klein[81] using the reaction of C_2H_4 and B_4H_{10}.

$$C_2H_4 + B_4H_{10} \xrightarrow[\text{reactor}]{\text{hot–cold}} B_4H_8C_2H_4 + H_2$$

The hot wall was 100°C, the cold was 0°. The reaction was stopped when pressure was a minimum and the yield was 70%. The preferred structure is one with a $-CH_2-CH_2-$ bridge across the top of the B_4H_8 unit (alkylated on 2 and 4 borons).

Dobson and Schaeffer[78] describe 2-bromotetraborane-10, B_4H_9Br, prepared by the reaction:

$$Br_2 + B_4H_{10} \xrightarrow[\text{12–18 hr}]{0-15°C} HBr + B_4H_9Br$$

The melting point of B_4H_9Br is given as $-37°C$ and its vapor pressure at 0°C is 9.9 mm.

C. The Pentaboranes (B_5H_9 and B_5H_{11}) and Their Alkyl and Halo Derivatives

1. The Synthesis of Pentaborane-11

a. Methods for B_5H_{11} Synthesis. The structure of pentaborane-11 and its planar projection are shown in Figure 3. The compound is best prepared from B_4H_{10} using the hot–cold reactor described earlier for use in B_4H_{10} synthesis.[72] The reaction is:

$$5B_4H_{10} \longrightarrow 4B_5H_{11} + 3H_2$$

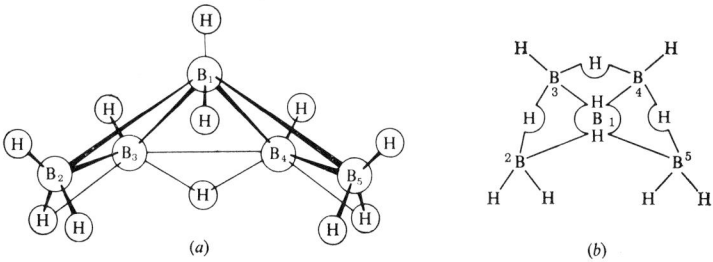

Fig. 3. (a) Pentaborane-11, perspective; (b) B_5H_{11}, planar projection.

Yields of 70% were obtained when the hot wall was at 120°C and the cold surface at −30°C. The mixture can be fractionated in the vacuum line as noted below.

Pentaborane-11 can also be obtained by holding equimolar amounts of purified B_2H_6 and B_4H_{10} in a one-liter bulb at 100°C (total press = 600 mm) for about 50–70 sec. The bulb is immediately quenched in liquid nitrogen. The reaction flask is then placed in a −95°C bath (toluene slush) and the volatile materials are passed through a trap at −126°C and through a trap cooled with liquid nitrogen. Tetraborane and a little B_5H_{11} are stopped at −126°C while B_2H_6 and entrained gases are stopped at −196°C. Further purification of material in the reaction flask can be achieved by fractionation through traps at −46, −78, and −196°C. The B_5H_{11} is stopped in the −78°C trap. Further fractionation of this material through a trap held at −64°C gives pure B_5H_{11}. In view of the extensive trap-to-trap distillation required to purify B_5H_{11}, a low-temperature, low-pressure fractionating column was recommended and described by Spielman and Burg.[87] A different low-temperature fractionating column operating at −98 to −101°C was recommended by Dobson and Schaeffer.[79,86]

The synthesis of B_5H_{11} directly from B_2H_6 without isolation of B_4H_{10} was effected by Norman and Schaeffer[79] using a hot–cold reactor.[72] The hot surface was held at 155 ± 5°C and the cold surface at −80°C. A B_2H_6 pressure of 1 atm and a reaction time of 6–7 hr converted 30–40% of the B_2H_6 to a mixture of B_4H_{10} and B_5H_{11} which was fractionated as described above.

b. Isotopically Labeled Pentaborane-11. $^{11}B^{10}B_4H_{11}$ may be prepared by the reaction between B_4H_8CO and $^{11}B_2H_6$ at 25°C for 2 min.[79] The preparation of B_4H_8CO and its reaction with B_2H_6 to give B_5H_{11} was first described by Spielman and Burg.[87] If $^{11}B_2H_6$ is used, the system is quenched in liquid nitrogen immediately after the initial reaction period, and low temperatures are maintained for fractionation. The ^{11}B NMR spectrum[79] indicated that ^{11}B was uniformly distributed over the four basal positions of B_5H_{11}. As the temperature was raised, ^{11}B became statistically distributed over the entire molecule.[79]

c. Alkyl-Substituted Pentaborane-11. Lutz and Ritter[77] prepared $B_5H_{10}CH_3$ by the low-yield reactions:

$$B_4H_{10} + \tfrac{1}{2}(BH_2CH_3)_2 \longrightarrow B_5H_{10}CH_3 + H_2$$
$$B_5H_{11} + B_2H_5CH_3 \longrightarrow B_5H_{10}CH_3 + B_2H_6$$

Products were separated by vapor phase chromatography (VPC) using helium as a carrier gas and standard white oil No. 9 on Johns-Manville firebrick, 32–65 mesh, as a column packing. (See also ref. 148 for VPC of boranes). A small amount of $B_5H_9(CH_3)_2$ was also separated from the first reaction. On the basis of NMR data, the methyl group in $B_5H_{10}CH_3$ was tentatively assigned to the frontal-pyramidal-base boron atom in B_5H_{11} (2 or 5 in Fig. 3), though no unequivocal structural data are available.

Macquire, Solomon, and Klein[102] reported that B_5H_{11} reacts with ethylene to give ethylpentaborane-11 and $C_2H_4B_4H_8$, reported earlier as a product of the ethylene B_4H_{10} reaction.

2. *The Synthesis of Pentaborane-9*

Pentaborane-9 is the most thermally stable of the fluid boron hydrides. Its structure and planar projection are shown in Figure 4.

a. The Direct Synthesis of B_5H_9 from B_2H_6. Like B_4H_{10} and B_5H_{11}, B_5H_9 is also prepared by the pyrolysis of B_2H_6. The relevant equation is:

$$5B_2H_6 \xrightarrow{\Delta} 2B_5H_9 + 6H_2$$

Competing reactions are:

$$5B_2H_6 \xrightarrow{\Delta} B_{10}H_{14} + 8H_2$$
$$xB_2H_6 \longrightarrow (BH)_{2x}(s) + 2xH_2$$

The conversion was studied very extensively in the U.S. high-energy fuel program. The results have been summarized in a monograph.[89] In the gas-phase pyrolysis, flow reactors, which are much superior to static reactors, can be used with reactor temperatures in the range 200–240°C; a residence time of a few seconds (depending on reactor size) and a 4:1 to 5:1 ratio of H_2 to B_2H_6 is recommended. Hydrogen helps to prevent the formation of

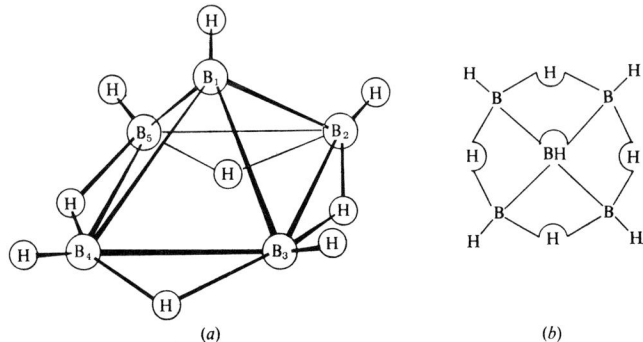

Fig 4. (*a*) Pentaborane-9, perspective; (*b*) B_5H_9, planar projection.

$B_{10}H_{14}$ and $(BH)_x$ solids by repressing the competing reactions shown above. Hydrogen also has the deleterious effect of retarding B_2H_6 conversion, in part because of its influence on gas molecule-wall contact. The temperature used in the pyrolysis is higher than that used in B_4H_{10} and B_5H_{11} production; at the higher temperature, B_4H_{10} and B_5H_{11} are converted to the more thermally stable B_5H_9. If the temperature is too high, excessive quantities of $(BH)_x$ solids are produced. Heating by adiabatic compression was also effective in this conversion and served as the basis for a plant design. In one of the most effective units for conversion, a 5:1 H_2/B_2H_6 mixture was passed at 400 cc/min through 12 reactor stages at a temperature of 240°C. Rapid quenching of the effluent gases retarded solid formation. About 57% of the boron was converted to pentaborane-9, 29% remained as B_2H_6, and the remainder was converted to solids. Pentaborane-11 could be obtained by operating the unit at 170°C. The details of operation have been carefully worked out. B_5H_9 could be an article of commerce if a demand were to be created.[84]

b. The Synthesis of B_5H_9 From B_4H_{10} through Base Catalysis. Edwards[88,150] and his colleagues found that the base adduct of B_3H_7 will react with excess B_4H_{10} to give B_5H_9 at temperatures near 100°C:

$$H_7B_3NR_3 + B_4H_{10} \xrightarrow[100°C]{near} (CH_3)_3NBH_3 + B_5H_9 + H_2 + \tfrac{1}{2}B_2H_6$$

Since the product $(CH_3)_3NBH_3$ will then react with more B_4H_{10} to regenerate $(CH_3)_3NB_3H_7$ and B_2H_6, $N(CH_3)_3$ can serve as a catalyst for the low-temperature synthesis of B_5H_9. The equation is:

$$(CH_3)_3NBH_3 + B_4H_{10} \longrightarrow \tfrac{1}{2}B_2H_6 + (CH_3)_3NB_3H_7$$

The equations indicating the catalytic and self-sustaining nature of the process are shown in Sequence 8. The overall process, as indicated by the sum of the equations between the lines in Sequence 8, can be summarized as:

$$4B_4H_{10} \xrightarrow{(CH_3)_3N} 3B_2H_6 + 2B_5H_9 + 2H_2$$

Since tetraborane can be formed from diborane by low-temperature pyrolysis, $(CH_3)_3N$ might well be a catalyst in the low-temperature conversion of B_2H_6 to B_5H_9. Indeed, when three moles of $(CH_3)_3NBH_3$ and one mole of B_2H_6 were heated to 125°C in a tube containing $\tfrac{1}{2}$ atm total pressure, almost 30% yield of B_5H_9 could be obtained in less than two hours.[150]

Initiation:

$2(CH_3)_3N + B_4H_{10} \longrightarrow (CH_3)_3NBH_3 + (CH_3)_3NB_3H_7$

Continuation No. 1

$(CH_3)_3NB_3H_7 + B_4H_{10} \longrightarrow (CH_3)_3NBH_3 + B_5H_9 + H_2 + \frac{1}{2}B_2H_6$
$2(CH_3)_3NBH_3 + 2B_4H_{10} \longrightarrow 2B_2H_6 + 2(CH_3)_3NB_3H_7$
$(CH_3)_3NB_3H_7 + B_4H_{10} \longrightarrow (CH_3)_3NBH_3 + B_5H_9 + H_2 + \frac{1}{2}B_2H_6$

Continuation No. 2

$(CH_3)_3NB_3H_7 + B_4H_{10} \longrightarrow (CH_3)_3NBH_3 + B_5H_9 + H_2 + \frac{1}{2}B_2H_6$
$2(3H_3)_3NBH_3 + 2B_4H_{10} \longrightarrow 2B_2H_6 + 2(CH_3)_3NB_3H_7$
$(CH_3)_3NB_3H_7 + B_4H_{10} \longrightarrow (CH_3)_3NBH_3 + B_5H_9 + H_2 + \frac{1}{2}B_2H_6$

etc.

Sequence 8

c. The Purification of B_5H_9. Pentaborane-9 may be purified by refluxing at 63°C to convert B_4H_{10} and B_5H_{11} to B_5H_9 and higher hydrides and then distilling the resulting mixture.[85a] Alternatively, Lewis bases may be added to react with other hydrides; then the B_5H_9 may be distilled away.[85b] Finally, B_5H_9 may be purified by gas chromatography.[85c,148]

d. The Synthesis of Pentaborane-9 by Specialized Reactions. A number of reactions involving difficultly available hydrides or hydride derivatives give B_5H_9, but these processes hold little attraction for use in general B_5H_9 synthesis. They may, however be of interest in the synthesis of specific B_5H_9 samples such as isotopically labeled molecules. These reactions may be summarized by the following equations:

$$B_6H_{12} \xrightarrow{(CH_3)_2O} B_5H_9 + \tfrac{1}{2}B_2H_6$$
$$2B_5H_{11} \longrightarrow B_5H_9 + H_2$$
$$2B_5H_{11} + H_2 \rightleftarrows B_5H_9 + 2\tfrac{1}{2}B_2H_6$$

e. Alkylated Pentaborane-9. The alkylation of B_5H_9 was extensively studied in the U.S. high-energy fuel program. A large quantity of commercial literature was developed; a portion of this has been reviewed in book form (ref. 4, p. 624 and ref. 89). A number of groups[90,91] demonstrated that B_5H_9 will react with alkyl halides or olefins in the presence of Lewis acid catalysts to give monoalkyl pentaborane-9. The alkyl group is attached to the apex boron of B_5H_9. $AlCl_3$ is a very effective catalyst when an alkyl halide is the alkylating agent and $FeCl_3$ is very effective when an olefin is the alkylating compound. Alkyl chlorides and bromides are more effective agents than alkyl iodides. Yields of alkyl pentaborane-9 ranged from 50 to 85% based on the B_5H_9 consumed. Only one ethyl or methyl group is added under these conditions. However, if the 1-alkyl pentaborane is allowed to rearrange to a 2-penta in the presence of a base catalyst,* the 2-alkyl pentaborane can be methylated further under Friedel-Crafts conditions to give $1,2(CH_3)_2B_5H_7$.

In the presence of 2,6-dimethylpyridine, rearrangement of 1,2-dimethylpentaborane-9 occurs over a 5-hr period at 25°C to give 2,3-dimethylpentaborane-9 with the methyl groups on adjacent basal borons of the B_5H_9 pyramid.[101] In general, ease of alkylation seems to increase in the order $CH_3—$, C_2H_5, $C_3H_7—$.

The polyethylated pentaboranes were prepared in very small yield by the pyrolysis of B_2H_6 in the presence of ethylene.[91b] The different products were separated by VPC and studied by IR and NMR spectroscopy. Compounds identified were: $B_5H_8(C_2H_5)$, $B_5H_7(C_2H_5)_2$, $B_5H_6(C_2H_5)_3$ (3 isomers), and $B_5H_5(C_2H_5)_4$ (2 isomers). A patent has been issued for the preparation of diethylpentaborane by the treatment of pentaborane with diethylborane at 25–65°C.[96]

The Friedel-Crafts alkylation, as well as apex deuteration and halogenation, appears to involve electrophilic attack on B_5H_9.

f. Deuterated or Halogenated Pentaboranes. Deuterium chloride exchanges the apex proton with B_5H_9 in the presence of $AlCl_3$.[95] The rearrangement at 25°C of 1-deuteropentaborane when catalyzed by 2,6-dimethylpyridine, was studied by Onak, Gerhart, and Williams.[95] They concluded that the base-catalyzed low-temperature rearrangement is an *intra*molecular process. In contrast, an *inter*molecular mechanism prevails when 1-deuteropentaborane undergoes deuterium–protium exchange at 145°C.

The 1-halopentaboranes can be prepared in yields above 80% by the direct halogenation of pentaborane in the presence of $AlCl_3$.[91c,97–100]

$$X_2 + B_5H_9 \xrightarrow{AlCl_3} B_5H_8X + HX$$

In the absence of $AlCl_3$, only traces of 1-chloropentaborane were obtained, but a 15% yield of 2-chloropentaborane was isolated. Gaines[97] suggested that in the absence of a strong Lewis acid, the chlorination of B_5H_9 is a radical reaction. By contrast, the halogenation with Br_2 or I_2 is visualized

---

* This arrangement is promoted by 2,6-dimethylpyridine[92a] and by $(CH_3)_3N$.[93] The same rearrangement is achieved by heating 1-alkyl pentaborane to 200°C.[92b] Considerable speculation on mechanism has been given.[92,94,101]

as a heterolytic cleavage of the halogen followed by displacement of an apex proton by X^+. The $AlCl_3$ presumably facilitates the halogenation process in the conventional fashion, i.e., by promoting the heterolytic cleavage of X_2 to give X^+ and $AlXCl_3^-$. Reaction of B_5H_9 with ICl or ICl_3 gives 1-iodopentaborane-9 in yields above 90%.

Burg and Sandhu[100] prepared 2-BrB_5H_8 from 1-BrB_5H_8 by using hexamethylenetetramine as the basic rearrangement catalyst. Dimethyl ether is also effective. Onak and Dunks[98] prepared 2-chloropentaborane from 1-bromopentaborane and $AlCl_3$.

Burg and Sandhu[100] found that 1-BrB_5H_8 reacts with $(CH_3)_2O$ at 38°C over a period of 30 hr to give a fair yield of 1-$CH_3B_5H_8$.

D. The Hexaboranes (B_6H_{10}, Isomeric B_6H_{10}, and B_6H_{12})

1. Hexaborane-10 (B_6H_{10})

The structure of hexaborane is shown in Figure 5. This hydride, first prepared by Stock in very small amounts from the acid hydrolysis of magnesium boride, is difficult to prepare. In a recent refinement of the magnesium boride procedure, Timms and Phillips[104] obtained 6% yields of B_6H_{10} by the action of $8M$ H_3PO_4 on Mg_3B_2. Except for this single low-efficiency process, B_6H_{10} is obtained from other hydrides which are themselves relatively unavailable. For example, B_5H_{11} decomposes in the presence of dimethyl ether to give hexaborane and diborane.[103] According to Edwards and co-workers, the equation given below is followed almost

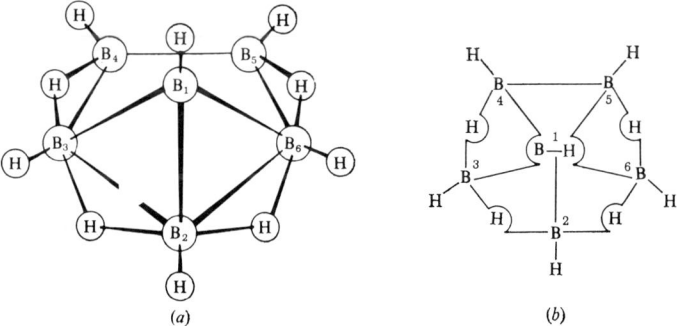

Fig. 5. (a) Hexaborane-10; (b) B_6H_{10}, planar projection. The boron framework bonds are strongly delocalized. The number of 2-center and 3-center bonds is taken to give the correct number of electrons only.

quantitatively* under appropriate conditions:

$$2B_5H_{11} \xrightarrow{(CH_3)_2O} B_6H_{10} + 2B_2H_6$$

Other weak bases also decompose B_5H_{11} to give yields of B_6H_{10} ranging from 17 to 30%.[76b] Bases used included $(CH_3)_3N$, $(CH_3)_2O$,* diglyme, and the residue left from the reaction of B_5H_{11} and $[(CH_3)_2N]_2BH$.

Very low yields of B_6H_{10} are obtained from the passage of B_2H_6 through a silent electric discharge.[105] This procedure is not recommended. A more recent method of B_6H_{10} synthesis involves the action of polyphosphoric acid on $[(C_2H_5)_3NH]B_9H_{14}$ at 25°C in a vacuum.[107] Yields of B_6H_{10} were only 2.5% based on the $[(C_2H_5)_3NH]B_9H_{14}$ used, but the method was suggested as a practical one since $[(C_2H_5)_3NH]B_9H_{14}$ can be made rather easily[106] from the action of aqueous KOH on $B_{10}H_{14}$:

$$B_{10}H_{14} + 2OH^- \longrightarrow [B_{10}H_{13}OH]^{-2} + H_2O$$
$$[B_{10}H_{13}OH]^{-2} + H_3O^+ + H_2O \longrightarrow B_9H_{14}^- + B(OH)_3 + H_2$$

In evaluating this process, it is important to note that the 2.5% yield is less than half of the 6% yield of B_6H_{10} isolated by Timms and Phillips[104] from the hydrolysis of magnesium boride. The choice between these two methods will obviously be dependent upon the relative availability of $[(C_2H_5)_3NH]B_9H_{14}$ and Mg_3B_2.

Hexaborane-10 is also observed as a product in the decomposition of B_8H_{12} [82] and in the reaction of B_8H_{12} with limited amounts of water.[151]

$$B_8H_{12} \longrightarrow B_6H_{10} + \frac{2}{x}(BH)_x(\text{solids})$$

$$B_8H_{12} + 3H_2O \longrightarrow B_6H_{10} + B_2O_3 + 4H_2$$

Dobson and Schaeffer[151] report yields of 98% for the B_8H_{12}–water reaction. This fact coupled with a direct route to B_8H_{12} from KB_9H_{14} [144] and the ready availability of KB_9H_{14} might make this route to B_6H_{10} acceptable under certain circumstances. Unfortunately, hexaborane-10 is still a difficult material to prepare.

2. Isomeric Hexaborane-10

An isomer of B_6H_{10} has recently been isolated by Shore and Geanangel[67] from the reaction of the newly prepared $[(CH_3)_4N]B_5H_8$ and B_2H_6:

---

* Burg[76b] did not verify the quantitative decomposition of B_5H_{11} reported by Edwards.[103] Under the conditions used by Burg [warming a $B_5H_{11}/(CH_3)_2O$ mixture (ratio 0.64) from -78 to $-20°C$], the product was a mixture consisting of 25% B_2H_6, 20% B_4H_{10}, 3% B_5H_9, 25–27% B_6H_{10}, 2% $B_{10}H_{14}$, and about 25% nonvolatiles.

$$[(CH_3)_4N]B_5H_8 + B_2H_6 \longrightarrow MBH_4 + i\text{-}B_6H_{10}$$

The isomeric B_6H_{10} converts to normal B_6H_{10} on standing at room temperature. It gives 20–30% yields of $B_{10}H_{14}$ when it is heated in diglyme. Shore and Geanangel suggest that the new hydride is formed by insertion of a $BH_2{}^+$ group into the open bridge of the $B_5H_8{}^-$ ion. The $BH_2{}^+$ results from nonsymmetrical cleavage of the B_2H_6 molecule. The process is presented in Figure 6.

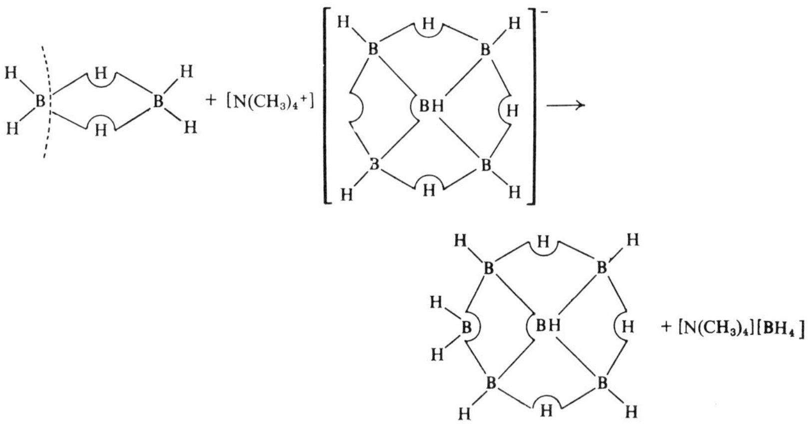

Fig. 6. Proposed mode of formation of $i\text{-}B_6H_{10}$.

3. Hexaborane-12 (B_6H_{12})

No x-ray data are available for B_6H_{12}, but the most probable structure based on spectroscopic evidence shows a striking resemblance to B_5H_{11}.[53] The proposed B_6H_{12} structure is shown in Figure 7.

Although B_6H_{12} was tentatively identified by Stock,[1] it was never isolated as a distinct pure species until 1963.[53,74] It was obtained in two ways. Gaines and Schaeffer[74] obtained B_6H_{12} in about 4% yield from the previously-described reaction of polyphosphoric acid with $[(CH_3)_4N]B_3H_8$. Lutz, Phillips, and Ritter,[53] on the other hand, used gas–liquid chromatography to separate very small amounts of B_6H_{12} (yields less than 1%) from a variety of boron hydride interactions such as: (a) the reaction of B_4H_{10} and B_2H_6 in a static system at 110°C for 10 min, (b) the reaction of B_4H_{10} and B_2H_6 in a hot–cold reactor (110 and 0°C), (c) the decomposition of liquid B_5H_{11} at 25°C. The last procedure was recommended as the best of the three. It was noted that B_6H_{12} seems to be reasonably stable when very pure; but it is highly reactive, particularly toward other boron hydrides. The compound reacts quantitatively with an excess of water to

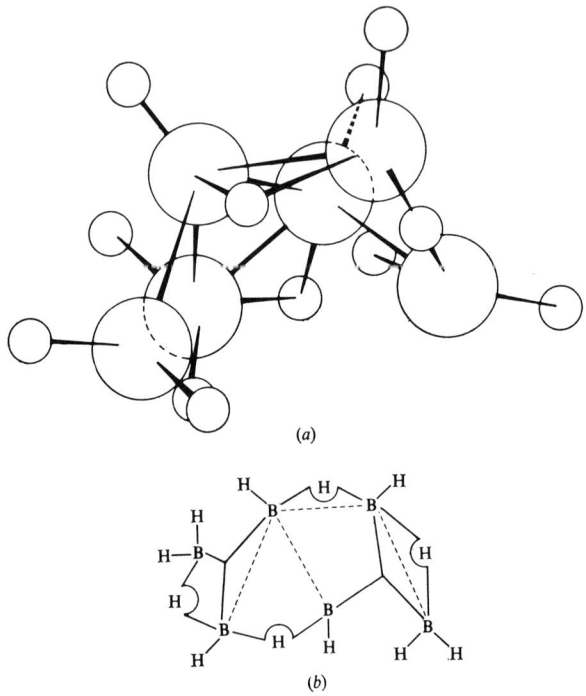

Fig. 7. (a) Proposed B_6H_{12} structure; (b) planar projection of B_6H_{12}. Dashed line outlines B_5H_{11} unit.

form B_4H_{10}[74] and with $(CH_3)_2O$ to form B_5H_9. With H_2, B_6H_{10} gives significant yields of B_5H_{11}, B_5H_9, B_4H_{10}, and B_2H_6.[53]

No derivatives of the hexaboranes have been reported.

E. Heptaboranes and Octaboranes (B_8H_{12}, B_8H_{14}, B_8H_{18})

1. Heptaboranes—Not Identified

Although at least four heptaboranes have been postulated from mass spectral data,[63,64] the most recent and most careful review of the evidence[82] suggests that mass spectral peaks in the heptaborane region were due to ethylpentaborane. At the present time, the statement of R. E. Williams provides the most defensible position on the state of the heptaboranes. He writes, "Since the only experimental evidence for heptaboranes to date has been simple mass spectral data, we feel that their existence has not been demonstrated and the evidence is open to serious doubt."

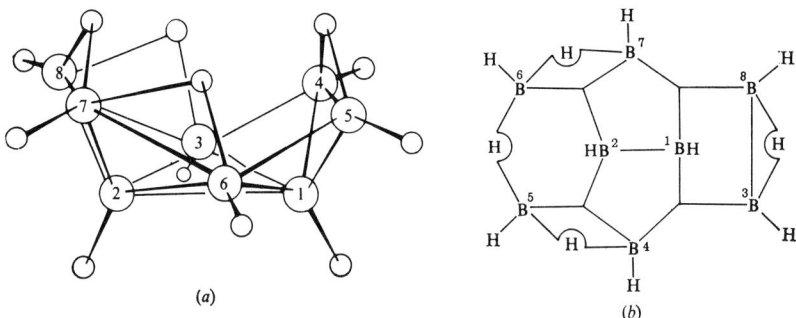

Fig. 8. (a) Octaborane-12, perspective; (b) B_8H_{12}, planar projection.

2. Octaborane-12 (B_8H_{12})

Although octaborane-12 is one of the least common and least stable of the boron hydrides, it has been unequivocally characterized and its structure established by a three-dimensional x-ray diffraction analysis. The structure is shown in Figure 8. The compound was suggested by Schlesinger and Burg and by Shapiro and Keilin,[108] but isolation and characterization of the species was first done by Enrione, Boer, and Lipscomb[55] in a difficult and ingenious study. A 2:1 mixture of B_2H_6 and B_5H_9 was swept through an electrical discharge using a low-pressure (12 mm) hydrogen stream as a carrier. The B_8H_{12} was obtained from the reaction mixture by a rather complex fractionation procedure.[55] Handling procedures were complicated by the fact that the compound is thermally unstable at temperatures much above $-20°C$.

More recently, Ditter, Spielman, and Williams[82] found that B_8H_{12} is a decomposition product of B_9H_{15} at very low pressures.

$$B_9H_{15} \underset{\text{excess diborane}}{\overset{\text{low pressure}}{\longleftrightarrow}} B_8H_{12} + \tfrac{1}{2}B_2H_6$$

Since B_9H_{15} is an unstable material and was prepared by the pressurization of B_5H_{11} with B_2H_6 (see section on nonaboranes), this route to B_8H_{12} also has limitations.

The best reported route to B_8H_{12} appears to be through isomeric B_9H_{15}. Dobson, Keller, and Schaeffer[144] report that isomeric B_9H_{15} (see section on nonaboranes for synthesis) decomposes above $-35°C$ to give 10–15% yields of B_8H_{12} which can be easily separated from the reaction mixture. The reactions suggested are:

$$i\text{-}B_9H_{15} \longrightarrow B_9H_{13} + H_2$$

$$2B_9H_{13} \longrightarrow B_8H_{12} + B_{10}H_{14}$$

Because KB_9H_{14} is available from KOH and $B_{10}H_{14}$, and because $i\text{-}B_9H_{15}$ is readily prepared from KB_9H_{14} and HCl, this may provide the most accessible route to B_8H_{12}.

In a typical experiment, a 3.0-g sample of KB_9H_{14} was treated with excess HCl at $-80°C$ in a 500 ml flask equipped with a magnetic stirrer. After excess HCl was removed, a 100 ml aliquot of dry n-pentane was condensed into the flask. The vessel was warmed to $-45°C$ and the contents stirred at this temperature for several minutes. The $-45°C$ bath was removed and replaced with one at $-30°C$ contained in an uninsulated vessel. The temperature of the system was then allowed to rise with constant stirring to about $-5°C$ over a 10-min period. The pressure of the evolved hydrogen was kept below one atmosphere by removal into the vacuum system. When the decomposition appeared to be nearly complete, as evidenced by only a slow pressure rise, the flask was cooled to $-78°C$, the cooling bath removed, and the major portion of the n-pentane distilled away. The last few milliliters of solvent contained most of the B_8H_{12} and all of the other products. To separate this mixture, the pentane was distilled away at $-78°C$, the trap was warmed to $-30°C$, and the B_8H_{12} was distilled away from the $B_{10}H_{14}$. A short distillation path for B_8H_{12} reduces decomposition. About 0.2–0.3 g of B_8H_{12} has been recovered using this procedure. Diethyl ether appears to stabilize B_8H_{12}.

Octaborane-12 is a strong[151] monobasic Lewis acid. Compounds of the type $B_8H_{12} \cdot L$ were prepared where L is $(C_2H_5)_2O$, $(CH_3)_3N$, H_3CCN. On the basis of NMR data, bonding to the 4-boron was suggested.[151] Hydrolysis with about three moles of H_2O per mole of B_8H_{12} gives a nearly quantitative yield of B_6H_{10}.

$$B_8H_{12} + 3H_2O \longrightarrow B_6H_{10} + B_2O_3 + 4H_2$$

Diborane reacts with B_8H_{12} to give an approximately 45% yield of $n\text{-}B_9H_{15}$.

$$2B_8H_{12} + \tfrac{3}{2}B_2H_6 \longrightarrow B_9H_{15} + B_{10}H_{14} + 2H_2$$

The reaction with C_2H_2 has not been resolved. Reactions of B_8H_{12} with NaH or Na (amalgam) gave a product which appears to be NaB_8H_{12}, NaB_8H_{13}, or a mixture. This was converted to $[(CH_3)_4N]B_8H_x$ by metathesis.

3. Octaborane-14 (B_8H_{14})

Small amounts of an unstable hydride, B_8H_{14}, have been obtained[151] from the reaction of HCl and the solid of empirical composition $[(CH_3)_4N]B_8H_x$ at $-78°C$. The hydride decomposes above $-30°C$. A sample warmed to $0°C$ for thirty seconds was 30% decomposed to octaborane-12. The structure *suggested* by Dobson and Schaeffer[151] for B_8H_{14} on the basis of NMR evidence can be represented in planar projection as shown here.

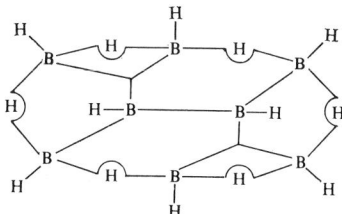

This structure would have hydrogen bridge bonds around the entire open periphery of the boron framework. It was suggested that such a structure might well give rise to hydrogen crowding and to the observed ease of hydrogen loss.

4. Octaborane-18 (B_8H_{18})

Dobson, Gaines, and Schaeffer[56] reported B_8H_{18} as a trace product of the reaction between $B_3H_8^-$ and polyphosphoric acid in a vacuum. The volatile products caught in a trap at $-65°C$ were subjected to continuous pumping. The more volatile hydrides were volatilized and B_6H_{12}, B_9H_{15}, and B_8H_{18} remained. In the low-temperature fractionation of this mixture, B_8H_{18} was retained at $-62°C$. A 10 g sample of $[(CH_3)_4N]B_3H_8$ gave about 30 mg of B_8H_{18}—a thermally unstable compound which was difficult to characterize. One structure proposed is that of two B_4H_{10} units linked by removing one H from a BH_2 unit on each B_4H_{10} molecule and then joining the two resulting $B_4H_9\cdot$ units with a boron–boron bond. Isomers would, of course, be possible; see Figure 9.

F. The Nonaboranes (B_9H_{15} and i-B_9H_{15})

1. Normal B_9H_{15}

Nonaborane-15, B_9H_{15}, was first* isolated by Kotlensky and Schaeffer[52] from the gases obtained by passing B_2H_6 through an electric

* The formula B_9H_{15} was first suggested by Norton[109] on the basis of mass spectral data, but definitive evidence for this hydride was not available until the work of Kotlensky and Schaeffer.

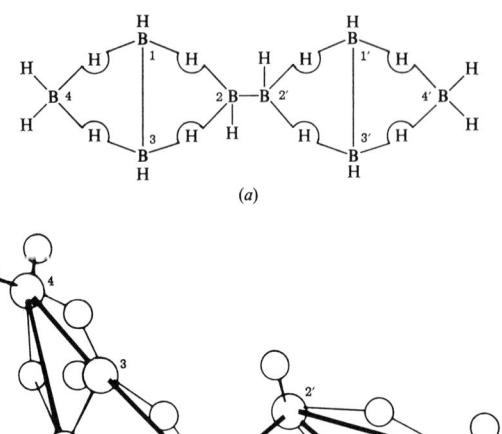

Fig. 9. (a) B_8H_{18}, planar projection; (b) proposed structure for B_8H_{18}, perspective.

discharge. Although yields were very low, a crystal was obtained which permitted a three-dimensional x-ray diffraction study.[110] The structure shown in Figure 10 has been established unequivocally.

Burg and Kratzer[111] improved the synthesis procedure dramatically. They obtained B_9H_{15} in 13% yields from the reaction between B_5H_{11} and a surface of hexamethylenetetramine, $(CH_2)_6N_4$. The B_9H_{15} was purified by a combination of high-vacuum distillation and low-temperature crystallization. Because the procedure is reasonably complex, the original reference[111] should be consulted.

Ditter, Spielman, and Williams[82] prepared B_9H_{15} by the reaction between B_5H_{11} and B_2H_6 at 25 atm and 25°C for several days. They note that B_9H_{15} is thermally unstable, decomposing at low pressures to give B_8H_{12} and B_2H_6

$$B_9H_{15} \underset{\text{excess } B_2H_6}{\xleftarrow{\text{low pressure}}} B_8H_{12} + \tfrac{1}{2}B_2H_6$$

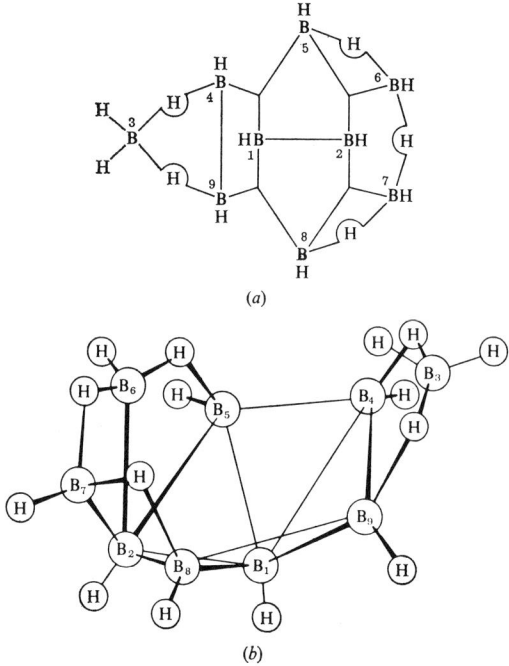

Fig. 10. (a) B_9H_{15}, planar projection; (b) nonaborane-15, B_9H_{15}, perspective.

2. Isomeric B_9H_{15}

An isomeric B_9H_{15} was prepared by Dobson, Keller, and Schaeffer[57,144] using the reaction between KB_9H_{15}[106] and excess liquid HCl at $-80°C$ in a sealed tube. The excess HCl was vaporized and the B_9H_{15} was extracted from the solid residue using dry, cold pentane. The compound decomposes above $-30°C$. Its structure is not known.

A number of reactions of i-B_9H_{15} have been recorded recently. When allowed to warm up fairly rapidly, i-B_9H_{15} gives respectable yields of B_8H_{12}, $B_{10}H_{14}$, and n-$B_{18}H_{22}$. When i-B_9H_{15} decomposes slowly, n-$B_{18}H_{22}$ is the major hydride isolated. It is suggested[144] that the decomposition process involves initial loss of H_2. The resulting product then either disproportionates (B_8H_{12} and $B_{10}H_{14}$) or condenses ($B_{18}H_{22}$). Ligand complexes of the form B_9H_{13} (ligand) could be isolated. Ligands were diethyl or di-n-butyl ethers, or triphenylphosphine. Refluxing $B_9H_{13}O(C_2H_5)_2$ gave n-$B_{18}H_{22}$ in 35% yield.

G. The Decaboranes

1. Decaborane-14 ($B_{10}H_{14}$)

a. Direct Synthesis of $B_{10}H_{14}$. The structure of decaborane was established by Kasper, Lucht, and Harker[117,152] in one of the classic studies of modern chemistry. The structure, based on the boron icosahedron found in $B_{12}C_3$ and in elemental boron, is shown in Figure 11.

Decaborane-14 was prepared and was well characterized by Stock.[1] Perturbations of his early methods of synthesis have been used for the large-scale preparation of $B_{10}H_{14}$. The procedures involve: (*a*) heating B_2H_6 to 115–120°C for 48 hr; (*b*) heating B_4H_{10} to 90–95°C for 5 hr; or (*c*) polymerizing B_5H_{11} at room temperature. Clearly, the lower hydrides go toward decaborane when heated or allowed to stand. Unfortunately for decaborane production, $B_{10}H_{14}$ itself will decompose on heating, to give undefined polymeric solids of approximate composition B_nH_n. In any synthesis operation, conditions must be carefully adjusted to maximize

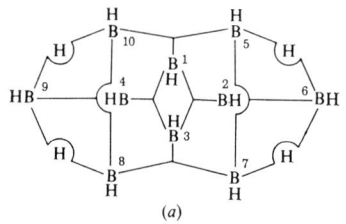

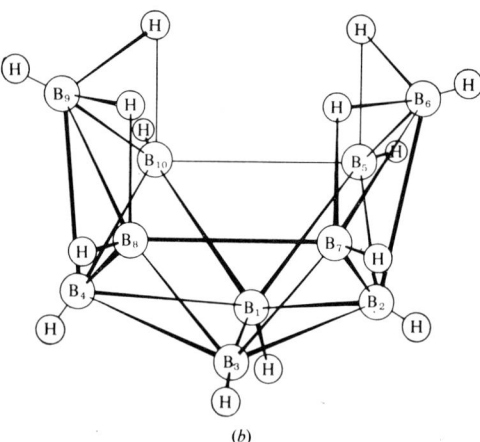

Fig. 11. (*a*) $B_{10}H_{14}$, planar projection; (*b*) decaborane-14, $B_{10}H_{14}$, perspective.

conversion of lower hydrides to $B_{10}H_{14}$ and to minimize the formation of polymeric solids. By using a pyrolysis temperature near 100°C and by recycling the lower unconverted hydrides, it is possible to obtain reasonable yields of $B_{10}H_{14}$ directly from B_2H_6. Decaborane can be separated from the complex reaction mixture by sublimation or by extraction with a hydrocarbon such as 2,2-dimethylbutane.

Hillman, Mangold, and Norman[112] prepared $B_{10}H_{14}$ by heating equimolar quantities of diborane and pentaborane-9. Pyrolysis of the mixture gave better yields than the pyrolysis of either component separately. The real nature of the interaction involving both B_5H_9 and B_2H_6 was demonstrated[112] when normal B_5H_9 was pyrolyzed in the presence of $^{10}B_2H_6$. The resulting decaborane had five ^{10}B atoms per molecule, indicating that *one* B_5H_9 combines with *two and one-half* B_2H_6 molecules. (The mechanism may involve five BH_3 units, or even a preformed intermediate generated from the initial pyrolysis of B_2H_6.) Further mechanistic speculation is considered later. Clearly, two B_5H_9 molecules do not combine preferentially to give $B_{10}H_{14}$. Copyrolysis of B_4H_{10} and B_5H_9 [113] and of B_2H_6 and B_4H_{10} [112,114] give $B_{10}H_{14}$ as does the straight pyrolysis of B_4H_{10}.[115]

Lewis bases are reported[116] to promote the formation of decaborane from the pyrolysis of diborane. The situation is analogous to that reported for B_5H_9 synthesis (see p. 70) and is probably attributable to the same type of reactions. Methyl ether is reported to give a yield of 50% $B_{10}H_{14}$ when B_2H_6 is pyrolyzed for 10 min at 150°C. CO was also listed as helpful in the pyrolysis.

b. Synthesis of Alkylated Decaboranes. The reaction between $B_{10}H_{14}$ and an alkyl halide in the presence of $AlCl_3$ (Friedel-Crafts conditions) was developed and patented as part of the U.S. high-energy fuel program.[118] Olefins may also be used as the carbon source.[119]

$$RX + B_{10}H_{14} \xrightarrow[CS_2]{AlCl_3} RB_{10}H_{13} + HBr$$

$$R-C(H)=CH_2 + B_{10}H_{14} \xrightarrow{AlCl_3} R-\underset{\underset{H}{|}}{\overset{\overset{H}{|}}{C}}-\underset{\underset{H}{|}}{\overset{\overset{H}{|}}{C}}-B_{10}H_{13}$$

Clearly, many alkylated decaboranes are possible even if one considers only one kind of alkyl group. The 2-methyl and 2-ethyl; 2,4- and 1,3-dimethyl or diethyl; 1,2,3- and 1,2,4-trimethyl; and 1,2,3,4-tetramethyl derivatives were separated from appropriate Friedel-Crafts mixtures by vapor-phase chromatography. The separated products were studied by

IR and NMR spectroscopy and structures were assigned.[120] Only the 1,2,3, and 4 positions were substituted under the conditions used in these studies. The order of probability of substitution under Friedel-Crafts conditions is: 2,4 > 1,3 > 5,7,8,10 > 6,9. Assuming that the process goes by electrophilic attack, this may be interpreted as the order of decreasing electron density in the transition state of $B_{10}H_{14}$. A charge of -0.254 has been assigned[121] to borons 2 and 4 and $+0.046$ to borons 1 and 3. All other borons are assumed to carry a higher positive charge. To the extent that ground-state molecular-orbital calculations can be used to estimate charge distribution in the transition state of $B_{10}H_{14}$, consistency is observed; however, as Hawthorne points out (ref. 5, p. 271), this correlation does not have to be one-to-one. A review of the extensive literature[89] of the high-energy fuel project indicates that alkylation of decaborane is best accomplished by the Friedel-Crafts method. In general, less polyalkylation occurs when an inert solvent such as pentane or *n*-hexane is used. If polychloroalkanes are used, some chlorination of the alkyldecaboranes results. Higher hydrocarbons such as kerosene inhibit the alkylation while a trace of water improves the effectiveness of the $AlCl_3$ catalyst. $AlCl_3$ seems to be the best catalyst, although $FeCl_3$ and $GaCl_3$ were also effective. Temperatures used ranged from 60–160°C.

Decaborane can also be alkylated by formation of a Grignard reagent.[122-125]

$$B_{10}H_{14} + CH_3MgI \longrightarrow B_{10}H_{13}MgI + CH_4$$
$$B_{10}H_3MgI + RX \longrightarrow RB_{10}H_{13} + MgXI$$

Alkylation under these conditions should presumably proceed by nucleophilic attack and the position of alkylation should be reversed from that observed under Friedel-Crafts conditions. It is not surprising then, that substitution is found chiefly in the 5 or 6 position. Some anomalous results relative to preferential substitution in either the 5 or 6 position have been recorded. Structural conclusions based on NMR[123] indicated that alkylation of the decaborane Grignard with *dimethyl sulfate* gave equal amounts of 5 and 6 methyldecaborane; if the alkylating agent were *diethyl sulfate*, it was reported that the product was primarily 5-ethyldecaborane, but if RX were benzylchloride, substitution in the 6-position was reported. These observations would not be consistent with an explanation based only on steric factors. Studies reported by Hawthorne (ref. 5, p. 269) have indicated that the nature of the leaving group is important in the displacement sequence. Decaborane Grignards react with alkyl fluorides, whereas in many cases corresponding reactions with alkyl chlorides, bromides, or iodides did not occur.[89] Fluoride, sulfate, and ethereal oxygens have generally been most effective. This fact gave rise to the suggestion (ref. 5, p. 269) that bonding of the leaving group to the magnesium in the transition state might aid in stabilization of the activated complex.

Decaborane-13 alkali metal "organometallics" give reactions with RX compounds similar to the reactions of Grignard reagents.

The sodium salt of decaborane-13 can be prepared by the reaction between NaH and $B_{10}H_{14}$:

$$NaH + B_{10}H_{14} \xrightarrow{(C_2H_5)_2O} NaB_{10}H_{13} + H_2$$

This product then reacts in ether with suitable RX compounds to give 5 or 6 substituted products:

$$NaB_{10}H_{13} + C_6H_5CH_2Br \xrightarrow{(C_2H_5)_2O} 6\text{-}C_6H_5CH_2\text{—}B_{10}H_{13} + NaBr$$

Reaction of *decaborane* with lithium ethyl gives alkylation principally in the 6 positions[123]; however, with methyl, some 5-substitution and some polymethyl substitution were observed.

$$LiR + B_{10}H_{14} \longrightarrow LiB_{10}H_{14}R$$
$$LiB_{10}H_{14}R + H_2O \longrightarrow LiOH + B_{10}H_{13}R + H_2$$

Diethyl mercury and diethyl zinc were ineffective (ref. 89, p. 158):

$$Zn(C_2H_5)_2 + B_{10}H_{14} \longrightarrow C_2H_6 + C_2H_5ZnB_{10}H_{13}$$
$$C_2H_5ZnB_{10}H_{13} + RX \longrightarrow \text{no reaction}$$

A rather complex mixture of alkylated deca- and other boranes was obtained by the copyrolysis of mixtures of diborane and ethyldiborane. Separation and purification problems inherent in this method render it of limited interest for specific synthesis problems.

A more detailed account of the alkylation of decaborane and a comprehensive tabulation of the properties of many boranes, alkyl boranes, and related compounds, is given in the summary of the U.S. high-energy fuel program reviewed by Holzman, Hughes, Smith, and Lawless.[89] Most of the alkylated decaboranes are either liquids or low-melting solids at room temperature. (See also ref. 4, p. 656.)

c. Synthesis of Halogenated Decaboranes. In general, halogens react very slowly with decaborane. A disubstituted compound seems to be the most stable product in each halogenation process (ref. 4, p. 61). Schaeffer[128] prepared $B_{10}H_{12}I_2$ by the direct reaction of $B_{10}H_{14}$ and I_2 and showed by x-ray methods that the iodines were in positions 2 and 4. The same decaborane–iodine reaction gives small amounts of a 2-monoiododecaborane plus a second monoiodo product now believed to be the 1-monoiodo isomer.[129] A second diiodo isomer is now believed to be the 1,2-diiodo product.[129] The reaction of Br_2 and $B_{10}H_{14}$ in the presence of $AlCl_3$ catalyst gives a monobromodecaborane.[130]

Patents[131] have been issued for the preparation of three isomeric monochlorodecaboranes from the reaction of $CHCl_3$ or CH_2Cl_2 with $B_{10}H_{14}$ in the presence of $AlCl_3$. Bromoform and decaborane gave dibromodecaborane if $AlCl_3$ was a catalyst.

Halogenation of decaborane, like Friedel-Crafts alkylation, appears to involve electrophilic attack on decaborane. It is then not surprising that $AlCl_3$ is an effective catalyst and that the 2,4 positions are easiest to halogenate and the 1,3 positions the next easiest. The process resembles alkylation and deuteration mechanistically.

2. Decaborane-16 ($B_{10}H_{16}$)

Decaborane-16, like B_8H_{18}, belongs to that group of hydrides formed by linking lower hydrides together through elimination of H_2. The compound $B_{10}H_{16}$ is made from two B_5H_8 units joined through the apex borons. The structure shown in Figure 12 was established unequivocally through x-ray diffraction.[132]

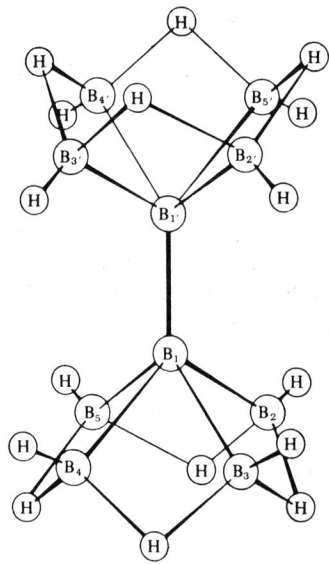

Fig. 12. Decaborane-16, perspective.

The structure really suggests the method of synthesis. Pentaborane-9 was carried through an electric discharge by an H_2 stream.[132] Preparation and characterization of the compound by Grimes, Wang, Lewis, and Lipscomb[132] was unequivocal. Hall and Koski[65] used neutron irradiation of B_5H_9 to form $B_{10}H_{16}$. Characterization was by mass spectroscopy.

$B_{10}H_{16}$ reacts with HI to give B_5H_9 and B_5H_8I. One of the rather unexpected features of $B_{10}H_{16}$ is the ease with which it is converted to $B_{10}H_{14}$ or its derivatives.[94] Iodination of $B_{10}H_{16}$ at 150°C produced $B_{10}H_{14}$ and 2HI. When $B_{10}H_{16}$ was treated with $AlCl_3$ in CS_2 solution, $B_{10}H_{14}$ was the principal product.[94] Pyridine converted $B_{10}H_{16}$ to the $B_{10}H_{14}$ derivative, $B_{10}H_{12}(Py)_2$. The mechanisms of these reactions have been related to the rearrangement of 1-alkyl to 2-alkyl pentaborane-9.[133]

3. Decaborane-9 Free Radical ($B_{10}H_9$)

Lewis and Kaczmarczyk[58] obtained good evidence for a decaborane-9 free radical. When $CuCl_2$ was treated with $K_2B_{10}H_{10}$ in ether, the reaction taking place was represented by the equations:

$$2CuCl_2(s) + K_2B_{10}H_{10}(s) \xrightarrow{\text{ether}} 2KCl(s) + CuCl(s) + HCl + CuB_{10}H_9$$

$$CuB_{10}H_9 + CuCl_2 \longrightarrow 2CuCl(s) + (B_{10}H_9 \cdot)$$

Little information on the chemistry of this radical is currently available.

H. Undecaboranes ($B_{11}H_x$)

Three distinct B_{11} hydrides have been reported. Hall and Koski[65] presented mass spectral evidence for a hydride $B_{11}H_{17}$, analogous to $B_{10}H_{16}$, which would be formed by linking a —B_5H_8 unit and a —B_6H_9 unit at the apex borons, H_8B_5—B_6H_9. Characterization is far from complete, but the structure and compound seem eminently reasonable in view of the known structure of H_8B_5—B_5H_8 and the assumed structure for the known compound, H_9B_4—B_4H_9. Two other 11-boron hydrides, $B_{11}H_{15}$ and $B_{11}H_{13}$, reported by Edwards and Mahklouf,[59] were prepared by the reactions:

$$NaB_{11}H_{14} \cdot 2.5C_4H_8O_2 + HCl \xrightarrow[\substack{6 \text{ hr} \\ 0°C}]{(CH_3)_2S} B_{11}H_{15} \cdot 2.0C_4H_8O_2 + NaCl$$

$$B_{11}H_{15} \cdot 2.0C_4H_8O_2 \longrightarrow B_{11}H_{13} \cdot 2.0C_4H_8O_2 + H_2$$

* $NaB_{11}H_{14}$ can be prepared[137] by the reaction of BH_4^- with decaborane in dioxane or dimethoxyethane at 90°C.

$$BH_4^- + B_{10}H_{14} \longrightarrow B_{11}H_{13}^- + 2H_2$$

Until the solvent is removed from these hydrides and complete characterization of the nonsolvated materials is obtained, they cannot be classed as new distinct hydrides but rather as Lewis base adducts comparable to $C_4H_8O_2BH_3$, $C_4H_8O_2B_3H_7$, and OCB_4H_8, etc.

Still, the recent discovery[151] that B_8H_{12} is a strong Lewis acid and readily adds bases to form compounds such as $B_8H_{12} \cdot$ ligand suggests that the border between "stable hydrides" and Lewis acid fragments may not be distinct and that compounds such as B_8H_{12} and perhaps $B_{11}H_{13}$ and $B_{11}H_{15}$ may lie in this twilight zone.

I. Unidentified Hydrides (Possible $B_{12}H_x$, $B_{13}H_x$, and $B_{14}H_x$)

No distinct, noncoordinated $B_{12}H_x$, $B_{13}H_x$, or $B_{14}H_x$ hydrides have yet been reported; but the linkage of simpler hydride units to give compounds comparable to B_8H_{18} and $B_{10}H_{16}$ would appear to be clear and reasonable possibilities. Compounds expected could include H_9B_6—B_6H_9, H_9B_6—B_8H_{11}, H_9B_4—B_9H_{14}, etc. In view of the instability of B_8H_{12} and the low stability of the hydride involving —B_4H_9 units, the latter two compounds would probably be very unstable. On the other hand, $B_{12}H_{18}$ might well be a reasonably stable compound which could be prepared by carrying B_6H_{10} through an electric discharge on a hydrogen stream.

J. Pentadecaborane ($B_{15}H_{22}$)

No clearcut characterization of any B_{15} hydride has been achieved; however, Hall and Koski[65] reported mass spectral evidence for a possible hydride, H_9B_5—$B_{10}H_{13}$.

K. The Octadecaboranes (n-$B_{18}H_{22}$ and i-$B_{18}H_{22}$)

1. Normal $B_{18}H_{22}$

Pitochelli and Hawthorne[134] have reported the preparation of $B_{18}H_{22}$ from the $B_{20}H_{18}^{-2}$ ion.* An ethanolic solution of the triethylammonium salt of $B_{20}H_{18}^{-2}$ was passed through an acidic ion exchange column and the resulting solution was concentrated.

---

* The $B_{20}H_{18}^{-2}$ can be prepared by the chemical oxidation of $B_{10}H_{10}^{-2}$ using a number of metal oxidizing agents with $E°$ values in the range 1.33–1.51. Useful agents are Ce^{+4}, PbO_2, MnO_4^-, and $Cr_2O_7^{-2}$. The $B_{10}H_{10}^{-2}$ can be prepared under appropriate conditions from the interaction of B_2H_6 with metal borohydrides.

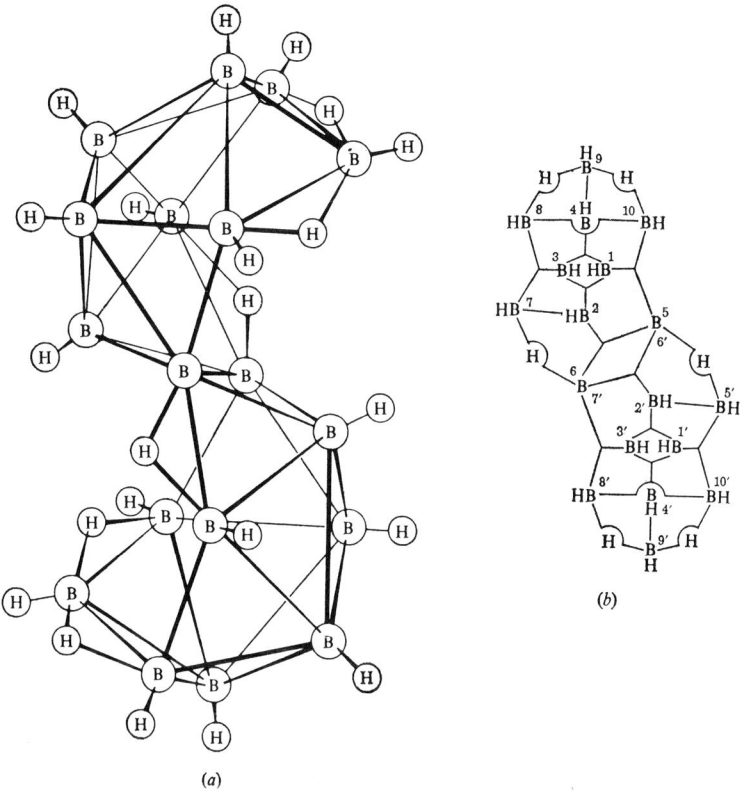

Fig. 13. (a) n-Octadecaborene-22, perspective; (b) n-$B_{18}H_{22}$, planar projection.

The concentrate was dissolved in diethyl ether and hydrolyzed. After the solvent was evaporated, the product was recrystallized from cyclohexane and sublimed. Yields exceeded 60% based on the $B_{20}H_{18}^{-2}$ used.

The structure of $B_{18}H_{22}$, as determined by Simpson and Lipscomb,[135] consists of two decaborane cages joined at the 5–6' and 6–5' positions. (See Fig. 13.)

The hydride $B_{18}H_{22}$ is a strong monoprotic acid, presumably losing H$^+$ from a bridge position.[136] $B_{18}H_{21}^-$ has a bright yellow color.

2. Isomeric-$B_{18}H_{22}$

Simpson, Folting, and Lipscomb[138] found that a lesser product of the reaction of $B_{20}H_{18}^{-2}$ with an acid was an *isomer* of $B_{18}H_{22}$. The structure

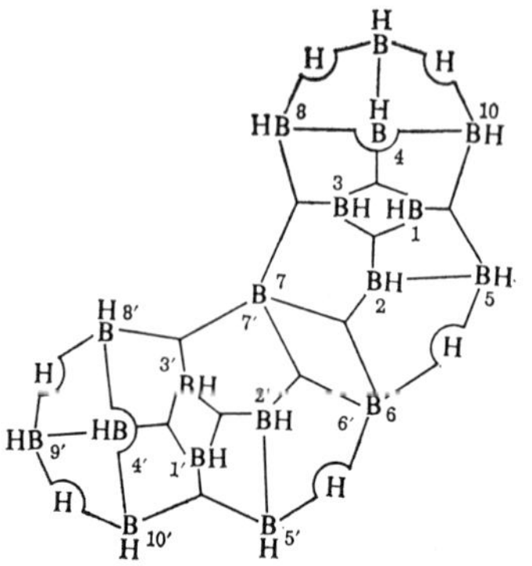

Fig. 14. *i*-Octadecaborane-22, planar projection.

which was determined by x-ray diffraction can be described as two $B_{10}H_{14}$ cages fused at the 5–6' and 6–7' positions (see Fig. 14)

L. The Icosaboranes ($B_{20}H_x$)

1. Icosaborane-16 ($B_{20}H_{16}$)

Icosaborane-16 was prepared and characterized independently in two separate laboratories. Miller and Muetterties[140] pyrolyzed $B_{10}H_{14}$ at 350°C and 1 mm pressure in the presence of catalytic amounts of $CH_3HNB(CH_3)_2$. The $B_{20}H_{16}$ which resulted was purified by vacuum sublimation and recovered in about 10% yield based on the reaction:

$$2B_{10}H_{14} \longrightarrow B_{20}H_{16} + 6H_2$$

The same compound was prepared independently by Friedman, Dobrott, and Lipscomb[141] by passing $B_{10}H_{14}$ through an electric discharge. Using x-ray diffraction methods, they showed that the structure can be represented as two $B_{10}H_{14}$ cages fused at the mouths. (See Fig. 15.)

2. Icosaborane-26 ($B_{20}H_{26}$)

By deuteron irradiation of $B_{10}H_{14}$, Hall and Koski[65] prepared a material whose mass spectrum suggested the formula, $H_{13}B_{10}$—$B_{10}H_{13}$. Characterization is still incomplete.

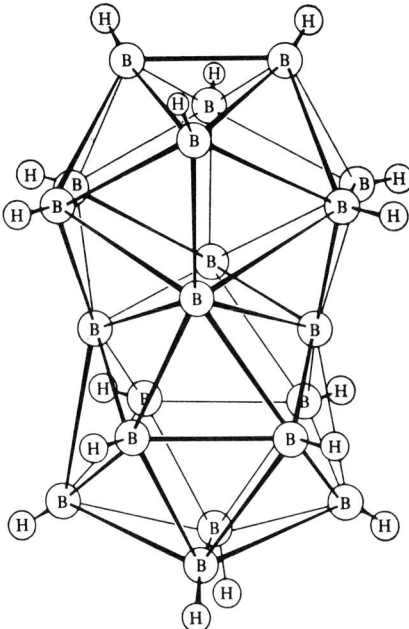

Fig. 15. Icosaborane-16, $B_{20}H_{16}$.

M. Higher Hydrides

Although no hydrides beyond $B_{20}H_{16}$ have been clearly characterized up to this time, the yellow solids obtained in boron hydride pyrolysis suggest strongly that other hydrides exist which consist of the simpler units linked together. Future developments will probably permit identification of these materials.

IV. THE MECHANISM OF BORON HYDRIDE CONVERSIONS

A. General Considerations

Because of the importance of borane pyrolysis in the synthesis of the higher hydrides, considerable effort has been devoted to an elucidation of the pyrolysis mechanism. Up to the present time such studies have frequently generated more heat than light.

Several types of reactions seem to dominate borane chemistry. One of these is loss or gain of a BH_3 group by a boron hydride fragment. The versatility of BH_3 loss or gain is emphasized by the fact that all but two of the known or probable hydrides listed in Table I can be represented[82,136] by the empirical formula $(BH)_n(BH_3)_x$.* Loss of a BH_3 would appear to provide an easy mode of interconversion. Another significant reaction would appear to be loss or gain of H_2 from the boron framework. These two processes seem to have dominated much of the thinking on mechanisms.

B. The Pyrolysis of Diborane

Early studies[154] on diborane pyrolysis indicated that the rate of disappearance of B_2H_6 was $\frac{3}{2}$ order with respect to diborane. An activation energy of 26 kcal/mole was reported; further, it was noticed that the addition of hydrogen reduced the rate of disappearance of diborane. These facts led to the following mechanistic proposal[154b]:

$$B_2H_6 \rightleftarrows 2BH_3$$

$$BH_3 + B_2H_6 \underset{k_3}{\overset{k_2}{\rightleftarrows}} B_3H_7 + H_2$$

$$B_3H_7 + B_2H_6 \longrightarrow B_4H_{10} + BH_3$$

Since the tetraborane recovered was only a small fraction of that expected, it was clear that further pyrolysis of the tetraborane was taking place. The foregoing would indeed provide a reasonable explanation except that several studies[154d,155] have shown rather clearly that the initial product *isolated* during pyrolysis is B_5H_{11}, not B_4H_{10}. B_4H_{10} appears later. This fact plus others led Clapper[156] and Stewart and Adler[154d] to propose the following for pyrolysis at 112°C:

$$B_2H_6 \rightleftarrows 2BH_3 \quad \text{(rapid)}$$
$$BH_3 + B_2H_6 \rightleftarrows B_3H_9 \quad \text{(rapid)}$$
$$B_3H_9 \rightleftarrows B_3H_7 + H_2$$
$$B_3H_7 + B_2H_7 \longrightarrow B_5H_{11} + H_2$$
$$B_3H_7 + B_2H_6 \longrightarrow B_4H_{10} + BH_3$$
$$B_5H_{11} \longrightarrow B_4H_8 + BH_3$$
$$B_4H_8 + H_2 \rightleftarrows B_4H_{10}$$
$$B_4H_8 + B_2H_6 \longrightarrow B_5H_{11} + BH_3$$

* For example, B_2H_6 is $(BH)_0(BH_3)_2$; B_4H_{10} is $BH(BH_3)_3$; B_5H_9 is $(BH)_3(BH_3)_2$, etc.

$$B_4H_{10} \longrightarrow B_3H_7 + BH_3$$
$$B_5H_{11} \longrightarrow B_5H_9 + H_2$$
$$B_5H_9 + B_3H_7 \longrightarrow B_8H_{14} + H_2$$
$$B_8H_{14} \longrightarrow B_8H_{12} + H_2$$
$$B_8H_{12} + B_2H_6 \longrightarrow B_{10}H_{16} + H_2$$
$$B_{10}H_{16} \longrightarrow B_{10}H_{14} + H_2$$
$$B_{10}H_{14} + BH_3 \longrightarrow B_{11}H_{15} + H_2$$
$$B_{11}H_{15} + B_2H_6 \longrightarrow B_{12}H_{12} + BH_3 + 3H_2 \quad \text{(rapid)}$$
$$B_{12}H_{12} \longrightarrow \text{polymer} + H_2$$

Several investigators[157,158] agree with the first three steps in the sequence shown above, but serious doubts have been raised over the interpretation of the time of B_5H_{11} appearance. Schaeffer[157] suggested that B_4H_{10} is the initial product but it is converted rapidly to B_5H_{11}. His process would be:

$$B_2H_6 \longrightarrow 2BH_3$$
$$BH_3 + B_2H_6 \longrightarrow B_3H_9$$
$$B_3H_9 \longrightarrow B_3H_7 + H_2$$
$$B_3H_7 + B_2H_6 \longrightarrow B_4H_{10} + BH_3$$
$$B_4H_{10} \longrightarrow B_4H_8 + H_2$$
$$B_4H_8 + B_2H_6 \longrightarrow B_5H_{11} + BH_3$$

Since Enrione and Schaeffer[157] found the rate of decomposition of B_2H_6 at 100°C to be five times faster than that of perdeuterodiborane, they suggested that the rate-controlling step must be

$$B_3H_9 \longrightarrow B_3H_7 + H_2$$

Fehlner and Koski[158] pyrolyzed B_2H_6 in a shock tube and suggested the following sequence in which B_4H_{10} precedes the formation of B_5H_{11}.

$$B_2H_6 \rightleftarrows 2BH_3$$
$$B_2H_6 + BH_3 \longrightarrow B_3H_9$$
$$B_3H_9 \rightleftarrows B_3H_7 + H_2$$
$$B_3H_7 + B_2H_6 \longrightarrow B_4H_{10} + BH_3$$
$$B_3H_9 + B_2H_6 \longrightarrow B_5H_9 + 3H_2$$
$$B_4H_{10} \longrightarrow B_3H_7 + BH_3$$
$$B_4H_{10} \rightleftarrows B_4H_8 + H_2$$
$$B_4H_8 + B_2H_6 \longrightarrow B_5H_{11} + BH_3$$
$$B_4H_8 + B_2H_6 \longrightarrow B_6H_{12} + H_2$$
$$B_6H_{12} \rightleftarrows B_6H_{10} + H_2$$

$$B_5H_{11} \rightleftarrows B_4H_8 + BH_3$$
$$B_3H_7 + B_4H_{10} \rightleftarrows B_7H_{13} + 2H_2$$
$$B_7H_{13} \longrightarrow B_7H_{11} + H_2$$
$$n\text{-}B_4H_8 \longrightarrow m(\text{solid}) + p\text{-}H_2$$

They reported that B_6H_{12} was the most abundant intermediate seen in their shock tube experiments.

The postulate that B_4H_{10} formation precedes the formation of B_5H_{11} was supported by Stafford[159] and his students in their mass spectral study of B_2H_6 pyrolysis. They reported direct evidence for a B_3H_x and a possible B_4H_8. More recent data[162] now suggest that the B_4H_8 may have been produced by the ionizing beam rather than by the diborane pyrolysis.

Although agreement on the cleavage of diborane to give $2BH_3$ groups seemed to be general and seemed to be demanded by the data, this point was brought into question in 1965 when Fehlner[160] carefully considered several anomalies in the data on B_2H_6 pyrolysis. He noted that the mechanism usually accepted did not account for the appearance of BH_2 units seen in mass spectral studies by Fehlner and Koski.[161] Further, it did not explain reported changes in reaction order with temperature,[160] and the large kinetic isotope effect in the production of H_2.[157] In view of these points, he proposed the following mechanism based on a nonsymmetrical cleavage of B_2H_6.

$$B_2H_6 \xrightarrow{1} BH_2 + BH_4$$
$$BH_4 \xrightarrow{2} BH_2 + H_2$$
$$BH_2 + B_2H_6 \xrightarrow{3} B_2H_5 + BH_3$$
$$B_2H_5 \xrightarrow{4} BH_3 + BH_2$$
$$B_2H_5 \xrightarrow{5} B_2H_4 + H$$
$$H + B_2H_6 \underset{7}{\overset{6}{\rightleftarrows}} B_2H_5 + H_2$$
$$2BH_3 \underset{9}{\overset{8}{\rightleftarrows}} B_2H_6$$
$$2BH_2 \xrightarrow{10} B_2H_4$$

It was then assumed that B_2H_4, the net product of the above sequence, condenses rapidly with itself and other species to give the observed products. He could rationalize reaction order and other known facts and anomalies in terms of this mechanism.

Even more recently in a mass spectral study of diborane pyrolysis, Baylis, Pressley, and Stafford[162] reported that the BH_3 unit was clearly identified in the pyrolysis but no BH_2 was found. They noted, however, that their result did not eliminate the BH_2 mechanism proposed by Fehlner since BH_2 could be preferentially destroyed in this system by wall reactions. Their work revealed fragments containing 3- and 4-, but no 5-boron atoms; however, it was established that these fragments arose from secondary processes involving neutral diborane and $B_2H_x{}^+$ generated in the ion source. It was noted that the B_4 fragment produced in this process is identical to the pattern attributed earlier[159] to B_4H_8 in the pyrolysis. It is thus doubtful that the B_4H_8 reported earlier has real significance in the absence of the ionizing beam.

In summary, it is quite clear that the pyrolysis of B_2H_6 is a very difficult problem and is not well understood. Although most investigators have favored the cleavage of B_2H_6 to give $2BH_3$ groups and the combination of BH_3 and B_2H_6 to give B_3H_9, even this concept has been questioned recently and a mechanism involving cleavage to BH_2 and BH_4 has been suggested. Data currently available to do not resolve the question. The so-called "first-intermediate question" also remains unanswered. Is the first stable hydride formed from B_2H_6, B_4H_{10}, or B_5H_{11}? Much of the current data point to B_4H_{10} as the first intermediate, but again, results are far from unequivocal. Mechanisms giving products above B_5H_{11} appear to be largely speculation.

C. Transitions and Exchanges Involving B_4H_{10} and B_5H_{11}

While the detailed data on diborane conversion are confusing, some of the chemistry and some of the exchange reactions for B_4H_{10} and B_5H_{11} offer strong mechanistic suggestion. Parry and Edwards[136] noted that the chemistry of tetraborane suggests two types of decomposition into neutral species. The first is loss of a BH_3 group to give a B_3H_7 unit (symmetrical bridge cleavage), and the second is loss of H_2 to give B_4H_8.

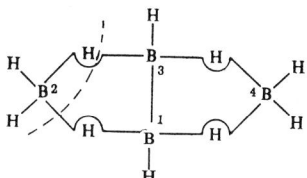

Clearly, borons 2 or 4 would be lost more easily than 1 or 3. A sizeable mass of evidence indicates that this is so. For example, the synthesis of $^{10}BB_3H_{10}$ with boron-10 in the 2 position and synthesis of monoalkylated

B_4H_{10} by reaction with $B_2H_4(CH_3)_2$ are consistent with this view as is the isolation of B_3H_7 complexes. Recently, Norman, Schaeffer, Baylis, Pressley, and Stafford[163] found by the use of ingenious labeling experiments that $B_3H_x^+$, $B_2H_x^+$, and BH_x^+ ions found in their mass spectrometer result from dissociation and/or ionization processes in which ion fragments are lost preferentially from the 2 or 4 positions of B_4H_{10}. They observed, however, that H or D was lost randomly and that there was no position preference.

Mechanisms involving loss of H_2 from B_4H_{10} are also supported by chemical arguments. Pearson and Edwards[164] suggested the following equilibria to explain the pyrolysis of tetraborane:

$$B_4H_{10} \rightleftarrows B_4H_8 + H_2$$
$$B_4H_{10} \rightleftarrows B_3H_7 + BH_3$$

Both reactions are assumed to proceed as competing operations. It is significant that B_4H_8CO can be obtained from B_4H_{10} and CO as would be expected from the first equation above. Reactions to produce base–B_3H_7 adducts are well known.

Cleavage of a BH_3 group from B_5H_{11} to give a B_4H_8 unit is also easily rationalized by structural arguments and is consistent with data given earlier from labeling experiments and the synthesis of $B_5H_{10}CH_3$. It is also consistent with the synthesis of $^{10}B^{11}B_4H_{11}$ from B_4H_8CO and $^{10}B_2H_6$.

D. Polyhedral Rearrangements and Exchange Reactions of Higher Hydrides

In an earlier review of boron hydride reactions,[136] it was noted that boron hydrides may lose H^+ by ionization from a bridge position. Further, the larger the boron framework, the greater the delocalization of electrons in the ion and the greater the degree of ion stabilization. Thus, the larger the boron framework, the more acidic the bridge hydrogens should be. This is clearly seen in $B_{10}H_{14}$ where four bridge hydrogens will exchange with D in basic solution and in $B_{18}H_{22}$ which is a strong acid. It is known that deuteriums originally in the bridge in $B_{10}H_{14}$ will undergo slow exchange and distribute to other positions in the molecule. Similarly, it was noted that alkyl groups, halo atoms, or deuterium atoms on the 1 position of B_5H_9 will undergo a base-catalyzed rearrangement to move the appropriate group to the 2 position away from the apex. These changes in both $B_{10}H_{14}$ and B_5H_9 appear to involve framework rearrangements which are of great interest in studies of the carboranes. Lipscomb[165] has treated the subject of framework rearrangements recently. The base-catalyzed rearrangement of B_5H_8R can be visualized by a distortion of the octahedral fragment of B_5H_8—R to give borons in an almost trigonal bipyramidal arrangement which then collapses to give the new structure. Hough, Edwards, and Stang[93] suggested that the base removes a proton from B_5H_8R to give the $B_5H_7R^-$ ion which can rearrange more readily. A more detailed discussion of framework rearrangements is appropriate for any discussion of the carboranes.

REFERENCES

1. A. E. Stock, *Hydrides of Boron and Silicon*, Cornell University Press, Ithaca, 1933.
2. H. I. Schlesinger, R. T. Sanderson, and A. B. Burg, **61**, 536 (1939); **62**, 3421 (1940); H. I. Schlesinger and H. C. Brown, *ibid.*, **62**, 3429 (1940); H. I. Schlesinger, H. C. Brown, R. Hoekstra, and L. R. Rapp, *ibid.*, **75**, 199 (1933).
3. W. N. Lipscomb, *Boron Hydrides*, Benjamin, New York, 1963.
4. R. M. Adams, *Boron, Metallo-Boron Compounds, and Boranes*, Interscience, New York, 1964.
5. E. L. Muetterties, *The Chemistry of Boron and Its Compounds*, Wiley, New York, 1967.
6. "Borax to Boranes," *Advan. Chem. Ser.*, **32** (1961).
7. A. E. Finholt, A. C. Bond, and H. I. Schlesinger, *J. Am. Chem. Soc.*, **69**, 1199 (1947).
8. H. I. Schlesinger, H. C. Brown, J. R. Gilbreath, and J. J. Katz, *J. Am. Chem. Soc.*, **75**, 195–9 (1953).
9. (a) R. M. Adams and R. K. Pearson, U.S. Pat. 2,968,531, Jan. 17, 1961; (b) F. H. May and H. Hammar, U.S. Pat. 2,013,863, Jan. 29, 1959.
10. (a) C. D. Good and H. D. Batha, U.S. Pat. 2,867,499; Jan. 6, 1959; (b) C. D. Good, U.S. Pat. 3,019,085; Jan. 30, 1962.
11. I. Shapiro, H. G. Weiss, M. Schmich, S. Skolnik, and G. B. L. Smith, *J. Am. Chem. Soc.*, **74**, 901 (1952).
12. H. I. Schlesinger and H. C. Brown, *J. Am. Chem. Soc.*, **75**, 219 (1953).
13. H. I. Schlesinger, H. C. Brown, R. Hoekstra, and L. R. Rapp, *J. Am. Chem. Soc.*, **75**, 199 (1953).
14. H. I. Schlesinger, University of Chicago, Signal Corps Contract W3434-sc-174 Final Report 1944.

15. H. I. Schlesinger and H. C. Brown, U.S. Pat. 2,543,511; Feb. 27, 1951.
16. J. D. Bush, R. A. Carpenter, and W. H. Schechter, U.S. Pat. 3,014,059; Dec. 19, 1961.
17. J. R. Elliott, E. M. Boldebuck, and C. F. Roedel, *J. Am. Chem. Soc.*, **74**, 5047 (1952); U.S. Pat. 2,658,815, 2,658,816; Nov. 10, 1953.
18. V. D. Aftandalian, H. C. Miller, and E. L. Muetterties, *J. Am. Chem. Soc.*, **83**, 2471 (1961).
19. C. B. Jackson, R. M. Bovard, and J. R. Taylor, U.S. Pat. 2,796, 328-9.
20. E. M. Fedneva, *Zh. Neorgan. Khim.*, **4**, 286 (1959); V. I. Mikheeva and E. M. Fedneva, *Bull. Acad. Sci. USSR, Div. Chem. Sci. (Engl. Trans.)*, **1956**, 925.
21. Metropolitan Vickers Electrical Co., Ltd., British Pat. 787,771; Dec. 18, 1957.
22. C. J. Smith, Jr., U.S. Pat. 2,983,582, May 9, 1961.
23. H. C. Brown and P. J. Tierney, *J. Am. Chem. Soc.*, **80**, 1522 (1958).
24. D. T. Hurd, *J. Am. Chem. Soc.*, **71**, 20 (1949).
25. V. I. Mikheeva and T. N. Dymova, *Zh. Neorgan. Khim.*, **2**, 2530 (1957).
26. E. L. Muetterties, U.S. Pat. 3,019,086; Jan. 30, 1962.
27. E. I. du Pont de Nemours and Co., British Pat. 832,049; 832,133; 832,134; April 6, 1960.
28. A. E. Stock, F. Kurzen and H. Laudenklos, *Z. Anorg. Allgem. Chem.*, **225**, 243 (1935); W. V. Hough, L. J. Edwards, and A. D. McElroy, *J. Am. Chem. Soc.*, **78**, 689 (1956); *ibid.*, **80**, 1838 (1958); D. L. Chamberlain, Jr., U.S. Pat. 3,029,128, April 10, 1962; H. C. Brown, E. J. Mead, and B. C. S. Rao, *J. Am. Chem. Soc.*, **77**, 6209 (1955).
29 (a) H. I. Schlesinger and H. C. Brown, *J. Am. Chem. Soc.*, **75**, 219 (1953); (b) H. I. Schlesinger, H. C. Brown, R. Hoekstra, and L. R. Rapp, *J. Am. Chem. Soc.*, **75**, 199 (1953); (c) H. I. Schlesinger, R. T. Sanderson, and A. B. Burg, *J. Am. Chem. Soc.*, **62**, 3421 (1940); (d) A. B. Burg and H. I. Schlesinger, *J. Am. Chem. Soc.*, **62**, 3425 (1940); (e) H. I. Schlesinger and H. C. Brown, *J. Am. Chem. Soc.*, **62**, 3429 (1940), etc.
30. H. I. Schlesinger, H. C. Brown, and A. E. Finholt, *J. Am. Chem. Soc.*, **75**, 205 (1953).
31. W. H. Schechter, C. B. Jackson, and R. M. Adams, *Boron Hydrides and Related Compounds*, Callery Chemical Co., Callery, Pa., 1954.
32. (a) D. Goerrig, W. Schlabacher, and F. Schubert, German Pat. 1,036,222; Aug. 14, 1958; (b) L. J. Edwards, U.S. Pat. 3,042,485; July 3, 1962.
33. E. C. Ashby and W. E. Foster, *J. Am. Chem. Soc.*, **84**, 3407 (1962); D. Goerrig, German Pat. 1,078,098, Aug. 9, 1958, **56**, 5634 (1962).
34. G. Broja and W. Schlabacher, German Pat. 1,108,670; Oct. 6, 1959; F. Schubert, K. Lang, and W. Schlabacher, German Pat. 1,067,005, Oct. 15, 1959; F. Schubert, K. Lang, W. Schlabacher, and D. Goerrig, German Pat. 1,053,746, March 26, 1959.
35. W. L. Jolly and T. Schmitt, *J. Am. Chem. Soc.*, **88**, 4282 (1966); *Inorg. Chem.*, **6**, 344 (1967).
36. R. M. Adams, *Advan. Chem. Ser.*, **32**, 60 (1961).
37. H. C. Brown, K. J. Murray, L. J. Murray, J. A. Snover, and G. Zweifel, *J. Am. Chem. Soc.*, **82**, 4233 (1960).
38. S. J. Chiras, U.S. Pat. 2,967,760, Jan. 10, 1961.
39. H. G. Weiss and I. Shapiro, *J. Am. Chem. Soc.*, **81**, 6167 (1959).
40. R. D. Schultz and C. L. Randolph, U.S. Pat. 3,020,127, Feb. 6, 1962.

41. B. J. Duke, J. R. Gilbert, and I. A. Read, *J. Am. Chem. Soc.*, **1964**, 540.
42. H. C. Brown and P. J. Tierney, *J. Am. Chem. Soc.*, **80**, 1522 (1958).
43. R. K. Pearson, L. L. Lewis, and L. J. Edwards, *Nucl. Sci. Abstr.*, **12**, 4069 (1958).
44. M. J. Nicole, French Pat. 1,181,025, June 11, 1959.
45. A. D. McElroy and R. M. Adams, U.S. Pat. 2,992,266, July 11, 1961.
46. R. Koester and K. Ziegler, *Angew. Chem.*, **69**, 94 (1957).
47. H. I. Schlesinger and A. B. Burg, *J. Am. Chem. Soc.*, **53**, 4321 (1931).
48. H. W. Myers and R. F. Putnam, *Inorg. Chem.*, **3**, 655 (1963).
49. T. A. Schegaleva and E. M. Belavskaia, *Dokl. Akad. Nauk SSSR*, **136**, 638 (1961).
50. R. E. Davis, A. E. Brown, R. Hopmann, and C. L. Kibby, *J. Am. Chem. Soc.*, **85**, 487 (1963).
51. A. I. Gorbunov and G. S. Solov'eva, *Zh. Neorgan. Khim.*, **12**, 3 (1967).
52. W. Kotlensky and R. Schaeffer, *J. Am. Chem. Soc.*, **80**, 4517 (1958).
53. C. A. Lutz, D. A. Phillips, and D. M. Ritter, *Inorg. Chem.*, **3**, 1191 (1964).
54. D. F. Gaines and R. W. Schaeffer, *Inorg. Chem.*, **3**, 438 (1964).
55. R. E. Enrione, F. P. Boer, and W. N. Lipscomb, *J. Am. Chem. Soc.*, **86**, 1451 (1964); *Inorg. Chem.*, **3**, 1659 (1964).
56. J. Dobson, D. Gaines, and R. Schaeffer, *J. Am. Chem. Soc.*, **87**, 4072 (1965).
57. J. Dobson, P. Keller, and R. Schaeffer, *J. Am. Chem. Soc.*, **87**, 3522 (1965).
58. J. S. Lewis and A. Kaczmarczyk, *J. Am. Chem. Sov.*, **88**, 1068 (1966).
59. L. J. Edwards and J. Makhlauf, *J. Am. Chem. Soc.*, **88**, 4728 (1966).
60. A. R. Pitochelli and M. F. Hawthorne, *J. Am. Chem. Soc.*, **84**, 3218 (1962).
61. N. E. Miller and E. L. Muetterties, *J. Am. Chem. Soc.*, **85**, 3506 (1963); N. E. Miller and J. A. Forstner, *Inorg. Chem.*, **3**, 1690 (1964).
62. L. B. Friedman, R. D. Dobrott, and W. N. Lipscomb, *J. Am. Chem. Soc.*, **85**, 3505 (1963).
63. R. W. Schaeffer, K. H. Ludlum, and S. E. Wiberley, *J. Am. Chem. Soc.*, **81**, 3157 (1959).
64. S. G. Gibbons and I. R. Shapiro, *J. Am. Chem. Soc.*, **82**, 2968 (1960).
65. L. H. Hall and W. S. Koski, *J. Am. Chem. Soc.*, **84**, 4205 (1962).
66. P. G. Simpson, K. Folting, and W. N. Lipscomb, *J. Am. Chem. Soc.*, **85**, 1879 (1963).
67. S. G. Shore and R. Geanangel, *J. Am. Chem. Soc.*, **89**, 6772 (1967).
68. M. F. Hawthorne, *The Chemistry of Boron and Its Compounds*, E. O. Muetterties, Ed., Wiley, New York, 1967, Chap. 5.
69. V. I. Mikheeva and V. Yu Markina, *Zh. Neorgan. Khim*, **1**, 619 (1956).
70. T. Wartik, T. Linevsky, and H. Bowkley, *Nucl. Sci. Abstr.*, **9**, 862 (1955); private communication 1955 and 1968.
71. R. M. Hunt, Callery Chemical Co., Callery, Pa., private communication, 1968.
72. M. J. Klein, B. C. Harrison, and I. J. Solomon, *J. Am. Chem. Soc.*, **80**, 4149 (1958).
73. D. F. Gaines, R. Schaeffer, and F. Tebbe, *Inorg. Chem.*, **2**, 526 (1963).
74. D. F. Gaines and R. Schaeffer, *Inorg. Chem.*, **3**, 438 (1964); *Proc. Chem. Soc.*, **1963**, 267.
75. R. Schaeffer and F. Tebbe, *J. Am. Chem. Soc.*, **84**, 3974 (1962).
76. (a) J. L. Boone and A. B. Burg, *J. Am. Chem. Soc.*, **80**, 1519 (1958); (b) **81**, 1766 (1959).
77. C. A. Lutz and D. M. Ritter, *Am. J. Chem.*, **41**, 1344 (1963).

78. J. Dobson and R. Schaeffer, *Inorg. Chem.*, **4**, 593 (1965).
79. A. D. Norman and R. Schaeffer, *J. Am. Chem. Soc.*, **88**, 1143 (1966); *Inorg. Chem.*, **4**, 1225 (1965).
80. (a) J. E. Todd and W. S. Koski, *J. Am. Chem. Soc.*, **81**, 2319 (1959); (b) T. P. Fehlner and W. S. Koski, *J. Am. Chem. Soc.*, **85**, 1905 (1963).
81. B. C. Harrison, I. J. Solomon, R. D. Hites, and M. J. Klein, *J. Inorg. Nucl. Chem.*, **14**, 195 (1960).
82. J. F. Ditter, J. R. Spielman, and R. E. Williams, *Inorg. Chem.*, **5**, 122 (1966).
83. L. V. McCarty and P. A. DiGiorgio, *J. Am. Chem. Soc.*, **73**, 3138 (1951).
84. C. S. Herrick, N. Kirk, T. L. Etherington, and A. E. Schubert, *Ind. Eng. Chem.*, **52**, 105 (1960).
85. (a) N. Kirk, U.S. Pat. 3,021,264, Feb. 13, 1962; (b) M. Ford, U.S. Pat. 3,032,480, May 1, 1962; (c) *Chem. Eng. News*, **40**, 74 (1962); C. A. Lutz and D. M. Ritter, *Am. J. Chem.*, **41**, 1347 (1963).
86. J. Dobson and R. Schaeffer, "Studies of Boron–Nitrogen Compounds X," to be published; private communication, 1968; J. Dobson, Ph.D. thesis, Indiana University, 1966.
87. J. R. Spielman and A. B. Burg, *Inorg. Chem.*, **2**, 1139 (1963).
88. L. J. Edwards, W. V. Hough, and M. D. Ford, *Congr. Intern. Chim. Pure Appl. 16ᵉ, Paris, 1957, Mem. Sect. Chim. Minerale*, pp. 475–481, 1958.
89. R. Holtzman, R. L. Hughes, I. C. Smith, and E. W. Lawless, *Production of the Boranes and Related Research*, Academic Press, New York, 1967.
90. G. E. Ryschkewitsch, S. W. Harris, E. J. Mezey, H. H. Sisler, E. A. Weilmuenster, and A. B. Garrett, *Inorg. Chem.*, **2**, 890 (1963).
91. N. J. Blay, I. Dunstan and R. L. Williams, *J. Chem. Soc.*, **1960**, 430; (b) N. Blay, J. Williams, and R. L. Williams, *J. Chem. Soc.*, **1960**, 424; (c) B. Figgis and R. L. Williams, *Spectrochim. Acta*, **1959**, 331.
92. (a) T. P. Onak, *J. Am. Chem. Soc.*, **83**, 2584 (1961); (b) T. P. Onak and F. J. Gerhart, *Inorg. Chem.*, **1**, 742 (1962).
93. W. V. Hough, L. J. Edwards, and A. F. Stang, *J. Am. Chem. Soc.*, **85**, 831 (1963).
94. R. Grimes and W. N. Lipscomb, *Proc. Natl. Acad. Sci. U.S.*, **48**, 496 (1962).
95. T. P. Onak, F. J. Gerhart, and R. E. Williams, *J. Am. Chem. Soc.*, **85**, 1754 (1963); T. P. Onak and R. E. Williams, *Inorg. Chem.*, **1**, 106 (1962).
96. M. Lichtenwalter and K. E. Harwell, U.S. Pat. 2,979,530, April 11, 1961.
97. D. F. Gaines, *J. Am. Chem. Soc.*, **88**, 4528 (1966).
98. T. P. Onak and G. B. Dunks, *Inorg. Chem.*, **3**, 1060 (1964).
99. I. Shapiro and H. Landesman, *J. Chem. Phys.*, **33**, 1590 (1960).
100. A. B. Burg and J. S. Sandhu, *J. Am. Chem. Soc.*, **87**, 3787 (1965).
101. T. Onak, L. B. Friedman, J. A. Hartsuk, and W. N. Lipscomb, *J. Am. Chem. Soc.*, **88**, 3439 (1966).
102. R. G. Macquire, I. J. Solomon, and M. J. Klein, *Inorg. Chem.*, **2**, 1133 (1963).
103. M. W. Forsythe, W. V. Hough, M. D. Ford, G. T. Hefferson, and L. J. Edwards, Report presented at 135th Am. Chem. Soc. Meeting, Boston, 1959.
104. P. L. Timms and C. S. G. Phillips, *Inorg. Chem.*, **3**, 297 (1964).
105. W. Kotlensky and R. Schaeffer, *J. Am. Chem. Soc.*, **80**, 4517 (1958).
106. L. E. Benjamin, S. F. Stafiej, and E. A. Takacs, *J. Am. Chem. Soc.*, **85**, 2674 (1963).
107. H. A. Beall and W. N. Lipscomb, *Inorg. Chem.*, **3**, 1783 (1964).
108. I. Shapiro and B. Keilin, *J. Am. Chem. Soc.*, **76**, 3864 (1959).
109. F. J. Norton, *J. Am. Chem. Soc.*, **72**, 1849 (1950).

110. (a) R. E. Dickerson, P. J. Wheatley, P. A. Howell, W. N. Lipscomb, and R. Schaeffer, *J. Chem. Phys.*, **25**, 606 (1956). (b) R. E. Dickerson, P. J. Wheatley, P. A. Howell, and W. N. Lipscomb, *J. Chem. Phys.*, **27**, 200 (1957). (c) P. G. Simpson and W. N. Lipscomb, *J. Chem. Phys.*, **35**, 1340 (1961).
111. A. B. Burg and R. Kratzer, *Inorg. Chem.*, **1**, 725 (1962).
112. M. J. Hillman, D. J. Mangold, and J. H. Norman, *Advan. Chem. Ser.*, **32**, 151 (1961); *J. Inorg. Nucl. Chem.*, **24**, 1565 (1963).
113. J. P. Faust and N. Goodspeed, U.S. Pat. 2,987,377, June 6, 1961.
114. G. F. Judd, U.S. Pat. 2,968,354, Jan. 17, 1961.
115. J. A. Neff, U.S. Pat. 2,989,374, June 20, 1961.
116. W. DeAcetis and S. I. Trotz, U.S. Pat. 2,983,581, Dec. 5, 1956.
117. J. S. Kasper, C. M. Lucht, and D. Harker, *Acta Cryst.*, **3**, 436 (1950); E. B. Moore, Jr., R. E. Dickerson, and W. N. Lipscomb, *J. Chem. Phys.*, **27**, 209 (1957).
118. A. B. Garrett, E. Weilmuenster, S. W. Harris, and E. R. Altwicker, U.S. Pat. 2,999,117, Sept. 5, 1961.
119. J. A. Neff and E. J. Wandel, U.S. Pat. 2,987,552, June 6, 1961.
120. R. L. Williams, I. Dunstan, and N. J. Blay, *J. Chem. Soc.*, **1960**, 5006; N. J. Blay, I. Dunstan, and R. L. Williams, *J. Chem. Soc.*, **1960**, 430.
121. E. B. Moore, Jr., L. L. Lohr, Jr., and W. N. Lipscomb, *J. Chem. Phys.*, **35**, 1329 (1961).
122. B. Siegal, J. L. Mack, J. V. Lowe, and J. Gallaghen, *J. Am. Chem. Soc.*, **80**, 4523 (1958).
123. I. Dunstan, N. J. Blay, and R. L. Williams, *J. Chem. Soc.*, **1960**, 5016; *ibid*, **1960**, 5012.
124. J. Callaghen and B. Siegal, *J. Am. Chem. Soc.*, **81**, 504 (1959).
125. M. S. Cohen and C. E. Pearl, U.S. Pat. 3,098,876, July 23, 1963.
126. R. J. F. Palchak, J. H. Norman, and R. E. Williams, *J. Am. Chem. Soc.*, **83**, 3380 (1961).
127. R. J. Palchak, U.S. Pat. 3,002,026, Sept. 26, 1961.
128. R. Schaeffer, *J. Am. Chem. Soc.*, **79**, 2726 (1957).
129. R. E. Williams, *J. Inorg. Nucl. Chem.*, **20**, 198 (1961).
130. M. S. Cohen and C. E. Pearl, U.S. Pat. 2,990,239, June 27, 1961.
131. P. R. Wunz, U.S. Pat. 3,046,086, July 24, 1962.
132. R. N. Grimes, F. E. Wang, R. Lewis, and W. N. Lipscomb, *Proc. Natl. Acad. Sci. U.S.*, **47**, 996 (1961).
133. R. N. Grimes and W. N. Lipscomb, *Proc. Natl. Acad. Sci. U.S.*, **48**, 496 (1962); T. P. Onak, *J. Am. Chem. Soc.*, **83**, 2584 (1961).
134. A. R. Pitochelli and M. F. Hawthorne, *J. Am. Chem. Soc.*, **84**, 3218 (1962).
135. P. G. Simpson and W. N. Lipscomb, *Proc. Natl. Acad. Sci. U.S.*, **48**, 1490 (1962).
136. R. W. Parry and L. J. Edwards, *J. Am. Chem. Soc.*, **81**, 3554 (1959).
137. V. D. Aftandilian, H. C. Miller, G. W. Parshall, and E. L. Muetterties, *Inorg. Chem.*, **1**, 734 (1962).
138. P. G. Simpson, K. Folting and W. N. Lipscomb, *J. Am. Chem. Soc.*, **85**, 1879 (1963).
139. N. E. Miller and E. L. Muetterties, *J. Am. Chem. Soc.*, **85**, 3506 (1963).
140. N. E. Miller and E. L. Muetterties, *Inorg. Chem.*, **3**, 1690 (1964).
141. L. B. Friedman, R. D. Dobrott, and W. N. Lipscomb, *J. Am. Chem. Soc.*, **85**, 3505 (1963).
142. (a) G. M. Almy and R. B. Harsfall, Jr., *Phys. Rev.*, **51**, 491 (1937); (b) G. Hertz-

berg and J. W. C. Johns, *Proc. Roy. Soc.*, **298A**, 142–159 (1967); J. R. Morrey, A. B. Johnson, Y. C. Fu, and G. R. Hill, *Advan. Chem. Ser.*, **32**, 157 (1961); J. S. Rigden and W. S. Koski, *J. Am. Chem. Soc.*, **83**, 3037 (1961).
143. H. I. Schlesinger and A. O. Walker, *J. Am. Chem. Soc.*, **57**, 621 (1935); H. I. Schlesinger, N. W. Flodin, and A. B. Burg, *J. Am. Chem. Soc.*, **61**, 1078 (1939); L. Van Alten, G. R. Seeley, J. Oliver, and D. M. Ritter, *Advan. Chem. Ser.*, **32**, 107 (1961).
144. J. Dobson, P. C. Keller, and R. Schaeffer, "The Chemistry of i-B_9H_{15}," *Inorg. Chem.* (1968), in press.
145. C. R. Dillard, Ph.D. thesis, University of Chicago, 1949.
146. G. F. Judd, Callery Chemical Co., Report No. 1024–TR–210, Jefferson Chemical Co., Nov. 29, 1956.
148. G. R. Seeley, J. P. Oliver, and D. M. Ritter, *Anal. Chem.*, **31**, 1993 (1959).
149. C. E. Nordman and W. N. Lipscomb, *J. Am. Chem. Soc.*, **75**, 4116 (1953); *J. Chem. Phys.*, **21**, 1856 (1953).
150. (a) M. D. Ford and L. J. Edwards, Callery Chemical Co., Report No. 1024-TR-199, Sept. 5, 1956; (b) W. V. Hough, M. D. Ford, G. T. Hefferan, and L. J. Edwards, Report No. CCC-1024-TR-274, Dec. 19, 1957; W. V. Hough, C. D. Park, M. D. Ford, and L. J. Edwards, CCC-1024-TR-297, Sept. 19, 1958.
151. J. Dobson and R. Schaeffer, "Some Chemistry of Octaborane-12 and Preparation of Octaborane-14," *Inorg. Chem.*, **7** (1968), in press.
152. J. S. Kasper, C. M. Lucht, and D. Harker, *J. Am. Chem. Soc.*, **70**, 881 (1948); *Acta Cryst.*, **3**, 436 (1950); C. M. Lucht, *J. Am. Chem. Soc.*, **73**, 2373 (1951).
153. J. K. Bragg, L. V. McCarty, and F. J. Norton, *J. Am. Chem. Soc.*, **73**, 2134 (1951); R. P. Clarke and R. Pease, *ibid.*, **73**, 2132 (1951).
154. (a) R. P. Clarke, Ph.D. thesis, Princeton University, 1949; (b) R. P. Clarke and R. N. Pease, *J. Am. Chem. Soc.*, **73**, 2132 (1951); (c) J. K. Bragg, L. V. McCarty, and F. J. Norton, *J. Am. Chem. Soc.*, **73**, 2134 (1951); (d) R. D. Stewart and R. G. Adler, Amer. Potash and Chem. Corp. Report No. WTR-6024 Contract No. AF 22(600)-35745, March 29, 1962.
155. K. Borer, A. B. Littlewood, and C. S. G. Phillips, *J. Inorg. Nucl. Chem.*, **15**, 316 (1960).
156. T. W. Clapper, A. F. N. Inc., Los Angeles, Calif., Tech. Report NR ASD-TDR-62-1025 Vol II, June, 1962.
157. R. Schaeffer, *J. Inorg. Nucl. Chem.*, **18**, 103 (1961); R. E. Enrione and R. Schaeffer, *J. Inorg. Nucl. Chem.*, **18**, 103 (1961).
158. T. P. Fehlner and W. S. Koski, *J. Am. Chem. Soc.*, **85**, 1905 (1963).
159. A. B. Baylis, G. A. Pressley, Jr., E. J. Sinke, and F. E. Stafford, *J. Am. Chem. Soc.*, **86**, 5358 (1966).
160. T. P. Fehlner, *J. Am. Chem. Soc.*, **87**, 4200 (1965).
161. T. P. Fehlner and W. S. Koski, *J. Am. Chem. Soc.*, **86**, 581 (1964).
162. A. B. Baylis, G. A. Pressley, Jr., and F. E. Stafford, *J. Am. Chem. Soc.*, **88**, 2428 (1966).
163. A. D. Norman, R. Schaeffer, A. B. Baylis, G. A. Pressley, Jr., and F. E. Stafford, *J. Am. Chem. Soc.*, **88**, 2151 (1966).
164. R. K. Pearson and L. J. Edwards, Abstr. Papers, 132nd meeting, *Am. Chem. Soc.*, New York, Sept., 1957, p. 15N.
165. W. N. Lipscomb, *Science*, **153**, 373 (1966).

Compounds Containing P—P Bonds

EKKEHARD FLUCK

Institut für Anorganische Chemie der Universität
Stuttgart, Germany

CONTENTS

I. Introduction	103
II. Compounds with P—P Bonds from Elemental Phosphorus	111
III. Formation of P—P Bonds	113
A. Reactions of Primary or Secondary Phosphines with Organohalophosphines	113
B. Reactions of Metal Phosphides with Phosphorus Halides or Halogens and of Phosphorus Halides with Metals	120
C. Reactions of Phosphonic or Thiophosphonic Anhydrides with Potassium	126
D. Reactions of Organophosphorus Halides with Metal Hydrides	127
E. Reactions of Thiophosphoryl Halides or Thiophosphonic Halides with Organomagnesium Halides	128
F. Reactions of Primary or Secondary Phosphines with Aminophosphines	132
G. Reactions of Phosphorus, Phosphonous, or Phosphinous Esters with Organophosphorus(III) Halides	133
H. Reactions of Organophosphorus Halides or Thiophosphonic Anhydrides with Tertiary Phosphines	136
I. Condensation of Primary Phosphine Oxides	138
J. Miscellaneous	138
1. Elimination of Iodine with Ether	139
2. Oxidation of Phosphides	139
3. Thermal Decomposition	139
4. Decomposition by Electric Discharge	140
5. Reaction of Organomercury Compounds with Phosphines	141
6. Rearrangement Reactions	141
7. Reactions of Phosphoryl Halides with Phosphines	142
IV. ^{31}P-NMR Data	143
General References	151
References	151

I. INTRODUCTION

Despite the enormous progress that has been made in the chemistry of phosphorus in the last 20 years, until a few years ago, relatively few compounds were known in which two or more phosphorus atoms were linked directly. In 1964,[1] it was still possible to collect all the previously known compounds of this type in a short table. More recently, however, numerous papers have appeared refuting the view, which is often put forward even today, that compounds with P—P bonds are unstable and that the small number of known compounds can be attributed to this. From the fact that the P—P and e.g., the S—S bond energies are similar (P—P 51.3 kcal/mole and S—S 50.9 kcal/mole), compounds with P—P bonds should not be uncommon. In fact, many recent attempts at the synthesis of such compounds have been successful, so that the number of known compounds with P—P bonds has grown by leaps and bounds. Table I gives a survey of

TABLE I
Compounds with P—P Bonds

Compound	Refs.
P—P	
P_2H_4	2,3
P_2Cl_4	4
P_2J_4	5
P_2R_4	6
$Me_2P_2R_2$	7
MeP_2R_3	7
$P_2R_2J_2$	8
$P_2(CF_3)_4$	8
$P_2(CF_3)_2H_2$	9,10
$P_2R_4O_2$	8
$P_2R_4S_2$	8,178
P_2R_4O	11
P_2R_4S	8
P_4S_7	12
$P_4S_3J_2$	13
$\begin{bmatrix} R\ R & & R\ R \\ P\!\!-\!\!P\!\!=\!\!N\!\!=\!\!P\!\!-\!\!P \\ R\ R & & R\ R \end{bmatrix}^+$	14
$\begin{matrix} R\ R & R \\ P\!\!-\!\!P\!\!=\!\!N\!\!-\!\!P \\ R\ R & R \end{matrix}$	14
$\begin{matrix} R\ R & R \\ S\!\!=\!\!P\!\!-\!\!P\!\!=\!\!N\!\!-\!\!P\!\!=\!\!S \\ R\ R & R \end{matrix}$	14
$\begin{matrix} S\ \ R \\ RP\!\!-\!\!PR \\ S\ \ R \end{matrix}$	15,16

(continued)

TABLE I (continued)

Compound	Refs.
P—P	
R_3P-PCF_3	17
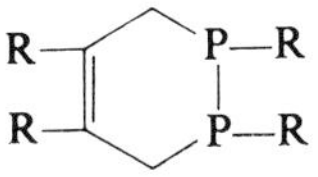 ![R,R cyclic structure with P-R, P-R groups] $\begin{array}{c} R \\ R \end{array} \diagdown \diagup \begin{array}{c} P-R \\ P-R \end{array}$	18
$\begin{array}{cc} R & OR' \\ P-P&=O \\ R & OR' \end{array}$	11
$\begin{array}{cc} R & OR' \\ P-P&=O \\ R & R \end{array}$	11
$\begin{array}{cc} R & OR' \\ P-P&=O \\ R'O & OR' \end{array}$	11
$\begin{array}{cc} R & R \\ P-P&=O \\ Cl & R \end{array}$	11
$\begin{array}{cc} O & O \\ RP\!\!-\!\!\!-\!\!PR \\ OH & OH \end{array}$	19,190
$\begin{array}{cc} S & S \\ RP-PR \\ SH & SH \end{array}$	19,190
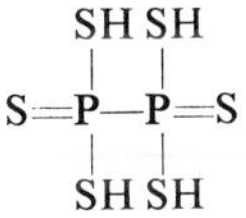 $\begin{array}{c} SH\ SH \\ \mid\ \ \mid \\ S=P-P=S \\ \mid\ \ \mid \\ SH\ SH \end{array}$	20
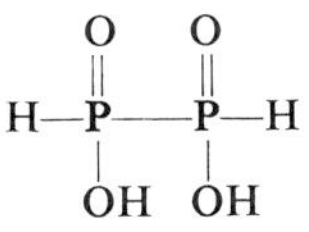 $\begin{array}{cc} O & O \\ \parallel & \parallel \\ H-P\!\!-\!\!\!-\!\!P-H \\ \mid & \mid \\ OH & OH \end{array}$	21
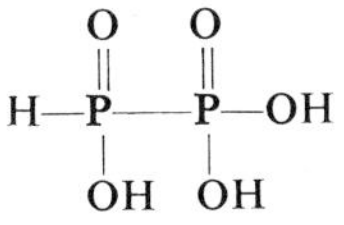 $\begin{array}{cc} O & O \\ \parallel & \parallel \\ H-P\!\!-\!\!\!-\!\!P-OH \\ \mid & \mid \\ OH & OH \end{array}$	22
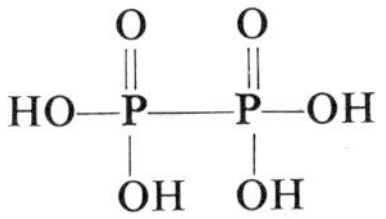 $\begin{array}{cc} O & O \\ \parallel & \parallel \\ HO-P\!\!-\!\!\!-\!\!P-OH \\ \mid & \mid \\ OH & OH \end{array}$	23

(continued)

TABLE I (*continued*)

Compound	Refs.
P—P	
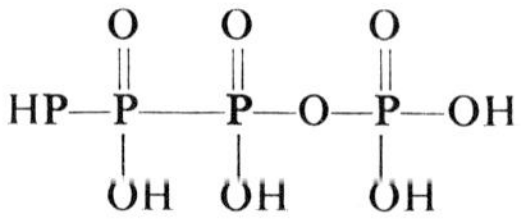 $\begin{array}{ccc} O & O & O \\ \| & \| & \| \\ HP-P\!-\!-\!-\!-\!P\!-\!O\!-\!P\!-\!OH \\ \| & \| & \| \\ OH & OH & OH \end{array}$	24
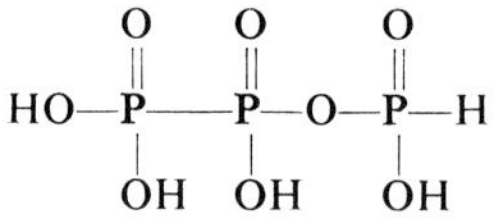 $\begin{array}{ccc} O & O & O \\ \| & \| & \| \\ HO-P\!-\!-\!-\!-\!P\!-\!O\!-\!P\!-\!H \\ \| & \| & \| \\ OH & OH & OH \end{array}$	27
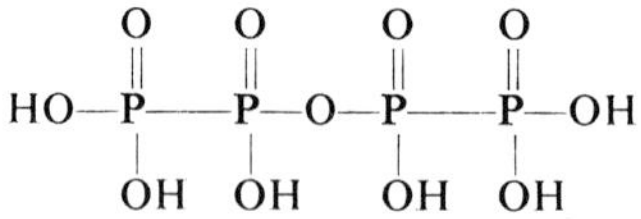 $\begin{array}{cccc} O & O & O & O \\ \| & \| & \| & \| \\ HO-P\!-\!-\!-\!P\!-\!O\!-\!P\!-\!-\!-\!P\!-\!OH \\ \| & \| & \| & \| \\ OH & OH & OH & OH \end{array}$	28
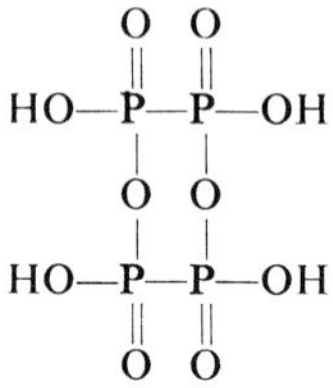 $\begin{array}{cc} O & O \\ \| & \| \\ HO-P-P-OH \\ \| & \| \\ O & O \\ \| & \| \\ HO-P-P-OH \\ \| & \| \\ O & O \end{array}$	28
P—P—P or 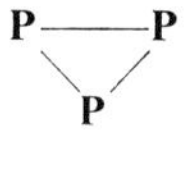 $\begin{array}{c} P\!-\!-\!-\!-\!P \\ \diagdown\;\diagup \\ P \end{array}$	
P_3H_5	29
P_4S_3	30
P_4Se_3	31
P_4S_5	32
$P_3[N(CH_3)_2]_5$	29
$CF_3P[P(H)CF_3]_2$	10,33
$RP[P(H)R]_2$	29
P_3R_5	29
$Me_2[PR]_3$	34
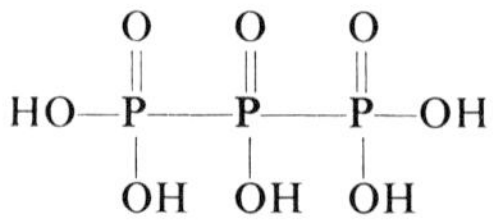 $\begin{array}{ccc} O & O & O \\ \| & \| & \| \\ HO-P\!-\!-\!-\!P\!-\!-\!-\!P\!-\!OH \\ \| & \| & \| \\ OH & OH & OH \end{array}$	35

(*continued*)

TABLE I (*continued*)

Compound	Refs.
P—P—P or a triangular P₃ ring	
$\begin{array}{c}\text{OR}' \ \ \text{R} \ \ \ \ \text{OR}' \\ \text{O}{=}\text{P}{-}\text{P}{-}\text{P}{=}\text{O} \\ \text{OR}' \ \ \ \ \ \ \ \text{OR}'\end{array}$	11
$\begin{array}{c}\text{R} \ \ \ \text{R} \ \ \ \text{R} \\ \text{O}{=}\text{P}{-}\text{P}{-}\text{P}{=}\text{O} \\ \text{OR}' \ \ \ \ \text{OR}'\end{array}$	11
P—P—P—P or a square P₄ ring	
P_4R_6	29
$Me_2P_4R_4$	7
$[RP]_4$	36,37
$[RPS]_4$	38
$[RPSe]_4$	39
$[CF_3P]_4$	191
$CF_3(H)P—P(CF_3)—P(CF_3)—P(H)CF_3$	33
$\begin{array}{c}CF_3P{-}{-}{-}PCF_3 \\ \ \ | \ \ \ \ \ \ \ \ \ \ \ | \\ CF_3P{\diagdown}_S{\diagup}PCF_3\end{array}$	40
P—P—P—P—P or a pentagonal P₅ ring	
$[RP]_5$	36,37,41
$[CF_3P]_5$	192

(*continued*)

TABLE I (continued)

Compound	Refs.
P—P—P—P—P—P or (hexagonal P₆ ring)	
(cyclic hexaphosphonic acid structure with six P atoms, each bearing =O and —OH groups)	35, 42
[RP]₆ (bicyclic/cage structure with RP units and OP=O groups)	29
(six-membered ring with alternating P—R, P—NR₂, and PR substituents)	37, 41
	43
[P—P]ₙ	
H₂P—[PH]ₙ—[P]ₘ—PH₂ [a,b]	44, 186–188
Cl₂P—[PCl]ₙ—PCl₂ [c]	45
[RP]ₙ	37, 41

[a] According to Van Wazer,[44] lower phosphorus hydrides with nonstoichiometric compositions which exist as amorphous, yellow-to-orange solids, contain three-dimensional networks built from the units H₂P—, HP=, and P≡.

[b] Recently, Baudler and co-workers were able to identify by mass spectrometry a series of polyphosphines having the general formulas P_nH_{n+2} (chain compounds) and P_nH_n (ring compounds). The oily products are formed by hydrolysis of calcium phosphide.[186–188]

[c] The yellow compounds formed by the reaction of PCl₃ with biphosphines, R₂PPR₂, and by many other reactions in which PCl₃ is involved (and produced also by the decomposition of P₂Cl₄) can be regarded as polyphosphorus chlorides, Cl₂P—[PCl]ₙ—PCl₂, where n is very large.

the well-characterized compounds with P—P bonds. The literature references in the table are, in general, references to work which gives evidence for the structure of the compound in question.

Two procedures are available, in principle, for the synthesis of compounds with P—P bonds:

1. The synthesis beginning with elementary phosphorus. In the α modification of white phosphorus, four atoms form a discrete tetrahedral molecule. α-Phosphorus forms a cubic crystal with an elementary cell built from 56 P_4 molecules. Despite the bond angle of 60° in the P_4 molecule, the model of which is shown in Figure 1, only the $3p$ orbitals of the phosphorus atom are used in the P—P bond, according to the most recent investigations of Hart, Robin, and Kuebler.[46,47]

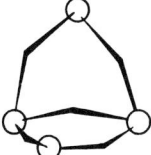

Fig. 1. Model of a P_4 molecule.

Examination of the electronic spectrum of white phosphorus showed, in the opinion of the authors that, contrary to previous hypothesis, no pd^2 hybrids are used in the formation of the σ-bonding system. Such a hybridization would, admittedly, give rise to a bond angle of 66°26′, and make possible an approach to the bond angle of 60°, i.e., the strain energy would fall almost to zero. On the other hand, a pd^2 hybridization would require an exceptionally large promotion energy. The bonds which are formed from pure p orbitals are therefore very considerably bent, as in the cyclopropane molecule. Commercial red phosphorus, commonly used for reactions in the laboratory, is almost completely amorphous. The bond angle P—P—P is about 99°.[48]

2. The synthesis beginning with mononuclear phosphorus compounds and forming the phosphorus–phorphorus bond by a chemical reaction. A number of different procedures are available, which can be characterized by the following reaction types:

A. $\diagdown$P—H + Hal—P$\diagup$ [M = metal, Hal = halogen]

B. 1. \P—M + Hal—P/

2. \P—Hal + M

3. \P—M + Hal$_2$

C. 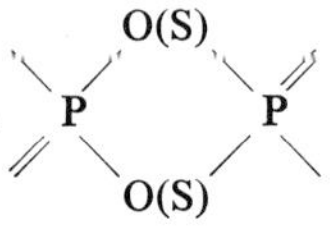 \P(=O(S))(O(S))=P/ + K

D. \P—Hal + H—M

E. SPCl$_3$ + RMgHal or RP(S)Hal$_2$ + RMgHal

F. \P—H + (CH$_3$)$_2$N—P/

G. \P—OR + Hal—P/

H. 1. \P—Hal + PR$_3$

2. \P—Hal + P(C$_4$H$_9$)$_3$

3. 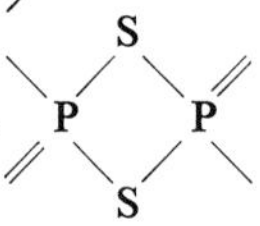 \P(=S)(S)=P/ + P(C$_4$H$_9$)$_3$

I. —\P=O + H$_2$P/

J. Miscellaneous
  1. Elimination of iodine with ether.
  2. Oxidation of phosphides.
  3. Thermal decomposition.
  4. Decomposition by electric discharge.
  5. Reaction of organomercury compounds with phosphine.
  6. Rearrangement reactions.
  7. Reactions of phosphoryl halides with phosphines.

Some of the known compounds listed in Table I have been made neither by cleavage reactions from elementary phosphorus nor from any of the listed reaction schemes. Their preparation is not described in detail in this chapter. They are obtained, for example, by the decomposition of cyclophosphines with alkali metals, by the hydrolysis of trifluoromethyl-cyclophosphines, or by the treatment of P$_4$S$_3$ with iodine, etc.—in other words, by reactions using compounds in which the P—P bonds are already present. The present chapter confines itself, apart from syntheses which start from elementary phosphorus, to those in which a new P—P bond is actually formed. The small amount of space taken up by the fission reactions of elementary phosphorus will, however, be used in emphasizing how few syntheses are known which start from the element. When one considers that, in most industrial syntheses of phosphorus compounds, white phosphorus is used in only the first of many steps, this should be regarded as a challenge to find more direct syntheses for compounds with P—P bonds as well as for mononuclear phosphorus compounds.

II. COMPOUNDS WITH P—P BONDS FROM ELEMENTAL PHOSPHORUS

Only a relatively small number of compounds with P—P bonds are synthesized directly from elementary phosphorus. A partial breakage of the P—P bonds of the molecules of white phosphorus can, in principle, be achieved by reduction as well as by oxidation reactions although, in practice, syntheses of compounds with P—P bonds from elementary phosphorus are confined to the latter. Although both phosphine and biphosphine[2,49] are produced by the treatment of white phosphorus with potassium hydroxide solution, the only suitable preparation of the latter compound known at present is the hydrolysis of calcium phosphide.[50,51] Reductive methods which start from red phosphorus to give compounds with P—P bonds which are useful for further synthetic steps have hitherto only been hinted at in the literature.[180] Various compounds with P—P bonds are obtained by such methods, e.g., by further reaction of the products obtained from red phosphorus and sodium in liquid ammonia: $(PNa)_n$ or $(PNa_2)_n$.

By contrast, a series of oxidation reactions of white and red phosphorus suitable for the preparation of compounds with P—P bonds have long been known and used. The most important are the syntheses of the phosphorus sulfides, P_4S_3 (see Fig. 2)[52,53] and P_4S_7 (see Fig. 4).[54] These

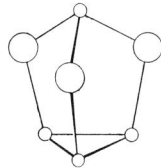

Fig. 2. Model of a P_4S_3 molecule.

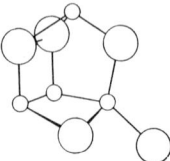

Fig. 3. Model of a P_4S_5 molecule.

are made by melting sulfur and red phosphorus together in the appropriate molecular proportions. The third phosphorus sulfide that has a phosphorus atom directly bound to another, P_4S_5 (see Fig. 3), is made by irradiation with diffuse daylight for several days of a solution of P_4S_3 and sulfur in carbon disulfide.[55,56] P_4Se_3 is made in a manner analogous to P_4S_3, by heating white phosphorus with selenium in tetralin.[57]

A partial breakage of the P—P bonds of the white phosphorus molecule occurs with elementary iodine, when this is used in the correct molecular proportions. In this way, diphosphorus tetraiodide, P_2I_4, is readily accessible by reaction between white phosphorus and iodine in carbon disulfide,[58-60]

$$P_4 + 4I_2 \longrightarrow 2P_2I_4 \qquad (1)$$

P_2I_4 has also been made by direct combination of iodine and red phosphorus, or by heating PI_3 and red phosphorus in butyl iodide.[61]

Another oxidation product of red phosphorus containing P—P bonds is hypophosphoric acid, $H_4P_2O_6$, and its salts. Various methods have been described for the preparation of disodium dihydrogen hypophosphates, all of which, however, depend on the oxidation of white or red phosphorus. The preparative method most frequently used in the laboratory starts with red phosphorus and uses sodium chlorite as the oxidizing agent.[62]

$$2P\,(red) + 2NaClO_2 + 8H_2O \longrightarrow Na_2H_2P_2O_6 \cdot 6H_2O + 2HCl \qquad (2)$$

Other oxidizing agents which lead to the same reaction product are hypochlorite,[63,64] permanganate,[65] hydrogen peroxide,[65] and iodine.[66]

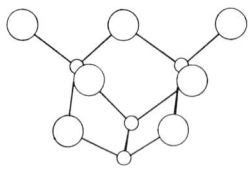

Fig. 4. Model of a P_4S_7 molecule.

White phosphorus also gives hypophosphate under careful oxidation with air[67] or with cupric nitrate in aqueous solution.[68,69]

The P—P bonds of white phosphorus are partially preserved when it is heated in a sealed, evacuated tube for about 12 hr at 235–250° with tetraphenyltin. Among other products, this reaction gives tetrakis(triphenylstannyl)biphosphine, $[(C_6H_5)_3Sn]_2P—P[Sn(C_6H_5)_3]_2$.[70]

Blaser and Worms[35] were able to isolate an exceptionally interesting oxidation product of red phosphorus after treating it with hypobromite, namely a salt of the cyclic acid.

Phosphorus–phosphorus bonds in elementary phosphorus also remain if white phosphorus is treated with butylmagnesium bromide and butyl bromide in the molecular proportions 1:2:2 in boiling tetrahydrofuran.[71,72] Tetrameric and pentameric butylcyclophosphine are produced in 42% yield by this reaction. The method appears generally suitable for the preparation of distillable cyclic polyphosphines.[73]

Finally, it may be mentioned that treatment of white phosphorus with CF_3 radicals, obtained from CHF_3 and benzoyl peroxide, gives tetrameric trifluoromethylcyclophosphine $(CF_3P)_4$.[74]

III. FORMATION OF P—P BONDS

A. Reactions of Primary or Secondary Phosphines with Organohalophosphines

$$\diagdown\!\!\!\!\diagup \text{P—H} + \text{Hal—P} \diagdown\!\!\!\!\diagup$$

The oldest known method of forming a P—P bond is the condensation of organophosphines with organohalophosphines. In 1888, Dörken[75] obtained by this method tetraphenylbiphosphine (1), from diphenylphosphine and diphenylchlorophosphine in boiling petroleum ether (at 100°).

$$(C_6H_5)_2PH + ClP(C_6H_5)_2 \longrightarrow (C_6H_5)_2P-P(C_6H_5)_2 + HCl \quad (3)$$
$$(1)$$

The preparation worked out by Kuchen and Buchwald[76] for **1**, which can be modified for other tetraarylbiphosphines, can appropriately be given here.

1. Preparation of Tetraphenylbisphosphine[76]

6 g of diphenylphosphine and 7.1 g of diphenylchlorophosphine are dissolved in 100 ml of petroleum ether (bp 90–100°, distilled from sodium). The solution is heated under reflux for $3\frac{1}{2}$ hours with stirring, in a stream of nitrogen. Then the solution is allowed to cool with continuous stirring, and the precipitated product, a finely crystalline powder, is filtered in the absence of air and washed repeatedly with petroleum ether. It is dried in a high vacuum and stored under nitrogen, mp 120.5°.

Tetraalkylbisphosphines may be made by the same process. For example, the reaction between $(CH_3)_2PCl$ and $(CH_3)_2PH$ follows eq. 4 with 81% yield.

$$(CH_3)_2PCl + 2(CH_3)_2PH \longrightarrow (CH_3)_2P-P(CH_3)_2 + (CH_3)_2PH_2Cl \quad (4)$$

Although other preparative methods, to be dealt with in the following section, are known for symmetrical biphosphines of the type R_2P-PR_2 or $RR'P-PR'R$, unsymmetrical biphosphines can be prepared only by reactions of the type discussed here, or by reaction 23 on p. 122. Thus, Grant and Burg[78] made the unsymmetrical dimethylbistrifluoromethylbiphosphine from bistrifluoromethylphosphine and dimethylchlorophosphine at $-78°$, with or without trimethylamine as HCl acceptor.

$$(CH_3)_2PCl + (CF_3)_2PH \longrightarrow (CH_3)_2P-P(CF_3)_2 + HCl \quad (5)$$

In an analogous manner, Issleib and Krech[79] made the unsymmetrical 1,1-diphenyl-2,2-dicyclohexylbiphosphine from $(C_6H_5)_2PCl$ and $(c-C_6H_{11})_2PH$ in ether in the presence of triethylamine, with 36.6% yield. The melting and boiling points of the numerous biphosphines are listed in Table II.

The synthesis of fully substituted triphosphines also was achieved by a hydrogen halide elimination process. Wiberg[29] describes the preparation of pentaphenyltriphosphine. A frozen ethereal solution of 2 moles of $(C_6H_5)_2PH$, 1 mole $C_6H_5PBr_2$, and 2 moles triethylamine is melted in a

TABLE II. Physical Properties of Biphosphines

Compound	mp, °C	bp, °C (mm Hg)	Refs.

$$\begin{array}{c} R R \\ \diagdown \diagup \\ P\!-\!P \\ \diagup \diagdown \\ R R \end{array}$$

R			
H	−99	66.7	80
CH_3	−2.25 to −2.15	140.2(calc.)	77,81
C_2H_5		220–222	82–84
$n\text{-}C_3H_7$		144–145(16)	81
		112–113(5)	77
$n\text{-}C_4H_9$		180–182(14)	81
C_6H_{11}	173		84–86
C_6H_5	120.5	258–260(1)	83
$p\text{-}CH_3C_6H_4$			83
CF_3		84	87
$N(CH_3)_2$	48	50(0.01)	88
$Sn(C_6H_5)_3$	95–110		70

$$\begin{array}{c} R_1 R_1 \\ \diagdown \diagup \\ P\!-\!P \\ \diagup \diagdown \\ R_2 R_2 \end{array}$$

R_1	R_2			
H	CF_3		69.5	9,10
CH_3	C_2H_5		188–190(740)	84
CH_3	C_4H_9		51–52(0.01)	89
CH_3	C_6H_5	73–76	128–130(0.5)	90
C_2H_5	$N(C_2H_5)_2$		142–145(12)	91
C_6H_5	C_6H_5CO	117–117.5		90
C_6H_5	$N(CH_3)_2$		150(0.5)	43
C_6H_5	$N(C_2H_5)_2$		236–238(15)	91
C_6H_{11}	$N(C_2H_5)_2$		234–235(15)	91

$$\begin{array}{c} R_1 R_2 \\ \diagdown \diagup \\ P\!-\!P \\ \diagup \diagdown \\ R_1 R_2 \end{array}$$

R_1	R_2			
CH_3	CF_3	−79.2 to −79.1	120	78
C_6H_5	$N(CH_3)_2$		137–140(0.001)	43

$$\begin{array}{c} C_6H_5 C_6H_5 \\ \diagdown \diagup \\ P\!-\!\!-\!\!-\!P \\ | | \\ CH_2CH_2CH_2 \end{array}$$

			184–190(4)	92

$$\begin{array}{c} CH_2\!-\!CH_2 CH_2\!-\!CH_2 \\ | \diagdown \diagup | \\ P\!-\!P \\ | \diagup \diagdown | \\ CH_2\!-\!CH_2 CH_2\!-\!CH_2 \end{array}$$

			50(0.05)	93

bath at $-30°$ and warmed to room temperature, whereupon 2 moles of $[(C_2H_5)_3NH]Br$ precipitates from the solution, and one obtains an ethereal solution of $(C_6H_5)_5P_3$, eq. 6.

$$(C_6H_5)PBr_2 + 2HP(C_6H_5)_2 + 2(C_2H_5)_3N \longrightarrow$$
$$(C_6H_5)_2P\text{—}P(C_6H_5)\text{—}P(C_6H_5)_2 + 2[(C_2H_5)_3NH]Br \quad (6)$$

The inverse method, namely the condensation of one mole of a primary phosphine with two moles of a monohalophosphine to give a triphosphine, was adopted by Burg[94]:

$$CH_3PH_2 + 2(CF_3)_2PCl + 2(CH_3)_3N \longrightarrow 2(CH_3)_3NHCl + CH_3P[P(CF_3)_2]_2 \quad (7)$$

This process obviously goes in two stages, which can also be separated, and then gives virtually quantitative yields, just as in the above-described reactions:

$$(CF_3)_2PI + CH_3PH_2 \longrightarrow CH_3PH_3I + CH_3(H)P\text{—}P(CF_3)_2 \quad (8)$$
$$CH_3(H)P\text{—}P(CF_3)_2 + (CF_3)_2PI + (CH_3)_3N \longrightarrow$$
$$(CH_3)_3NHI + CH_3P[P(CF_3)_2]_2 \quad (9)$$

As well as bi- and triphosphines, biphosphine disulfides may be synthesized by the reaction scheme discussed above. Niebergall and Langenfeld[83] carried out the reactions described in eqs. 10–12, in which they heated the reactants to $100°$ without solvent:

$$(C_6H_5)_2P(S)Cl + HP(S)(C_6H_5)_2 \longrightarrow (C_6H_5)_2P(S)\text{—}P(S)(C_6H_5)_2 + HCl \quad (10)$$
(Yield 79%)

$$(CH_3C_6H_4)_2P(S)Cl + HP(S)(CH_3C_6H_4)_2 \longrightarrow$$
$$(CH_3C_6H_4)_2P(S)\text{—}P(S)(CH_3C_6H_4)_2 + HCl \quad (11)$$
(Yield 90%)

$$(c\text{-}C_6H_{11})_2P(S)Cl + HP(S)(c\text{-}C_6H_{11})_2 \longrightarrow$$
$$(c\text{-}C_6H_{11})_2P(S)\text{—}P(S)(c\text{-}C_6H_{11})_2 + HCl \quad (12)$$

Even before Dörken used the HCl elimination to form P—P bonds (see above) and thus prepared the first tetraorganobiphosphine, Köhler and Michaelis[95] had employed this type of reaction. These authors treated primary phenylphosphine with phenyldichlorophosphine and obtained "phosphobenzene," $(C_6H_5P)_n$. As later work showed,[37,41] pentaphenylcyclopentaphosphine (2) and hexaphenylcyclohexaphosphine (3) are mainly produced in this reaction.

In the same way, other aryl- and alkylcyclopolyphosphines may be prepared. Primary arylphosphines react readily even at room temperature with aryldichlorophosphines, whereas with the primary alkyl phosphines

(2)   (3)

and the corresponding chlorophosphines, higher temperatures are necessary. In general, the reactions are carried out in a solvent such as benzene, toluene, or xylene. Listed in Table III are the cyclophosphines which have been made by the types of reactions discussed here. The melting and boiling points of the known cyclophosphines are shown in Table IV.

Finally, in this section, two more reactions which give compounds with P—P bonds should be mentioned, although it is not certain that the bonds are formed by hydrogen halide elimination. They are the hydrolyses of phosphorus halides.

Blaser[22] was able to isolate the trisodium salt of the acid from the

$$\begin{array}{cc} O & OH \\ \| & | \\ H-P-\!\!\!-P=O \\ | & | \\ OH & OH \end{array}$$

hydrolysis products of phosphorus tribromide in sodium–bicarbonate solution, while Falius[20] obtained, by the reaction of thiophosphoryltrichloride with aqueous sulfide solutions, the dipotassium salt of hexathiohypophosphoric acid.

$$\begin{array}{cc} SK & SK \\ | & | \\ S=P-\!\!\!-P=S \\ | & | \\ SK & SK \end{array}$$

Nothing is known of the mechanism which leads to the formation of each of these compounds. In principle, it would be conceivable that the

TABLE III

Preparation of Cyclophosphines

$$RPH_2 + Cl_2PR \longrightarrow (2/n)(PR)_n + 2HCl$$

Reaction product	Reaction medium	Temp., °C	Yield, %	Refs.
$(C_2H_5P)_{4,5}$	Toluene	110	81.8	96,97
$(n\text{-}C_3H_7P)_{4,5}$	Benzene	50	92	98
$(NCCH_2CH_2P)_4$	$CHCl_3$/pyridine	0	43	98
$(n\text{-}C_4H_8P)_{4,5}$	Benzene	78	82	98
$(i\text{-}C_4H_9P)_4$	Xylene	140	78	98
$(c\text{-}C_6H_{11}P)_4$	Toluene	110	79	98,99
$(n\text{-}C_8H_{17}P)_{4,5}$	Toluene	110	92	98,100
$(C_6H_5P)_5$	Ether	35	84–93	41,98,101–104
$(C_6H_5P)_5$ [a]	No solvent	20		90,95
$(C_6H_5P)_6$	Benzene	—	77	98
$(C_6H_5P)_6$	Benzene	20	81	104,105
$(C_6H_5P)_6$	Benzene/triethylamine	20	96.5	104,106
$(C_6H_5P)_6$ [a]	No solvent	20		90
$(C_6H_5P)_n$ [b]	No solvent	20	100	90,104
$(C_6H_5P)_n$ [b]	Benzene	20	100	98
$(C_6H_5P)_n$ [b]	Cyclohexane	20	96	98
$(o\text{-}CH_3C_6H_4P)_4$	Ether	25	86	107
$(o\text{-}CH_3C_6H_4P)_5$	Benzene	20	65.5	107
$(o\text{-}CH_3C_6H_4P)_5$	No solvent	20	86	107
$(p\text{-}ClC_6H_4P)_5$	Ether	35	97	107
$(p\text{-}ClC_6H_4P)_5$	Benzene	20		107
$(p\text{-}ClC_6H_4P)_6$	Benzene	20		107
$(p\text{-}ClC_6H_4P)_n$	No solvent	20	95	107
$(C_6F_5P)_4$	Petroleum ether	40	94	108

[a] Large batches yield mainly $(C_6H_5P)_n$.
[b] n is still unknown.

formation of the P—P bond in the first case should take place by elimination of hydrogen bromide from two molecules:

$$\text{\textbackslash P—H} + \text{Br—P/} \longrightarrow \text{\textbackslash P—P/} + HBr$$

but it cannot be disregarded as yet that the P—P bond is formed by elimination of water:

$$\text{\textbackslash P—H} + \text{HO—P/} \longrightarrow \text{\textbackslash P—P/} + H_2O$$

Even more complex is the course of the second reaction, for here at the same time that the hydrolysis occurs, there is a reduction reaction binding the sulfide ions.

TABLE IV. Physical Data of the Cyclophosphines $[RP]_n$ (from ref. 37)

Compound	mp, °C	bp, °C (mm Hg)	Refs.
$(CH_3P)_5$		86(0.0002)	96,98,102,
		99(0.2)	109,110
		110–111(1)	111
		123(2)	
$(CF_3P)_4$	66.4	135(760 calc.)	9,10,74,112,113
$(CF_3P)_5$	−33	190(760 ber.)	9,113,114,115
$(C_2H_5P)_{4,5}$		112–114(0.0002)	96–98,102,116
		123(0.2)	
		124–129(0.05)	
		168–170(15)	
$(n\text{-}C_3H_7P)_4$ [a]		140–145(0.03)	98,100,116
		120–124(1.0)	
$(i\text{-}C_3H_7P)_4$	23–24	110–114 (0.04)	98
$(NCCH_2CH_2P)_4$	87–89		98
$(n\text{-}C_4H_9P)_4$ [a]		170 (0.02)	71,98,116
		136–140(0.007)	
$(i\text{-}C_4H_9P)_4$		140(0.1)	98,100
		145–148(1.0)	
$(tert\text{-}C_4H_9P)_4$	167–169		117
$[(C_2H_5)_2CHP]_4$	91–92		98,100
	92–93		
$(c\text{-}C_6H_{11}P)_4$	219–220		98,99
	222–224		
$(n\text{-}C_8H_{17}P)_4$ [a]		230 (0.1)	98
$(C_6H_5P)_5$	150		15,16,95
	151–152		
	153–155		29,90,98,101–104,118,
	154–156		119
$(C_6H_5P)_6$	188–192		90,98,104,106
	193		
$(C_6H_5P)_n$ [b]	252–256		90,98,104
	260–280		
	240–275		
	289–305		
$(C_6H_5P)_n$ [b]	260–285		98
$(o\text{-}CH_3C_6H_4P)_4$	198–202		37
$(o\text{-}CH_3C_6H_4P)_5$	290–299		37
$(p\text{-}ClC_6H_4P)_5$	187–1290		37
$(p\text{-}ClC_6H_4P)_6$	194–198		37
$(p\text{-}ClC_6H_4P)_{4,5}$	161–165		37
$(p\text{-}FC_6H_4P)_n$	167–169		120
$(p\text{-}CH_3OC_6H_4P)_5$	188–192		15,16
$(p\text{-}C_2H_5OC_6H_4)_5$	188		15,16
$(C_6F_5P)_5$	156–161		121
$(C_6F_5P)_5$	145		121
$(\beta\text{-}C_{10}H_7P)_n$ [a]			15,16

[a] Probably a mixture of 4- and 5-membered ring compounds.
[b] Molecular weight unknown.

B. Reactions of Metal Phosphides with Phosphorus Halides or Halogens and of Phosphorus Halides with Metals

(a) $\quad >\!\!P\!-\!M + Hal\!-\!P\!<$

(b) $\quad >\!\!P\!-\!Hal + M$

(c) $\quad >\!\!P\!-\!M + Hal_2$

The formation of a P—P bond by an HCl elimination is analogous to the formation of P—P bonds by elimination of alkali-metal halides from an alkali-metal phosphide and an organohalophosphine according to reaction a above. In most cases, where organohalophosphines are reduced by metals (especially alkali metals) to compounds with P—P skeletons, the same reaction mechanism prevails.

Of the two reaction schemes (eqs. 13–16),

$$R_2PCl + M \longrightarrow R_2P\cdot + MCl \qquad (13)$$

$$2R_2P\cdot \longrightarrow R_2P\!-\!PR_2 \qquad (14)$$

and

$$R_2PCl + 2M \longrightarrow R_2PM + MCl \qquad (15)$$

$$R_2PM + R_2PCl \longrightarrow R_2P\!-\!PR_2 + MCl \qquad (16)$$

the latter is the more probable, i.e., just as in reaction b it is a question, in the final stage where the P—P bond is formed, of the elimination of a metal halide. The same probably applies to the reaction c. Here, in all probability, a metal–halogen exchange takes place first, and then the unchanged halophosphine reacts with the alkali-metal phosphide so formed:

$$R_2PM + Hal_2 \longrightarrow R_2P\!-\!Hal + MHal \qquad (17)$$

$$R_2PM + R_2P\!-\!Hal \longrightarrow R_2P\!-\!PR_2 + MHal \qquad (18)$$

Both diaryl- and dialkylchlorophosphines react with sodium to give the corresponding tetraorganobiphosphines. For example, diphenylchlorophosphine, with sodium in boiling diisopropyl or diiosbutyl ether or in dioxane, gives tetraphenylbiphosphine[122,123]:

$$2(C_6H_5)_2PCl + 2Na \longrightarrow (C_6H_5)_2P\!-\!P(C_6H_5)_2 + 2NaCl \qquad (19)$$

Tetraalkylbiphosphines are made under the same conditions,[82,124] from dialkylchlorophosphines and sodium:

$$2(alkyl)_2PCl + 2Na \longrightarrow (alkyl)_2P\!-\!P(alkyl)_2 + 2NaCl \qquad (20)$$

Here, a general scheme may be given for the preparation of tetraalkylbiphosphines after eq. 20.

1. Preparation of Tetraalkylbiphosphines[82]

Dioxane, freshly distilled over benzophenone-sodium serves as reaction medium. The dialkylchlorophosphine is dissolved in dioxane under nitrogen and the equivalent amount of sodium is added to the solution. Finally, the reaction mixture is heated, with vigorous stirring, to the boiling point of dioxane. It is heated under reflux until the sodium has almost completely reacted. Then the precipitate is filtered off, washed with dioxane, and the filtrate fractionally distilled. In the case of the tetracyclohexylbiphosphine, part of the reaction product remains in the filtration residue and is isolated by extraction with toluene.

Lithium may replace sodium in many cases. Lithium flakes react with diethylchlorophosphine in cold tetrahydrofuran with 78% yield of tetraethylbiphosphine and lithium chloride.[122]

N-Substituted aminophosphines are reduced with alkali metals in the same way as aryl- and alkylhalophosphines. Bis-dimethylaminochloro- and bis-dimethylaminobromophosphines were reduced with lithium, sodium, and potassium.[88,125] While lithium reacted only very slowly, and potassium gave many side-products in a very vigorous reaction, sodium proved suitable for the reaction.

$$2[(CH_3)_2N]_2PCl + 2Na \longrightarrow 2NaCl + [(CH_3)_2N]_2P—P[N(CH_3)_2]_2 \quad (21)$$

Tetrakis-dimethylaminobiphosphine results in 25% yield. In addition, compounds with longer phosphorus skeletons resulted, e.g., pentakis-dimethylaminotriphosphine, $[(CH_3)_2N]_2P—P[N(CH_3)_2]—P[N(CH_3)_2]_2$, and dimethylaminopolyphosphine, $P_x[N(CH_3)_2]_y$.

Aminoorganohalophosphines react with alkali metals analogously to diaminohalophosphines. Seidel and Issleib[91] obtained, e.g., from diethylaminoorgano chlorophosphines in ether at -10 to $-20°C$ with a sodium–potassium alloy (according to eq. 22) symmetric aminoorganobi-

$$2(C_2H_5)_2NPR'Cl + 2(Na,K) \longrightarrow (C_2H_5)_2NR'P—PR'N(C_2H_5)_2 + 2(Na,K)Cl \quad (22)$$

$$[R' = C_6H_5, c\text{-}C_6H_{11}, C_2H_5]$$

phosphines. Good yields of the phenyl and cyclohexyl compounds were obtained, but the ethyl compound was formed only in small yields.

The solvents used for the previously-mentioned reactions, the reaction temperatures, and the yields are listed in Table V. Unsymmetrical biphosphines are produced by reaction between organohalophosphines and alkalimetals, so that the two steps (eqs. 15 and 16) are carried out separately, and the alkali-metal phosphide R_2PM from eq. 15 is then reacted with the halophosphine R_2PCl as in eq. 16. 1,1-diphenyl-2,2-bis(dimethylamino)biphosphine was prepared in this way[43]:

$$(C_6H_5)_2PNa + ClP[N(CH_3)_2]_2 \longrightarrow (C_6H_5)_2P-P[N(CH_3)_2]_2 + NaCl \quad (23)$$

P—P bonds are formed also by treatment of organodichlorobiphosphines with metals. Lithium[98,100,109,129] sodium,[103,130] magnesium,[98,104,131,132] zinc,[133] mercury,[9,10,29,87,121] and antimony[134] have been described as having been used as halogen acceptors. The dehalogenation of the organo-dichlorophosphines may go via the phosphinidene $R\bar{p}$, analogous to methylene, which probably exists in melts of cyclo phosphines.[135] For the reduction of the organodichlorophosphine, lithium or magnesium is preferred. A survey of the reactions of this type described in the literature is given as Table VI. The reaction formally resembles the Wurtz-Fittig reaction in organic chemistry.

The use of mercury has so far been confined to fluoroalkyl- and fluoroaryldihalophosphines, although mercury was used at a very early

TABLE V

Preparation of Tetraorganobiphosphines According to the Equation
$2R_2PCl + 2M \longrightarrow R_2P-PR_2 + 2MCl$

R	Metal	Solvent	Temp., °C	Yield, %	Refs.
C_2H_5	Na	Dioxane	101	65.8	82
C_2H_5	Na	Dibutyl ether	110–120	91	83,124
t-C_4H_9	Na	Dioxane	101	52	117
n-C_4H_9	Na	Dioxane	101	77.8	82
c-C_6H_{11}	Na	Dioxane	101	67.7	82
C_6H_5	Na	Diisopropyl ether	69		123
C_6H_5	Na	Diisobutyl ether			123
C_6H_5	Na	Dioxane	101		123
$[(C_2H_5)_2N], C_6H_5$	Na-K alloy	Ether	−10 to −20	66	91
$[(C_2H_5)_2N], C_6H_{11}$	Na-K alloy	Ether	−10 to −20	73	91
$[(C_2H_5)_2N], C_2H_5$	Na-K alloy	Ether	−10 to −20	46	91
C_6F_5 [a]	Mg	No solvent			108
C_6F_5 [a]	Hg	No solvent	20		126
CF_3 [b]	Hg	No solvent	20	95	127
CF_3 [c]	Hg	No solvent	100	80	128
$(C_6H_5), (C_6F_5)$	Mg	No solvent			108

[a] $(C_6F_5)_2PBr$ was used for the reaction.
[b] $(CF_3)_2PI$ was used for the reaction.
[c] $(CF_3)_2PBr$ was used for the reaction.

TABLE VI
Preparation of Cyclophosphines

$$RPHal_2 + 2M \longrightarrow (1/n)(RP)_n + 2MHal$$

Reaction product	RPHal$_2$	Metal	Reaction medium	Temp., °C	Yield, %	Refs.
(CH$_3$P)$_5$	CH$_3$PCl$_2$	Li	THF	−40	63	98,109
(C$_2$H$_5$P)$_{4,5}$	C$_2$H$_5$PCl$_2$	Li	THF	78	81	98,116
(i-C$_4$H$_9$P)$_4$	i-C$_4$H$_9$PCl$_2$	Li	THF	78	60	98,100
(C$_6$H$_5$P)$_5$ [a]	C$_6$H$_5$PCl$_2$	Li	THF	−40		129
(tert-C$_4$H$_9$P)$_4$	tert-C$_4$H$_9$PCl$_2$	Na	Dioxane	100	63	117
(C$_6$H$_5$P)$_5$	C$_6$H$_5$PCl$_2$	Na	Octane	105	72	103
(CH$_3$P)$_5$	CH$_3$PCl$_2$	Mg	THF	−30	19	98
(CH$_3$P)$_5$	CH$_3$PBr$_2$	Mg	Ether/benzene	50	53.5	132
(C$_2$H$_5$P)$_{4,5}$	C$_2$H$_5$PCl$_2$	Mg	THF	50	70	98,116
(C$_2$H$_5$P)$_{4,5}$	C$_2$H$_5$PBr$_2$	Mg	Ether/benzene	50	77	132
(i-C$_3$H$_7$P)$_4$	i-C$_3$H$_7$PCl$_2$	Mg	THF	50	80	98
(i-C$_4$H$_9$P)$_4$	i-C$_4$H$_9$PCl$_2$	Mg	THF	50	60	98
(n-C$_8$H$_{17}$P)$_{4,5}$	n-C$_8$H$_{17}$PCl$_2$	Mg	THF	78	46	98,116
(C$_6$H$_5$P)$_5$	C$_6$H$_5$PCl$_2$	Mg	THF or ether	30–50	84	98,104
(C$_6$H$_5$P)$_5$	C$_6$H$_5$PBr$_2$	Mg	Ether/benzene	50	76	131,132
(C$_6$H$_5$P)$_5$	C$_6$H$_5$PCl$_2$	Zn	THF	25		133
(C$_6$H$_5$P)$_5$	C$_6$H$_5$PBr$_2$	Hg	Benzene	80	91	29
(CF$_3$P)$_{4,5}$	CF$_3$PI$_2$	Hg		20	100	9,10,112
(CF$_3$P)$_{4,5}$	CF$_3$PBr$_2$	Hg		20	70	128
(C$_6$F$_5$P)$_5$	C$_6$F$_5$PBr$_2$	Hg	Ether	20	91	29
(C$_6$F$_5$P)$_5$	C$_6$F$_5$PI$_2$	Hg	No solvent			121
(CF$_3$P)$_{4,5}$	CF$_3$PI$_2$	Sb			70	128

[a] For molecular weight see ref. 104.

date by Besson to prepare diphosphorus tetraiodide from PI_3.[136] The reaction of (trifluoromethyl)diiodophosphine with a large excess of mercury proceeds at room temperature according to eq. 24,

$$CF_3PI_2 + Hg \longrightarrow HgI_2 + (1/n)(CF_3P)_n \qquad (24)$$

to give trifluoromethylcyclophosphine. The yield of this reaction is quantitative and consists of about 60% $(CF_3P)_4$ and 40% $(CF_3P)_5$.[9,10,112] The reduction of CF_3PI_2 with antimony powder at 100°[134] is exactly analogous:

$$3nCF_3PI_2 + 2nSb \longrightarrow 2nSbI_3 + 3(CF_3P)_n \qquad (25)$$

Finally, the corresponding bromo compound can be used instead of trifluoromethyldiiodophosphine, and reacts at room temperature with mercury.

$$nCF_3PBr_2 + nHg \longrightarrow nHgBr_2 + (CF_3P)_n \qquad (26)$$

The main product of eq. 26 was a mixture of equal weights of $(CF_3P)_4$ and $(CF_3P)_5$. The yields amount to only about 70%.[128]

Besides organohalophosphines, halogen derivatives of phosphoric acid may be reduced with sodium to give compounds with P—P bonds. Baudler[137,138] obtained reaction products by the action of sodium on phosphorohalidic acid dialkyl esters which consisted of 65% of hypophosphoric acid ester. The author considers that the first stage consists in the formation of sodium dialkylphosphite after eqs. 27 and 28 or eq. 29, and then reaction of the latter with unchanged phosphorohalidic acid dialkyl ester follows as in eq. 30. CH_3, C_2H_5, and n-C_3H_7 were used as

$$(RO)_2P(O)Hal + Na \longrightarrow (RO)_2PO + NaHal \qquad (27)$$

$$(RO)_2PO + Na \longrightarrow (RO)_2P(O)Na \qquad (28)$$

$$(RO)_2P(O)Hal + 2Na \longrightarrow (RO)_2P(O)Na + NaHal \qquad (29)$$

$$(RO)_2P(O)Na + HalP(O)(OR)_2 \longrightarrow (RO)_2P(O)-P(O)(OR)_2 \qquad (30)$$

alkyl groups, and the longer the alkyl chain, the lower the temperature at which the reaction commences. Hal can be Cl, Br, or I.

As was said at the beginning of this section, the reactions between alkali-metal phosphides and halogen which give biphosphines probably depend finally on the elimination of alkali-metal halide as in eq. 31. The first stage of the reaction may be considered to be a partial exchange of halogen for alkali–metal as in eq. 15. The formation of tetraphenylbiphosphine from potassium diphenylphosphide and elementary iodine follows the reaction sequence:

$$(C_6H_5)_2PK + I_2 \longrightarrow (C_6H_5)_2PI + KI \qquad (31)$$

$$(C_6H_5)_2PK + (C_6H_5)_2PI \longrightarrow (C_6H_5)_2P-P(C_6H_5)_2 + KI \qquad (31a)$$

Analogous reaction mechanisms are considered to hold for the reaction of alkali-metal dialkylphosphides with dihaloalkanes. While the reactions between alkali-metal diarylphosphides and dihaloalkanes give the corresponding ditertiary phosphines, $R_2P(CH_2)_nPR_2$, the reaction between lithium dialkylphosphide or lithium di(cyclohexyl)phosphide and 1,2-dibromo- or dichloromethane consists firstly of a partial exchange of halogen for lithium, giving dialkylchlorophosphine or di(cyclohexyl)-chlorophosphine, which then reacts with more alkali-metal phosphide to give the biphosphine (eqs. 32–35).[151] This reaction leads to an intramolecular cyclization when $C_6H_5P(Li)-(CH_2)_3-P(Li)C_6H_5$ is treated with dibromoethane.[96]

$$Br-CH_2-CH_2-Br + LiP(C_6H_{11})_2 \longrightarrow Br-CH_2-CH_2-Li + (C_6H_{11})_2PBr \quad (32)$$

$$Br-CH_2-CH_2-Li \longrightarrow CH_2=CH_2 + LiBr \quad (33)$$

$$(C_6H_{11})_2PBr + LiP(C_6H_{11})_2 \longrightarrow (C_6H_{11})_2P-P(C_6H_{11})_2 + LiBr \quad (34)$$

$$2LiP(C_6H_{11})_2 + C_2H_4Br_2 \longrightarrow (C_6H_{11})_2P-P(C_6H_{11})_2 + C_2H_4 + 2LiBr \quad (35)$$

$$C_6H_5P(Li)-(CH_2)_3-P(Li)C_6H_5 + C_2H_4Br_2 \longrightarrow$$

$$\begin{array}{c} C_6H_5P\!-\!-\!-\!H_2C \\ | \qquad\qquad\quad \diagdown \\ \qquad\qquad\qquad\qquad CH_2 + C_2H_4 + 2LiBr \quad (36) \\ | \qquad\qquad\quad \diagup \\ C_6H_5P\!-\!-\!-\!H_2C \end{array}$$

The same metal–halogen exchange takes place if primary alkali-metal phosphides of the type KPHR are reacted with methylene chloride or 1,2-dibromoethane.

The scheme in eqs. 37–40 is suggested for the reaction.[152–154]

$$KHPR + BrCH_2CH_2Br \longrightarrow BrCH_2CH_2K + BrPHR \quad (37)$$

$$BrCH_2CH_2K \longrightarrow KBr + CH_2CH_2 \quad (38)$$

$$nBrPHR \longrightarrow (RP)_n + nHBr \quad (39)$$

$$KPHR + HBr \longrightarrow H_2PR + KBr \quad (40)$$

A similar scheme is suggested for the reaction of primary alkylphosphides with bromine, leading also to cyclophosphines.[152–154]

The reaction of ethyl iodoacetate with lithium di(cyclohexyl) phosphide, which gives tetra(cyclohexyl)biphosphine, probably also depends on a partial exchange of halogen for lithium,[86] eq. 44.

$$2ICH_2CO_2C_2H_5 + 2LiP(C_6H_{11})_2 \longrightarrow$$
$$(C_6H_{11})_2P-P(C_6H_{11})_2 + H_5C_2O_2CCH_2CH_2CO_2C_2H_5 + 2LiI \quad (44)$$

The reaction between lithium di(cyclohexyl)phosphide and phosgene probably takes a different course, also giving tetra(cyclohexyl)bisphophine in benzene with evolution of CO. The authors assume that $R_2P-COCl$ is first formed here, and then reacts further with more lithium di(cyclohexyl)-phosphide, eqs. 45–47.[86]

$$COCl_2 + MPR_2 \longrightarrow R_2P-COCl + MCl \quad (45)$$

$$R_2P-COCl + MPR_2 \longrightarrow R_2P-CO-PR_2 + MCl \quad (46)$$

$$R_2P-CO-PR_2 \longrightarrow CO + 2R_2P\cdot \longrightarrow R_2P-PR_2 \quad (47)$$

$$R = C_2H_5, c\text{-}C_6H_{11}, C_6H_5 \quad M = Li, Na, K$$

P—P bonds probably exist in solid phosphorus monosulfides $(PS)_n$ formed by the reaction of $SPBr_3$ with magnesium in absolute ether as eq. 48, but nothing further is known about the structure of this compound.[139]

$$2n SPBr_3 + 3n Mg \longrightarrow 2(PS)_n + 3n MgBr_2 \quad (48)$$

An analogy exists here with the compound $(PO)_n$, produced when metallic magnesium reacts with a solution of $OPBr_3$ in ether.[140]

$$2n OPBr_3 + 3n Mg \longrightarrow 2(PO)_n + 3n MgBr_2 \quad (49)$$

C. Reactions of Phosphonic or Thiophosphonic Anhydrides with Potassium

A reaction recently discovered by Fluck and Binder[19] applies a process analogous to the reduction of phosphorus halides with metals to compounds with P—P bonds, in that cyclic acid anhydrides are reduced with elementary potassium, with the formation of a P—P bond.

A solution of potassium in liquid ammonia is best for the reduction of phenylphosphonic anhydride, eq. 50, whereas the reduction of phenyl-tetrathiophosphonic anhydride must be carried out with a solution of

$$\text{Ph-P(=O)-O-P(=O)-Ph (cyclic)} + 2K \longrightarrow \text{Ph-P(=O)(OK)-P(=O)(OK)-Ph} \quad (50)$$

potassium in phosphoric tris-dimethylamide, since the anhydride reacts very rapidly with liquid ammonia, eq. 51.

$$\text{Ph-P(=S)(S-)(S)P(=S)-Ph (cyclic)} + 2K \longrightarrow \text{Ph-P(=S)(SK)-P(=S)(SK)-Ph} \quad (51)$$

(5)

1. Preparation of Potassium Diphenylhypophosphonate (4)

14.0 g (0.05 mole) phenylphosphonic anhydride is added to a solution of potassium (3.9 g, 0.1 g-atom) in 100 ml liquid ammonia, whereupon the blue color rapidly disappears. After removal of the solvent, the salt 4 can be extracted from the residue with 80% ethanol. On concentration of the solution, 4 is precipitated as colorless needles that are readily soluble in water (yield 47.5%).

2. Preparation of Potassium Diphenyltetrathiohypophosphonate (5)

17.2 g (0.05 mole) of tetrathiophenylphosphonic anhydride are added to a solution of 3.9 g (0.1 g-atom) potassium in 50 ml of $OP[N(CH_3)_2]_3$. Heat is evolved, and within about 30 min the initially blue solution becomes colorless. After the amide has been distilled off in a vacuum, a pale yellow product is found which can be recrystallized from 80% ethanol; yield 35%.

D. Reactions of Organophosphorus Halides with Metal Hydrides

$$\text{\textbackslash P-Hal + H-M}$$

Organodihalophosphines may be reduced to cyclophosphines, with formation of P—P bonds, not only by metals but also by metal hydrides. Preparations of cyclophosphines of this type which are described in the literature are collected in Table VII. To date, lithium hydride and lithium alanate have been used as reducing agents, and tetrahydrofuran and ether have been used as reaction media.

Without doubt, the formation of the cyclophosphines takes place in stages. In one case, a triphosphine derivative could be isolated from the

TABLE VII

Preparation of Cyclophosphines

$$RPCl_2 + LiH \longrightarrow [RP] + LiCl + HCl$$

Reaction product	RPCl$_2$	Metal hydride	Medium	Temp., °C	Yield, %	Refs.
(CH$_3$P)$_5$	CH$_3$PCl$_2$	LiH	THF	78	66	96
(C$_2$H$_5$P)$_{4,5}$	C$_2$H$_5$PCl$_2$	LiH	THF	78	58	96
(C$_6$H$_5$P)$_5$	C$_6$H$_5$PCl$_2$	LiH	THF	78	80	103,118
(C$_6$H$_5$P)$_5$	C$_6$H$_5$PCl$_2$	LiAlH$_4$	Ether	0–5	82	37,90,98, 103,118, 120,141
(p-FC$_6$H$_4$P)$_n$ [a]	p-FC$_6$H$_4$PCl$_2$	LiAlH$_4$	Ether		20	37
(o-CH$_3$C$_6$H$_4$P)$_5$	o-CH$_3$C$_6$H$_4$PCl$_2$	LiAlH$_4$	Ether	0–6	87	37

[a] n is unknown.[37]

reduction of organodihalophosphines with lithium hydride. Wiberg et al.[29] obtained 1,2,3-triphenyltriphosphine by the reaction of phenyldibromophosphine with LiH in benzene solution at 5°:

$$3C_6H_5PBr_2 + 6LiH \longrightarrow \underset{H}{\overset{C_6H_5}{P}}-\underset{}{\overset{C_6H_5}{P}}-\underset{H}{\overset{C_6H_5}{P}} + 6LiBr + 2H_2 \quad (52)$$

E. Reactions of Thiophosphoryl Halides or Thiophosphonic Halides with Organomagnesium Halides

$$SPHal_3 + RMgHal \text{ or}$$
$$RP(S)Hal_2 + RMgHal$$

For the preparation of tetraorganobiphosphine disulfides, a reaction is of interest in which the phosphorus atoms of thiophosphoryl halides are linked together and at the same time, the remaining halogen atoms are replaced by organic groups. According to eq. 53, tetraorganobiphosphine

$$2SPHal_3 + 6RMgHal' \longrightarrow R_2P(S)P(S)R_2 + 6MgHalHal' + R-R \quad (53)$$
$$(Hal \text{ and } Hal' = Cl, Br, I)$$

disulfides result from the reaction of organomagnesium halides with thiophosphoryl halides in good yields. The reaction of SPCl$_3$ with organomagnesium halides was investigated at an early stage but the authors only partially understood the nature of the reaction products.[142] This type of reaction was first used by Kabachnik for the preparation of a biphosphine derivative when he made tetramethylbiphosphine disulfide.[143]

Following this, many other members of this class of compounds were prepared by the same method. Issleib and Tzschach[144] describe, for example, the preparation of tetraethylbiphosphine disulfide. Further compounds have been isolated by Niebergall and Langenfeld.[83]

Whether the reaction does, in fact, go according to eq. 53 is not yet known, for Pollart and Harwood[179] isolated ethylene and ethane from the preparation of tetraethylbiphosphine disulfide instead of the butane expected from eq. 53.

The best yields, over 90%, are obtained by taking alkylmagnesium bromide and thiophosphoryl chloride in the molar ratio 3.2:1. The most favorable reaction temperatures for the preparation of the methyl-, ethyl-, *n*-propyl-, and *n*-butyl compounds lie in the range 0–20°.

Because of the tendency towards polymerization, the alkyl compound should be prepared at about $-20°$. Use of alkyl magnesium chlorides or iodides instead of bromides leads to substantially smaller yields.

Benzyl magnesium chloride reacts with $SPCl_3$ only in very poor yield to give tetrabenzylbiphosphine disulfide. The reactions with vinyl-, phenyl-, and cyclohexylmagnesium halides take another course.

As an example of the preparation of a tetraorganobiphosphine disulfide according to eq. 53, the method of preparing tetraethylbiphosphine disulfide will be given here.

1. Preparation of Tetraethylbiphosphine Disulfide[144]

A solution of 86 g $SPCl_3$ in 100 ml ether is added dropwise over 4–5 hr to a Grignard solution prepared from 36 g magnesium and 163.5 g ethyl bromide in 500 ml absolute ether at 22–23° (with vigorous stirring and cooling with cold water). The slurry formed towards the end of the reaction is warmed for 1 hr on the water bath and finally decomposed with enough 10% sulfuric acid to form two clear layers. The ethereal layer is separated, dried over sodium sulfate, and the ether then distilled off on the water bath. The residual oil solidifies on cooling almost completely, to colorless leaflets. For purification, these are suspended in 50–100 ml methanol and after filtration, crystallized from acetone–water; mp 76–77°; yield 40 g (62.5% of theory).

It should be noted that the biphosphine disulfides readily accessible in this way can serve as starting materials for the preparation of biphosphines, to which they may be reduced with metals like sodium, potassium, zinc, copper, and others[83,145] and in many cases with tri-*n*-butylphosphine.[84,170]

TABLE VIII

Preparation of Tetraorganobiphosphine Disulfides
$2SPHal_3 + 6RMgHal' \longrightarrow R_2P(S)P(S)R_2 + 6MgHalHal' + R-R$

Hal	R	Hal'	Reaction medium	Temp., °C	Yield, %	Refs.
Cl	CH_3	Cl	Ether	3–5	44	83
Cl	CH_3	Br	Ether	3–5	82	83
Cl	CH_3	I	Ether	0–5	58	143,146
Cl	C_2H_5	Cl	THF	65	22	83
Cl	C_2H_5	Br	THF	65	75	83,145
Cl	C_2H_5	Br	THF/ether	0–5	90	83
Cl	C_2H_5	Br	THF/ether	20	85	83
Cl	C_2H_5	Br	Ether	0–5	46	146
Cl	$n\text{-}C_3H_7$	Br	Ether	0–5	46	146
Cl	$n\text{-}C_3H_7$	Br	Ether	0–5	86	83,145
Cl	$CH_2=CH-CH_2$	Cl	Ether	−20 to −25	40	83
Br	$CH_2=CH-CH_2$	Cl	Ether	−20 to −25	12	83
Cl	$n\text{-}C_4H_9$	Br	Ether	0–5	70	83
Cl	$n\text{-}C_4H_9$	Br	Ether	0–5	26	145,146

The tetraorganobiphosphine disulfides described in the literature and prepared by the method described above are collected in Table VIII. Table IX lists the melting points of numerous biphosphine disulfides.

Maier[147] subsequently found that organobiphosphine disulfides resulted not only from the reaction of $SPCl_3$ with Grignard reagents, but also from the reaction of alkyl- and arylphosphonothioic dihalides with the

$$2RP(S)Hal_2 + 4R'MgBr \longrightarrow R'RP(S)-P(S)RR' + R'-R' + 4MgBrHal \quad (54)$$

latter (eq. 54). These compounds have two asymmetric phosphorus atoms in the molecule, so that optical isomers are possible. Maier[147] was able to separate the products of several reactions in two isomeric forms. As Wheatley[177] was able to show by x-ray structural investigations on 1,2-dimethyl-1,2-diphenylbiphosphine, the high-melting form (cf. Table IX) has the *meso* structure (**A**), while the lower-melting form is the racemic mixture (**B**).

$$\begin{array}{ccc}
\overset{\overset{\displaystyle S}{\|}}{R_1-P-R_2} & \overset{\overset{\displaystyle S}{\|}}{R_1-P-R_2} & \overset{\overset{\displaystyle S}{\|}}{R_2-P-R_1} \\
R_1-P-R_2 & R_2-P-R_1 & R_1-P-R_2 \\
\underset{\displaystyle S}{\|} & \underset{\displaystyle S}{\|} & \underset{\displaystyle S}{\|} \\
\text{(A)} & \text{(B)} &
\end{array}$$

TABLE IX

Physical Properties of the Biphosphine Disulfides

$$\begin{array}{c} R\diagdown \overset{S}{\underset{\|}{}} \quad \overset{S}{\underset{\|}{}} \diagup R \\ P\!-\!P \\ R\diagup \qquad \diagdown R \end{array}$$

R	mp, °C	Ref.
CH_3	228–229	83
C_2H_5	77–78	83
$n\text{-}C_3H_7$	145, 147–148	83
$n\text{-}C_4H_9$	74.5–75	83
$i\text{-}C_4H_9$	92–93	
C_5H_{11}	43.5	
$c\text{-}C_6H_{11}$	205	83
$CH_2\!=\!CH\!-\!CH_2$	58–59	83
C_6H_5	168–169	83
$p\text{-}CH_3C_6H_4$	183–184	83
$C_6H_5CH_2$	145–150	
$N(CH_3)_2$	227	88

$$\begin{array}{c} R_1\diagdown \overset{S}{\underset{\|}{}} \quad \overset{S}{\underset{\|}{}} \diagup R_1 \\ P\!-\!P \\ R_2\diagup \qquad \diagdown R_2 \end{array}$$

R_1	R_2	mp, °C	Ref.
CH_3	C_2H_5	159–160; 103–104	147
CH_3	$n\text{-}C_3H_7$	155–156; 92–94	147
CH_3	$n\text{-}C_4H_9$	126–128; 47–50	147
CH_3	C_6H_5	206–208, 145–146	147
CH_3	$C_6H_5CH_2$	188–189, 120–123	147
C_2H_5	C_6H_5	156–157, 85–87	148
C_6H_{11}	$N(C_2H_5)_2$	128, 123	92
C_6H_5	$N(C_2H_5)_2$	191, 158–160	91

$$\begin{array}{c} C_6H_5\diagdown \overset{S}{\underset{\|}{}} \quad \overset{S}{\underset{\|}{}} \diagup C_6H_5 \\ P\!-\!P \\ \diagdown CH_2CH_2CH_2 \diagup \end{array}$$

	mp, °C	Ref.
	178–180	92

$$\begin{array}{c} CH_2\!-\!CH_2 \quad S \quad S \quad CH_2\!-\!CH_2 \\ |\qquad\qquad \|\quad\| \qquad\qquad | \\ \quad\quad P\!-\!P \\ |\qquad\qquad\qquad\qquad\qquad | \\ CH_2\!-\!CH_2 \qquad\qquad CH_2\!-\!CH_2 \end{array}$$

	mp, °C	Ref.
	185	93

$$\begin{array}{c} CH_2\!-\!CH_2 \quad S \quad S \quad CH_2\!-\!CH_2 \\ |\qquad\qquad \|\quad\| \qquad\qquad | \\ CH_2 \qquad P\!-\!P \qquad CH_2 \\ |\qquad\qquad\qquad\qquad\qquad | \\ CH_2\!-\!CH_2 \qquad\qquad CH_2\!-\!CH_2 \end{array}$$

	mp, °C	Ref.
	185–225	93

Table X shows the yields from the reaction of methylphosphonothioic dibromide with various alkyl and aryl magnesium bromides.

TABLE X

Preparation of Tetraorganobiphosphine Disulfides
$2R'P(S)Hal_2 + 4RMgBr \longrightarrow R'RP(S)P(S)RR' + 4MgBrHal + R-R$

R'	R	Hal	Yield, %	Ref.
CH_3	CH_3	Br	34.4	147
CH_3	C_2H_5	Br	75.5	147
CH_3	$n\text{-}C_3H_7$	Br	53.2	147
CH_3	$n\text{-}C_4H_9$	Br	52.6	147
CH_3	C_6H_5	Br	77.0	147
CH_3	$C_6H_5CH_2$	Br	14.8	147

F. Reactions of Primary or Secondary Phosphines with Aminophosphines

$$\text{>P-H} + (CH_3)_2N-P<$$

An interesting possibility for the formation of a P—P bond arose from Burg's hypothesis that in the biphosphines, two $P_{3p}-P_{3d}$ π bonds could be formed, as opposed to only one $N_{2p}-P_{3d}$ π bond in aminobiphosphines. Since the N—H bond is more stable than the P—H bond, aminophosphine and phosphine should react to give biphosphine and amine. This expectation was borne out by experiment. Equimolar amounts of dimethylphosphine and (dimethylamino)dimethylphosphine are brought to reaction, and 85% of the theoretical weight of tetramethylbiphosphine appears in a few minutes at 100°.[77,149]

$(CH_3)_2PH + (CH_3)_2N-P(CH_3)_2 \rightleftharpoons (CH_3)_2P-P(CH_3)_2 + (CH_3)_2NH$ (55)

The back-reaction is stopped by the presence of a molar amount of hydrogen chloride, and the yield then becomes nearly quantitative (eq. 56).

$(CH_3)_2NP(CH_3)_2 + (CH_3)_2PH_2Cl \longrightarrow (CH_3)_2NH_2Cl + (CH_3)_2P-P(CH_3)_2$ (56)

By the same method, Maier[104] achieved the preparation of a triphosphine derivative in over 50% yield. A mixture of diphenylphosphine and bis(dimethylamino)methylphosphine in xylene is heated, and the reaction goes according to eq. 56a.

$CH_3P[N(CH_3)_2]_2 + 2HP(C_6H_5)_2 \longrightarrow CH_3P[P(C_6H_5)_2]_2 + 2(CH_3)_2NH$ (56a)

The same reaction may be used in the preparation of "phosphobenzene."[104,150] Phenylphosphine reacts with dialkylaminophosphines at higher temperatures with the elimination of dimethylamine, giving good yields of "phosphobenzene," eq. 57.

$$(C_6H_5)PH_2 + [(CH_3)_2N]_2PC_6H_5 \longrightarrow [C_6H_5P]_n + HN(CH_3)_2 \quad (57)$$

G. Reactions of Phosphorus, Phosphonous, or Phosphinous Esters with Organophosphorus(III) Halides

$$\ce{>P-OR + Hal-P<}$$

Fluck and Binder[11] found a new method of forming P—P bonds. Trialkyl phosphites react with diphenylchlorophosphine when heated, preferably in an inert solvent such as benzene, by elimination of alkyl chloride to give compounds in which two phosphorus atoms with different coordination numbers link directly. The first step in the reaction probably consists of a Lewis acid–base addition, which is then followed by the elimination of alkyl halide, eq. 58.

As well as phosphorous acid alkyl esters, esters of other acids with a lone electron pair on the phosphorus atom react with the phosphorus atom of diphenylchlorophosphine, as long as the electron-donor power is sufficiently strong. Conversely, such esters can be coupled

$$\text{R'O-P(C}_6\text{H}_5)_2 + (\text{C}_6\text{H}_5)_2\text{P-Cl} \longrightarrow (\text{C}_6\text{H}_5)_2\text{P-P(=O)(C}_6\text{H}_5)(\text{OR'}) + \text{R'Cl} \quad (59)$$

$$\text{R'O-P(C}_6\text{H}_5)(\text{OR'}) + (\text{C}_6\text{H}_5)_2\text{P-Cl} \longrightarrow (\text{C}_6\text{H}_5)_2\text{P-P(=O)(C}_6\text{H}_5)(\text{OR'}) + \text{R'Cl} \quad (60)$$

with other halogen-containing phosphorus(III) compounds, as long as these have enough Lewis-acid character. For example, the alkyl esters of phenylphosphonous acid, $C_6H_5P(OR')_2$, and the alkyl ester of diphenylphosphinous acid, $(C_6H_5)_2P(OR')$ react in this way with diphenylchlorophosphine according to eqs. 59 and 60.

The reactions with phenyldichlorophosphine go in the same way as with diphenylchlorophosphine. In this case, the two-phosphorus reaction product first formed which has an extra chlorine atom as ligand on the phosphorus atom with coordination number three, can react further with

$$\text{R'O-P(OR')}_2 + \text{C}_6\text{H}_5\text{P(Cl)}_2 \longrightarrow \text{C}_6\text{H}_5(\text{Cl})\text{P-P(=O)(OR')}_2 + \text{R'Cl} \quad (61)$$
$$\text{(6)}$$

$$\text{C}_6\text{H}_5(\text{Cl})\text{P-P(=O)(OR')}_2 + \text{R'O-P(OR')}_2 \longrightarrow (\text{R'O})_2\text{(O=)P-P(C}_6\text{H}_5)\text{-P(=O)(OR')}_2 + \text{R'Cl} \quad (62)$$
$$\text{(7)}$$

a molecule of phosphorous acid ester or phosphonous acid ester, so that finally a three-phosphorus compound of type **7** results.

This is illustrated by eqs. 61 and 62, taking the reaction between trialkylphosphite and phenyldichlorophosphine as example.

It is not always necessary to use ester and halide components as starting materials for the preparation of compounds with P—P and P—P—P skeletons. For example, compounds of type **8** may be made by the reaction of phenyldichlorophosphine with aliphatic alcohols in presence of pyridine, after eq. 63.

$$3\, C_6H_5\text{—}PCl_2 + 4HOR' + 4C_5H_5N \longrightarrow$$

$$\underset{(8)}{O{=}\underset{OR'}{P}\text{—}\underset{C_6H_5}{P}\text{—}\underset{OR'}{\underset{C_6H_5}{P}}{=}O} + 2R'Cl + 4[C_5H_5NH]Cl \qquad (63)$$

Compounds of the types **9–14** and also **7** may be made by the reaction schemes discussed here.

$$\underset{(9)}{\underset{\underset{R}{|}\ \underset{OR'}{|}}{P\text{—}P{=}O}}\qquad \underset{(10)}{\underset{\underset{R}{|}\ \underset{OR'}{|}}{P\text{—}P{=}O}}\qquad \underset{(11)}{\underset{\underset{R}{|}\ \underset{R}{|}}{P\text{—}P{=}O}}\qquad \underset{(12)}{\underset{\underset{OR'}{|}\ \underset{OR'}{|}}{P\text{—}P{=}O}}\qquad \underset{(13)}{\underset{\underset{Cl}{|}\ \underset{R}{|}}{P\text{—}P{=}O}}$$

(with top substituents R,R / R,R / R,R / R,OR' / R,R respectively)

$$\underset{(14)}{O{=}\underset{|}{P}\text{—}\underset{|}{P}\text{—}\underset{|}{P}{=}O}\qquad \underset{(7)}{O{=}\underset{|}{P}\text{—}\underset{|}{P}\text{—}\underset{|}{P}{=}O}$$

The directions for the preparation of compounds of type **9** are given here [R = C₆H₅, R' = CH₃] and in appropriately modified form, these may be used for the preparation of the other compounds of types **7–14**.

1. Preparation of **9** [R = C₆H₅, R' = CH₃]

22 g (0.1 mole) diphenylchlorophosphine, (C₆H₅)₂PCl, is boiled under reflux for 2 hr with 12.4 g (0.1 mole) trimethyl phosphite, P(OCH₃)₃, in

100 ml absolute benzene in a slow stream of nitrogen. Then the benzene is removed in a rotary evaporator. $(C_6H_5)_2P-P(O)(OCH_3)_2$ remains behind as a colorless oil, which the ^{31}P-NMR spectrum shows to contain no other phosphorus compounds. The compound cannot be distilled in vacuum without decomposition.

2. *Preparation of* **11** [R = C_6H_5], *Tetraphenylbiphosphine Monoxide*

22 g (0.1 mole) $(C_6H_5)_2$PCl is refluxed in 100 ml absolute benzene for 2 hr with 21.6 g (0.1 mole) diphenylphosphinous acid methyl ester, $(C_6H_5)_2P(OCH_3)$, in a slow stream of nitrogen. The benzene is distilled off in a rotary evaporator and tetraphenylbiphosphine monoxide is left in the form of colorless crystals, which can be crystallized from a small amount of benzene; yield 37 g (95% of theory); mp 158–161° (*in vacuo*).

H. Reactions of Organophosphorus Halides or Thiophosphonic Anhydrides with Tertiary Phosphines

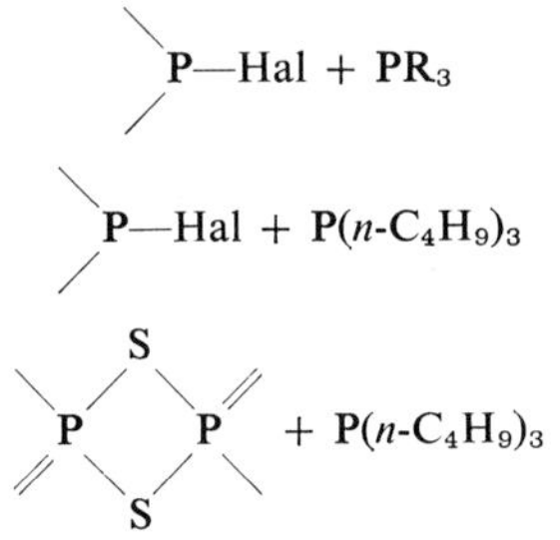

$$\diagdown\!\!\!\text{P}\!\!\!\diagup\text{—Hal} + \text{PR}_3$$

$$\diagdown\!\!\!\text{P}\!\!\!\diagup\text{—Hal} + \text{P}(n\text{-}C_4H_9)_3$$

(thiophosphonic anhydride) $+ \text{P}(n\text{-}C_4H_9)_3$

Tertiary aliphatic phosphines react with diorganohalophosphines forming pentaalkylbiphosphonium salts:

$$R_3P + R'_2PHal \longrightarrow [R_3P-PR'_2]Hal \qquad (63a)$$

According to this type of reaction, the following compounds were prepared with high yields[189]:

$[(C_2H_5)_3P-P(C_2H_5)_2]$Hal (Hal = Cl,I)

$[(C_2H_5)_3P-P(C_4H_9)_2]$Br

$[(C_2H_5)_3P-P(C_6H_5)_2]$Cl

Whether the addition reaction occurs depends highly on the base strength of the tertiary phosphine. While triethyl phosphine is reacting with aliphatic as well as with aromatic chlorophosphines, triphenylphosphine shows no reaction with diethyl- or diphenylchlorophosphine.

Tributylphosphine, on the other hand, seems to have exceptional properties in its behavior against halophosphines, as shown in the following paragraph.

In some cases, tri-n-butylphosphine proves a suitable reducing agent for reducing halophosphines with formation of a P—P bond. Tetraphenylbiphosphine[76] can be made similarly, for example, from diphenylchlorophosphine and tri-n-butylphosphine,

$$2(C_6H_5)_2PCl + (n\text{-}C_4H_9)_3P \longrightarrow (C_6H_5)_2P\text{—}P(C_6H_5)_2 + (n\text{-}C_4H_9)_3PCl_2 \quad (64)$$

Phenyldichlorophosphine gives "phenylphosphorus" on complete dehalogenation[21]:

$$nC_6H_5PCl_2 + n(C_4H_9)_3P \longrightarrow (C_6H_5P)_n + n(C_4H_9)_3PCl_2 \quad (65)$$

PCl$_3$ reacts exothermally when mixed with tributylphosphine. The amorphous red-yellow precipitate formed in this reaction consists probably of a three-dimensional network of phosphorus with alkyl and chloro endgroups. Tributylphosphine is oxidized to give tributyldichlorophosphorane, $(C_4H_9)_3PCl_2$[195]. "Phenylphosphorus," i.e., phenylcyclophosphine is also the reduction product from 1,2-diphenyl-1,2-diiodobiphosphine with diethylphenylphosphine.[8] In boiling benzene solution, phenylcyclophosphine is produced in 64% yield.

$$n(C_6H_5)(I)P\text{—}P(I)(C_6H_5) + nC_6H_5(C_2H_5)_2P \longrightarrow$$
$$2(C_6H_5P)_n + nC_6H_5(C_2H_5)_2PI_2 \quad (66)$$

The symproportionation reactions, resulting in the formation of a P—P bond, that occur when phosphonotrithioic acid anhydrides, i.e., compounds of the type [RPS$_2$]$_2$, are treated with 2 moles of tri-n-butylphosphine are to be regarded in the same light.[15,16] The compounds produced are to be considered as inner phosphonium phosphonates, stabilized by resonance (eq. 67).

$$\underset{(15)}{R\text{—}P\underset{S\diagdown \diagup S}{\overset{S\diagup \diagdown S}{\diagdown \diagup}}P\text{—}R} + 2(n\text{-}C_4H_9)_3P \longrightarrow 2\ R\text{—}\underset{S_\ominus}{\overset{S}{\overset{\|}{P}}}\text{—}\overset{\oplus}{P}(n\text{-}C_4H_9)_3 \quad (67)$$

The inner phosphonium phosphonates (15) react with more tributylphosphine according to eq. 68 with the formation of cyclophosphines and

$$R\text{—}\underset{S_\ominus}{\overset{S}{\overset{\|}{P}}}\text{—}\overset{\oplus}{P}(n\text{-}C_4H_9)_3 + P(n\text{-}C_4H_9)_3 \longrightarrow (1/n)(RP)_n + 2(n\text{-}C_4H_9)_3PS \quad (68)$$

tri-n-butyl phosphine sulfide.

If the reaction is used to prepare cyclophosphines, one can of course combine the partial reactions in eqs. 67 and 68. The total reaction, eq. 69,

$$R-P(S)(S)_2P-R + 4P(n\text{-}C_4H_9)_3 \longrightarrow (2/n)(RP)_n + 4(n\text{-}C_4H_9)_3PS \qquad (69)$$

then shows the analogy with the reaction in eq. 65. Just as in eq. 65 where the tri-n-butyl phosphine completely dehalogenates the organodichlorophosphine, in this case it completely desulfurizes the phosphonotrithioic acid anhydride, if used in excess.

Phosphinodithioic acid anhydride is fully desulfurized with tributylphosphine in the same way as phosphonotrithioic anhydride (eq. 69a).[6] This reaction is valuable for the practical synthesis of tetraorganobiphosphines,

$$R_2P(S)-S-P(S)R_2 + 3P(n\text{-}C_4H_9)_3 \longrightarrow R_2P-PR_2 + 3(n\text{-}C_4H_9)_3PS \qquad (69a)$$

I. Condensation of Primary Phosphine Oxides

$$>\!\!P=O + H_2P\!\!<$$

The conversion of primary phosphine oxides to cyclophosphines, described by Henderson et al.,[98,100] forms the first example of the linking of phosphorus atoms by elimination of water.

$$4RP(O)H_2 \longrightarrow (RP)_4 + 4H_2O \qquad (70)$$

It is true that the yields are not especially good. 3-pentyl- and cyclohexylphosphine oxides, $(C_2H_5)_2CH-P(O)H_2$ and $(c\text{-}C_6H_{11})P(O)H_2$, gave, after heating for 6 hr at 60° and 1 mm Hg pressure, 15 and 22% cyclotetraphosphines, respectively.

Similarly, poor yields of phenylcyclophosphine (10%) are obtained by heating phenylphosphinous acid, $C_6H_5P(O)(OH)H$.

J. Miscellaneous

A number of syntheses are assembled in this section which have not found wide applicability, although they appear to be capable of at least partial extension.

1. Elimination of Iodine with Ether

P—P bonds are formed by the treatment of phosphorus triiodide[157] or phenylphosphorus diiodide[158] with ether, thereby eliminating elementary iodine.

$$2PI_3 \xrightarrow{(C_2H_5)_2O} I_2 + P_2I_4 \qquad (71)$$

$$2C_6H_5PI_2 \xrightarrow{(C_2H_5)_2O} I_2 + C_6H_5P(I)-P(I)C_6H_5 \qquad (72)$$

Diphosphorus tetraiodide and diphenyldiiodobiphosphine are synthesized as in eqs. 71 and 72.

2. Oxidation of Phosphides

Oxidation of phosphides involving formation of P—P bonds can take place chemically or electrolytically.

Lithium ethylphenylphosphide is converted by iodine in good yield to 1,2-diethyl-1,2-diphenylbiphosphine:

$$2LiP(C_2H_5)(C_6H_5) + I_2 \longrightarrow (C_2H_5)(C_6H_5)P-P(C_6H_5)(C_2H_5) + 2LiI \qquad (73)$$

Similarly, ketyl-forming ketones such as benzophenone oxidize alkali-metal diphenylphosphides to tetraphenylbiphosphine.[159]

Finally, $TiCl_3 \cdot 3THF$ can oxidize alkali-metal phosphides. From $TiCl_3 \cdot 3$ THF and $LiP(c-C_6H_{11})_2$ in benzene/THF, a dark-brown solution is obtained containing tetracyclohexylbiphosphine[160]:

$$TiCl_3 \cdot 3THF + 3LiP(c-C_6H_{11})_2 \xrightarrow{-3LiCl} Ti[P(c-C_6H_{11})_2]_2 + (c-C_6H_{11})_2P-P(c-C_6H_{11})_2 \qquad (74)$$

The electrolytic oxidation of phosphides of copper, silver, or nickel can be used to obtain hypophosphites.[69]

3. Thermal Decomposition

On heating diphenyl(phenylcarbamoyl)phosphine under nitrogen at 180° decomposition, with formation of CO, tetraphenylbiphosphine, and diphenylurea, takes place[161]:

$$2(C_6H_5)_2P-CO-NHC_6H_5 \longrightarrow (C_6H_5)_2P-P(C_6H_5)_2 + CO + OC(NHC_6H_5)_2 \qquad (75)$$

The reaction may involve $(C_6H_5)_2P\cdot$ radicals. The thermal decomposition of phenylbis(phenylcarbamoyl)phosphine, which, under the same conditions, gives phenylcyclophosphine (40% yield), CO, and diphenylurea according to eq. 76, is similar in that it probably involves the intermediate formation of phenylphosphinidene, $C_6H_5\underline{P}$.

$$nC_6H_5P[CONHC_6H_5]_2 \longrightarrow (C_6H_5P)_n + nCO + nC_6H_5NHCONHC_6H_5 \quad (76)$$

Allied to the decomposition of organophenylcarbamoylphosphines is the thermal decomposition of phenylbis(alkylamino)phosphines which, as well as free amine and various phosphorus-nitrogen compounds, also gives phenylcyclophosphines.[162]

The thermal decomposition of methyldifluorophosphine appears usable for purposes of synthesis.[163] Good yields of methylcyclophosphine are obtained by prolonged heating at 35–40° in a sealed bomb tube:

$$10CH_3PF_2 \longrightarrow (CH_3P)_5 + 5CH_3PF_4 \quad (77)$$

The thermal decomposition of the recently prepared dimethylfluorophosphine is parallel, going according to eq. 78.[164]

$$3(CH_3)_2PF \longrightarrow (CH_3)_2PF_3 + (CH_3)_2P-P(CH_3)_2 \quad (78)$$

New P—P bonds are also prepared by the thermal decomposition of the biphosphine, $(CF_3PH)_2$, which gives chiefly $(CF_3P)_4$ and CF_3PH_2 at 225°.[9] Similarly, the catalytic decomposition of 1,2,3-tris(trifluoromethyl)-triphosphine, $CF_3(H)P-PCF_3-P(CF_3)H$, with nickel at room temperature, gives mixtures of $(CF_3P)_4$ and $(CF_3P)_5$ as well as CF_3PH_2.[9] On the other hand, the same triphosphine in the presence of mercury or a phosphine base such as trimethylphosphine gives the biphosphine $CF_3(H)P-P(H)CF_3$ and trifluoromethylcyclophosphine, $(CF_3P)_n$, a ring of unknown size. Likewise, the catalytic decomposition of 1,2,3,4-tetrakis(trifluoromethyl)tetraphosphine $CF_3(H)P-P(CF_3)-P(CF_3)-P(H)CF_3$ gives a mixture of tetrameric and pentameric trifluoromethylcyclophosphines as well as CF_3PH_2 and $[(CF_3)PH]_2$.[37] Also, the thermal decomposition of the unsymmetrical biphosphine $CF_3(H)P-P(CF_3)_2$ (see page 116) and pentakis(trifluoromethyl)triphosphine gives, as well as $(CF_3)_2PH$ and tetrakis(trifluoromethyl) biphosphine, $P_2(CF_3)_4$, a mixture of $(CF_3P)_4$, and $(CF_3P)_5$.[79,113]

Finally, formation of cyclophosphines, i.e., with formation of P—P bonds, takes place in the thermal decomposition of various other bi- and triphosphines.[9,29,94,154,160] The stable yellow products formed by the thermal decomposition of P_2Cl_4 [4] can be similarly regarded as polyphosphorus chlorides, $Cl_2P(PCl)_nPCl_2$, where n is very large.[45]

4. Decomposition by Electric Discharge

PCl_3 vapor is reduced by a high-voltage mercury arc to yield P_2Cl_4.[166] Better yields were obtained by conduction of an electric discharge in gaseous mixtures of PCl_3 and H_2 or in PCl_3 vapor above solutions of PCl_3 and white phosphorus.[182-184]

5. Reaction of Organomercury Compounds with Phosphines

The reaction described in eq. 79 appears especially worthy of a thorough investigation and generalization; namely the reaction of phenylphosphine with organomercury compounds which gives phenylcyclophosphine, mercury, and the aliphatic or aromatic hydrocarbon corresponding to the mercury ligands.[163]

$$4(C_6H_5CH_2)_2Hg + 4C_6H_5PH_2 \longrightarrow (C_6H_5P)_4 + 8C_6H_5CH_3 + 4Hg \qquad (79)$$

6. Rearrangement Reactions

Unlike tri(dimethylphosphino)amine, $[(CH_3)_2P]_3N$, tris(diphenylphosphino)amine is not stable, but rearranges to $(C_6H_5)_2P-P(C_6H_5)_2=N-P(C_6H_5)_2$ (**16**) which, as the hydrochloride (**17**), is formed also as the product of the reaction between bis(diphenylphosphino)amine and diphenylchlorophosphine,[14] eq. 80. The same compound (**17**) is also formed

$$(C_6H_5)_2P-NH-P(C_6H_5)_2 + (C_6H_5)_2PCl \longrightarrow$$
$$[(C_6H_5)_2P-P(C_6H_5)_2=NH-P(C_6H_5)_2]Cl \qquad (80)$$
$$(\mathbf{17})$$

by fission of hexamethldisilazane with diphenylchlorophosphine in pentane after eq. 81. $(C_6H_5)_2P-P(C_6H_5)_2=N-P(C_6H_5)_2$, itself, results from the deprotonation of **17** with triethylamine, eq. 82.

$$[(CH_3)_2Si]_2NH + 3(C_6H_5)_2PCl \longrightarrow$$
$$[(C_6H_5)_2P-P(C_6H_5)_2=NH-P(C_6H_5)_2]Cl + 2(CH_3)_3SiCl \qquad (81)$$
$$(\mathbf{17})$$

$$[(C_6H_5)_2P-P(C_6H_5)_2=NH-P(C_6H_5)_2]Cl + N(C_2H_5)_3 \longrightarrow$$
$$(C_6H_5)_2P-P(C_6H_5)_2=N-P(C_6H_5)_2 + [(C_2H_5)_3NH]Cl \qquad (82)$$
$$(\mathbf{16})$$

If the silazane fission reaction is carried out with diphenylchlorophosphine in acetonitrile, it goes according to eq. 83.

$$4[(CH_3)_3Si]_2NH + 12(C_6H_5)_2PCl \longrightarrow NH_4Cl + 8(CH_3)_3SiCl$$
$$+ 3[(C_6H_5)_2P-P(C_6H_5)_2=N=P(C_6H_5)_2-P(C_6H_5)_2]Cl \qquad (83)$$
$$(\mathbf{18})$$

A salt is produced, the cation of which contains two P—P bonds and is also formed by the reaction of di(trimethylsilyl)-aminodiphenylphosphine with diphenylchlorophosphine, eq. 84. A Michaelis-Arbusov type reaction has been discussed for the formation of **18**.

$(C_6H_5)_2P-N[Si(CH_3)_3]_2 + 3(C_6H_5)_2PCl \longrightarrow$
$[(C_6H_5)_2P-P(C_6H_5)_2=N=P(C_6H_5)_2-P(C_6H_5)_2]Cl + 2(CH_3)_3SiCl$ (84)
(18)

$$\begin{array}{c}(C_6H_5)_2P\\ \diagdown\\ (C_6H_5)_2P\end{array} N-P(C_6H_5)_2 + (C_6H_5)_2P'Cl \longrightarrow \begin{array}{c}\overset{\ominus}{P'(C_6H_5)_2Cl}\\ \overset{\oplus}{(C_6H_5)_2P}\diagup\\ (C_6H_5)_2P\diagdown\ N\diagup\ P(C_6H_5)_2\end{array}$$

$$\downarrow -(C_6H_5)_2PCl$$

$$+ (C_6H_5)_2PCl \quad\quad (C_6H_5)_2P-N=P(C_6H_5)_2-P'(C_6H_5)_2$$
(16)

$$\downarrow$$

$[(C_6H_5)_2P-P(C_6H_5)_2=N=P(C_6H_5)_2-P'(C_6H_5)_2]Cl$
(18)

7. Reactions of Phosphoryl Halides with Phosphines

When triphenylphosphine is added to excess boiling phosphoryl chloride, the chloride containing cation **19** is produced, in which there is a

$$(C_6H_5)_3P + OPCl_3 \longrightarrow (C_6H_5)_3P-PCl_2(O)]Cl \quad (85)$$
(19)

P—P bond.[181] Accordingly, the solution shows two doublets, among other lines, in its P-NMR spectrum.[45] For the mechanism of the reaction, it is assumed that the $OPCl_2^+$ ion, resulting from the dissociation of $OPCl_3$, acts as a Lewis acid, and the nucleophilic phosphine is then attached to it, eq. 86.

$$\left[O=P\diagup^{Cl}_{\diagdown Cl}\right]^+ + P(C_6H_5)_3 \longrightarrow [(C_6H_5)_3P-PCl_2(O)]^+ \quad (86)$$

The reaction of $(C_6H_5)_3P$ with $OPBr_3$ at temperatures above 193° goes in the same way to **20**:

$$[(C_6H_5)_3P-PBr_2(O)]Br$$
(20)

The reaction described by eq. 86 is comparable to the reactions of eq. 58 and 67 mentioned earlier.

IV. ^{31}P-NMR DATA

The nuclear magnetic ^{31}P resonance spectrum is invaluable for the characterization of phosphorus compounds. With its aid, the course of syntheses may easily be followed qualitatively and quantitatively. Spectra give further information on questions of structure. For general information on ^{31}P-NMR spectroscopy, see references 196 and 197.

Chemical shifts of compounds with P—P bonds are collected in Table XI. Although the influence on the chemical shift of the combination of σ-bond hybrids, the double-bond contribution, the electronegativity of the ligands etc., is not known in detail as yet, the chemical shifts of biphosphines may be broken down into empirically designated increments which can then be assigned to various ligands. For a P—P bond in a biphosphine, an increment of +18 ppm must be assumed. (All data given here refer to 85% aqueous orthophosphoric acid as standard.) The contributions of the biphosphine ligands to the chemical shift are indicated in Table XII. Chemical shifts for biphosphines calculated with the aid of these values fall for most compounds within ± 2 ppm of the observed values, and always within ± 10 ppm.

In compounds which contain linked phosphorus atoms which are not chemically equivalent, the resonance lines of the phosphorus atoms are in general considerably split by spin–spin coupling. The coupling constants J_{PP} can, of course, take very different values according to the nature of the compound. Exceptionally small coupling constants are observed for tetraorganobiphosphine disulfides. Thus, the coupling constant J_{PP} in the compounds $(C_6H_5)_2P(S)$—$P(S)(C_6H_{11})_2$ and $(C_6H_5)_2P(S)$—$P(S)(C_2H_5)_2$ is less than 30 cps; in $(C_2H_5)_2P(S)$—$P(S)(C_6H_{11})_2$ it is 69 cps. In unsymmetrical biphosphines, the coupling constants are, in general, substantially larger and lie between 200 and 300 cps.

The ^{31}P nuclear magnetic resonance spectra of symmetrical biphosphines in which the phosphorus atoms have two different ligands (i.e., compounds of the type **21**) in general show two resonance lines (cf. how-

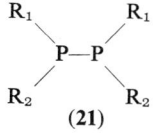

(21)

ever ref. 7, p. 185). This phenomenon was first observed by Maier.[147] It is explained by the hypothesis that the *meso* form and the racemic

TABLE XI

Chemical Shifts δ_{31P} and Coupling Constants J_{PP} of Compounds with P—P bonds
(Referred to 85% aqueous orthophosphoric acid)

Compound	Solvent	δ, ppm	J_{PP}, cps	Refs.
Biphosphines				
$(CH_3)_2P—P(CH_3)_2$	$(C_4H_9)_3PS$	59.5 ± 0.5		77,167,168
$(C_2H_5)_2P—P(C_2H_5)_2$	$(C_4H_9)_3PS$	34.2 ± 0.5		167,169
$(t-C_4H_9)_2P—P(t-C_4H_9)_2$		-40.0 ± 0.5		206
$(CH_3)(C_2H_5)P—P(CH_3)(C_2H_5)$				
Isomer 1	Pure	44.7 ± 0.5		84,147,170
Isomer 2	Pure	46.2 ± 0.5		84,147,170
$(CH_3)(C_6H_5)P—P(CH_3)(C_6H_5)$				
Isomer 1	$(C_4H_9)_3PS$	38.2 ± 0.5		84,167,170
Isomer 2	$(C_4H_9)_3PS$	41.7 ± 0.5		167
$(C_2H_5)(C_6H_5)P—P(C_2H_5)(C_6H_5)$				
Isomer 1	Benzene	21.5 ± 0.5		168
Isomer 2	Benzene	28.3 ± 0.5		168
$(C_2H_5)(n-C_4H_9)P—P(C_2H_5)(n-C_4H_9)$	Pure	37.5		168
$(C_6H_5)_2P—P(C_6H_5)_2$	Benzene	15.2		168
$(c-C_6H_{11})_2P—P(c-C_6H_{11})_2$	Toluene	21.5		168
$(C_2H_5)_2\underline{P}—P(c-C_6H_{11})_2$	Pure	42.2	282	168
$(C_2H_5)_2P—\underline{P}(c-C_6H_{11})_2$	Pure	13.8	282	168
$(C_6H_5)_2\underline{P}—P(c-C_6H_{11})_2$	Pure	28.8	224	168
$(C_6H_5)_2P—\underline{P}(c-C_6H_{11})_2$	Pure	8.2	224	168
$(C_4H_9)(C_6H_5)P—P(C_6H_5)(C_4H_9)$	Benzene	26.3 ± 0.5		45
		33.1 ± 0.5		

$(C_2H_5)P\text{———}P(C_2H_5)$
    $\|\quad\quad\quad\quad\quad\quad\|$
    $CH_2CH_2CH_2$
	46.0 ± 1.0		168

$(C_6H_5)P\text{———}P(C_6H_5)$
    $\|\quad\quad\quad\quad\quad\quad\|$
    $CH_2CH_2CH_2$
	34.3		45

$F_3C\quad\quad CF_3$
  $\diagdown\quad\diagup$
    $P\text{——}P$
  $\diagup\quad\diagdown$
$F_3C\quad\quad CF_3$
	40	55	204

Biphosphine monoxides			
$(C_6H_5)_2\underline{P}\text{—}P(O)(C_6H_5)_2$	21.6	224	
$(C_6H_5)_2P\text{—}\underline{P}(O)(C_6H_5)_2$	−36.9	224	
Biphosphine monosulfides			
$(CH_3)_2\underline{P}\text{—}P(S)(CH_3)_2$	54.7, 69.5		167
$(CH_3)_2P\text{—}\underline{P}(S)(CH_3)_2$	−43.9, −30.3		167
$(C_2H_5)_2\underline{P}\text{—}P(S)(C_2H_5)_2$	29.5, 44.6		167
$(C_2H_5)_2P\text{—}\underline{P}(S)(C_2H_5)_2$	−62.8, −47.8		167
$(CH_3)(C_2H_5)\underline{P}\text{—}P(S)(CH_3)(C_2H_5)$	40.9, 55.8		167
$(CH_3)(C_2H_5)P\text{—}\underline{P}(S)(CH_3)(C_2H_5)$	−53.9, 40.2		167
$(CH_3)(C_6H_5)\underline{P}\text{—}P(S)(CH_3)(C_6H_5)$	37.5, 47.4		167
$(CH_3)(C_6H_5)P\text{—}\underline{P}(S)(CH_3)(C_6H_5)$	−43.7, −33.9		167
Biphosphine dioxides			
$(C_4H_9)_2P(O)\text{—}P(O)(C_4H_9)_2$	-52.5 ± 1.5		

(*continued*)

TABLE XI (continued)

Compound	Solvent	δ, ppm	J_{PP}, cps	Refs.
Biphosphine disulfides				
$(CH_3)_2P(S)P(s)(CH_3)_2$	Benzene	−34.7		168
$(C_2H_5)_2P(S)-P(S)(C_2H_5)_2$	Benzene	−49.4		168
$(c-C_6H_{11})_2P(S)-P(S)(c-C_6H_{11})_2$	Benzene	−56.1		168
$(C_6H_5)_2P(S)-P(S)(C_6H_5)_2$	Toluene	−37.9		168
$(C_6H_5)_2(S)\underline{P}-P(S)(c-C_6H_{11})_2$	Toluene	−20.9		168
$(C_6H_5)_2P(S)-\underline{P}(S)(c-C_6H_{11})_2$	Toluene	−64.0		168
$(C_6H_5)_2P(S)-P(S)(C_2H_5)_2$	Toluene	−26.0		168
$(C_6H_5)_2P(S)-P(S)(C_2H_5)_2$	Toluene	−57.7		168
$(C_2H_5)_2P(S)-P(S)(c-C_6H_{11})_2$	Toluene	−47.6, −57.1	69	168
$(CH_3)(C_2H_5)P(S)-P(S)(CH_3)(C_2H_5)$	Toluene	−44.5		147
$(CH_3)(n-C_3H_7)P(S)-P(S)(CH_3)(n-C_3H_7)$	Toluene	−40.3		147
$(CH_3)(C_6H_5)P(S)(CH_3)(C_6H_5)$	Toluene	−37.0		147
$(C_2H_5)(n-C_4H_9)P(S)-P(S)(C_2H_5)(n-C_4H_9)$	Toluene	−47.6		168
![structure with (C₂H₅)P-P(C₂H₅), S, (CH₂)₄]	Benzene	−39.4		45
![cyclopentane-like P-P with S]		−61.2		209
![cyclohexane-like P-P with S]		−37.5		209

COMPOUNDS CONTAINING P—P BONDS

Compound	Solvent	δ	Ref.
Cyclophosphines			
$(CH_3P)_5$	CS_2	−21	98
$(C_2H_5P)_{4,5}$	CS_2	−17	98
	Pure	−15.8	168,171,172
		−16	116
$(n-C_3H_7P)_4$ [a]		−13	98
$(n-C_3H_7P)_n$ [b]		−53, −16, −12	116
$(i-C_3H_7P)_4$		66	98
$(NCCH_2CH_2P)_4$		−2	98
$(n-C_4H_9P)_4$		−14	98
$(n-C_4H_9P)_n$ [c]		−13, −17	116
		−10	71
$(i-C_4H_9P)_4$		−13	98
$(tert-C_4H_9P)_4$		57.8 ± 0.5	130
$[(C_2H_5)_2CHP]_4$		70	98
$(c-C_6H_{11}P)_4$		70	98
$(n-C_8H_{17}P)_4$ [a]		−13	98
$(C_6H_5P)_5$	CS_2	9	98
	Benzene	4.4	168
	Benzene	4.6	173
$(C_6F_5P)_4$		67.0	108
$(p-CH_3OC_6H_4P)_5$		11.8 ± 1.0	16
$(p-C_2H_5OPC_6H_4P)_5$		12.0 ± 1.0	16
$(\beta-C_{10}H_7P)_n$ [d]		−3.3	16
Other compounds			
P_4	Pure	461	174
P_2H_4		108.2 ± 0.2	201,202

(continued)

TABLE XI (continued)

Compound	Solvent	δ, ppm	J_{PP}, cps	Refs.
Other Compounds				
P_2F_4		−226		198
P_2Cl_4		−115		176
P_2I_4	CS_2	−106.9 ± 1.0		193,194
$P_2I_4S_2$		−9		207
P_4S_3	CS_2	120 ± 1 (d) −71 ± 1 (quad.)	86	175,199 207
![structure] P_A		−120		199
P_B		−20		199
P_C		120		199
P_D		−200		199
P_4Se_3		106		200
		−38		200
$K_2[P(C_6H_5)]_3$		49.8 ± 0.5		203
(CF₃-substituted cyclic structure with P_A, P_B) P_A		−55	220	204
P_B		41		
$(C_6H_5)_2\underline{P}—P(C_6H_5)—P(C_6H_5)_2$		15.9		205
$(C_6H_5)_2P—\underline{P}(C_6H_5)—P(C_6H_5)_2$		3.9		205
$Na_3[O_3\underline{P}—PO_2(H)]$		−7	480	207
$Na_3[O_3P—\underline{P}O_2(H)]$		−22.5		207

COMPOUNDS CONTAINING P—P BONDS

Compound	Solvent	Value		Ref
$Na_4P_2S_6 \cdot 6H_2O$	Water	-110.7 ± 1.0		45
$Na_2H_2P_2O_6 \cdot 6H_2O$	Water	-11.3 ± 0.5		45
$K_2[(C_6H_5)_2P_2S_4]$	Water/ethanol	-95.2 ± 1.0		45
$K_2[(C_6H_5)_2P_2O_4]$	Water/ethanol	-5.2 ± 1.0		45
$[(C_6H_5)_2P—P(C_6H_5)_2=N=P(C_6H_5)_2—P(C_6H_5)_2]^+Cl^-$		$-28.3, -21.8, 16.5;$ 23.0	270	14,98,102, 168
$[(C_6H_5)_3P—PCl_2(O)]Cl$		9.9, 17.7		45
$(C_6H_5)_2P_A—P_B(C_6H_5)_2=N—P_C(C_6H_5)_2$		$-41.3, -39.0$	AB 149	98
		$-21.3, -19.0$	BC 93	
		$-15.3, -13.0$		
		$7.5, 13.7^e$		
$(C_6H_5)_2P—P(O)(OCH_3)_2$		34.1(d)	192	11
		-35.7(d)	192	11
$(C_6H_5)_2P—P(O)(OC_2H_5)_2$		33.5(d)	178	11
		-31.8(d)	178	11
$(C_6H_5)_2P—P(O)(C_6H_5)(OC_2H_5)$		26.8(d)	205	11
		-47.0(d)	205	11
$(C_6H_5)(CH_3O)P—P(O)(OCH_3)_2$		50.6(d)	202	11
		-46.4(d)	202	11
$(C_6H_5)(Cl)P—P(O)(C_6H_5)_2$		26.6(d)	202	11
		-50.3(d)	202	11
$(C_6H_5)(C_2H_5O)P—P(O)(OC_2H_5)_2$		51.1(d)	201	11
		-43.4(d)	201	11
$(CH_3O)_2P(O)—P(C_6H_5)—P(O)(OCH_3)_2$		64.9(t)	168	11
		-33.7(d)	168	11
$(C_2H_5O)_2P(O)—P(C_6H_5)—P(O)(OC_2H_5)_2$		63.3(t)	157	11
		-26.7(d)	157	11
$(C_6H_5)(CH_3O)P(O)—P(C_6H_5)—P(O)(OCH_3)(C_6H_5)$		45.6(t)	210	11
		-45.7(d)	210	11

(continued)

TABLE XI (continued)

Compound	Solvent	δ, ppm	J_{PP}, cps	Refs.
Other Compounds				
$(C_6H_5)(C_2H_5O)P(O)$—$P(C_6H_5)$—$P(O)(OC_2H_5)(C_6H_5)$		45.5(t)	204	11
		−43.5(d)	204	11
Alkalipolyphosphides				
$Li(C_2H_5)\underline{P}$—$P(C_2H_5)(C_6H_5)$		112.8 ± 1.0	396	7
$Li(C_2H_5)P$—$\underline{P}(C_2H_5)(C_6H_5)$		17.2 ± 1.0	396	7
$K(C_2H_5)P$—$P(C_2H_5)K$		79.6 ± 1.0		7
$K(C_2H_5)\underline{P}$—$P(C_2H_5)$—$\underline{P}(C_2H_5)K$		78.5 ± 1.0	306	7
$K(C_2H_5)P$—$\underline{P}(C_2H_5)$—$P(C_2H_5)K$		23.9 ± 1.0		7
$Li(C_6H_5)\underline{P}$—$P(C_6H_5)$—$P(C_6H_5)Li$		86.0 ± 1.0	216	7
$Li(C_6H_5)P$—$\underline{P}(C_6H_5)$—$P(C_6H_5)Li$		8.2 ± 1.0	216	7

[a] Possibly as a mixture with $(RP)_5$.
[b] According to ref. 116, the 3 signals should be produced by different ring sizes.
[c] According to ref. 116, the 2 signals should be produced by different ring sizes.
[d] n is unknown.
[e] Lines of different intensities.

mixture of the *d* and *l* forms of the biphosphine have different chemical shifts.

$$\begin{array}{cc} R_2\diagup\!\!\!\stackrel{P-P}{}\!\!\!\diagdown R_2 \\ R_1 \quad R_1 \end{array} \qquad \begin{array}{cc} R_1\diagup\!\!\!\stackrel{P-P}{}\!\!\!\diagdown R_2 \\ R_2 \quad R_1 \end{array} \quad \begin{array}{cc} R_2\diagup\!\!\!\stackrel{P-P}{}\!\!\!\diagdown R_1 \\ R_1 \quad R_2 \end{array}$$

(*meso* form)   (*d,l* form)

TABLE XII
Contributions in ppm to the Chemical Shift from the Ligands of Phosphines and Biphosphines

Ligand	δ, ppm
CH_3	21
C_2H_5	7
n-C_4H_9	11
c-C_6H_{11}	−2
C_6H_5	3
−PR_2	18

GENERAL REFERENCES

The review articles in which the chemistry of P—P bonds is discussed are listed below.

A. H. Cowley, "The Chemistry of the Phosphorus–Phosphorus Bond," *Chem. Rev.*, **65**, 617 (1965).

A. H. Cowley and R. P. Pinnell, "The Structures and Reactions of Cyclophosphines," in *Topics in Phosphorus Chemistry*, Vol. 4, M. Grayson, and E. J. Griffith, Eds., Interscience, New York, 1966, p. 1.

E. Fluck, *Die Bindungseigenschaften des Phosphors*, *Chemiker-Zeitung*, **88**, 951 (1964).

J. E. Huheey, "Chemistry of Diphosphorus Compounds," *J. Chem. Educ.*, **40**, 153 (1963).

L. Maier, *Struktur, Darstellung und Reaktionen von Cyclopolyphosphinen*, *Fortschritte der chemischen Forschung*, Vol. 8, Springer-Verlag, Berlin, Heidelberg, New York, 1967, p. 1.

L. Maier, "Preparations and Properties of Primary, Secondary, and Tertiary Phosphines," in *Progress in Inorganic Chemistry*, Vol. 5, F. A. Cotton, Ed., Interscience, New York, 1963, p. 27.

E. Wiberg, "Neues aus der Chemie der Polyphosphane," *Angew. Chem.*, **75**, 814 (1963).

REFERENCES

1. E. Fluck, *Chemiker Ztg.*, **88**, 951 (1964).
2. P. Royen and K. Hill, *Z. Anorg. Allgem. Chem.*, **229**, 97 (1936).
3. E. C. Evers and E. H. Street, *J. Am. Chem. Soc.*, **78**, 5726 (1956).

4. A. Finch, *Can. J. Chem.*, **37**, 1793 (1959).
5. Y. C. Leung and J. Waser, *J. Phys. Chem.*, **60**, 539 (1956).
6. L. Maier, in *Progress in Inorganic Chemistry*, Vol. 5, F. A. Cotton, Ed., Interscience, New York, 1962, p. 69.
7. E. Fluck and K. Issleib, *Z. Anorg. Allgem. Chem.*, **339**, 274 (1965).
8. H. Hoffmann and R. Grünewald, *Chem. Ber.*, **94**, 186 (1961).
9. W. Mahler and A. B. Burg, *J. Am. Chem. Soc.*, **80**, 6161 (1958).
10. W. Mahler and A. B. Burg, *J. Am. Chem. Soc.*, **79**, 251 (1957).
11. E. Fluck and H. Binder, *Inorg. Nucl. Chem. Letters*, **3**, 307 (1967).
12. A. Vos and E. H. Wiebenga, *Acta Cryst.*, **8**, 217 (1955); **9**, 92 (1956).
13. S. van Houten and E. H. Wiebenga, *Acta Cryst.*, **10**, 156 (1957).
14. H. Nöth and L. Meinel, *Z. Anorg. Allgem. Chem.*, **349**, 225 (1967).
15. E. Fluck and H. Binder, *Angew. Chem.*, **78**, 677 (1966).
16. E. Fluck and H. Binder, *Z. Anorg. Allgem. Chem.*, **354**, 113 (1967).
17. A. B. Burg and W. Mahler, *J. Am. Chem. Soc.*, **83**, 2388 (1961).
18. U. Schmidt and I. Boie, *Angew. Chem.*, **78**, 1061 (1966).
19. E. Fluck and H. Binder, *Angew. Chem.*, **79**, 903 (1967).
20. H. Falius, *Z. Anorg. Allgem. Chem.*, **356**, 189 (1968).
21. M. Baudler, *Z. Naturforsch.*, **14b**, 464 (1959).
22. B. Blaser, *Chem. Ber.*, **86**, 563 (1953).
23. See J. R. Van Wazer, *Phosphorus and Its Compounds*, Vol. 1, Interscience, New York, 1958, p. 403.
24. B. Blaser and K.-H. Worms, *Z. Anorg. Allgem. Chem.*, **300**, 250 (1959).
27. B. Blaser and K.-H. Worms, *Z. Anorg. Allgem. Chem.*, **312**, 146 (1961).
28. B. Blaser and K.-H. Worms, *Z. Anorg. Allgem. Chem.*, **311**, 313 (1961).
29. E. Wiberg, M. Van Ghemen, and H. Müller-Schiedmayer, *Angew. Chem.*, **75**, 814 (1963).
30. D. A. Wright and B. R. Penfold, *Acta Cryst.*, **12**, 455 (1959).
31. E. Keulen and A. Vos, *Acta Cryst.*, **12**, 323 (1959).
32. S. Van Houten and E. H. Wiebenga, *Acta Cryst.*, **10**, 156 (1957).
33. A. B. Burg and L. K. Peterson, *Inorg. Chem.*, **5**, 943 (1966).
34. K. Issleib and E. Fluck, *Angew. Chem.*, **78**, 597 (1966).
35. B. Blaser and K.-H. Worms, *Z. Anorg. Allgem. Chem.*, **300**, 237 (1959).
36. L. Maier, in *Progress in Inorganic Chemistry*, Vol. 5, F. A. Cotton, Ed., Interscience, New York, 1962, p. 83.
37. L. Maier, *Fortschritte der Chemischen Forschung*, Vol. 8, Springer-Verlag, Berlin, Heidelberg, New York, 1967; p. 1.
38. W. Kuchen and H. Buchwald, *Chem. Ber.*, **91**, 2296 (1958).
39. H.-L. Krauss and H. Jung, *Z. Naturforsch.*, **15b**, 545 (1960).
40. A. B. Burg, *J. Am. Chem. Soc.*, **88**, 4298 (1966).
41. J. J. Daly and L. Maier, *Nature*, **203**, 1167 (1964).
42. J. Weiss, *Z. Anorg. Allgem. Chem.*, **306**, 30 (1960).
43. H.-J. Vetter and H. Nöth, *Chem. Ber.*, **96**, 1816 (1963).
44. J. R. Van Wazer, *Phosphorus and Its Compounds*, Vol. 1, Interscience, New York, 1958, p. 217.
45. E. Fluck, unpublished work.
46. R. R. Hart, M. B. Robin, and N. A. Kuebler, *Chem. Eng. News*, **1964**, 39.
47. R. R. Hart, M. B. Robin, and N. A. Kuebler, *J. Chem. Phys.*, **42**, 3631 (1965).
48. C. D. Thomas and N. S. Gingrich, *J. Chem. Phys.*, **6**, 659 (1938).

49. H. Rose, *Pogg. Ann.*, **6**, 199 (1826); **46**, 633 (1839).
50. P. Royen and K. Hill, *Z. Anorg. Allgem. Chem.*, **229**, 112 (1936).
51. M. Baudler and L. Schmidt, *Z. Anorg. Allgem. Chem.*, **289**, 219 (1957).
52. A. Stock, *Ber. Dtsch. Chem. Ges.*, **43**, 150 (1910);
53. F. C. Frary, German Pat. 309,618 (1918); *Chem. Zentr.*, **2**, 55 (1919).
54. A. Stock, *Ber. Dtsch. Chem. Ges.*, **43**, 414 (1910).
55. R. Boulouch, *Compt. Rend.*, **138**, 363 (1904).
56. W. D. Treadwell and Ch. Beeli, *Helv. Chim. Acta*, **18**, 1161 (1935).
57. J. Mai, *Ber. Dtsch. Chem. Ges.*, **61**, 1807 (1928).
58. A. Michaelis and M. Pitsch, *Liebigs Ann. Chem.*, **310**, 45 (1900).
59. F. E. E. Germann and R. N. Traxler, *J. Am. Chem. Soc.*, **49**, 307 (1927).
60. M. Baudler, *Z. Naturforsch.*, **13b**, 266 (1958).
61. E. S. Levchenko, I. E. Sheinkman, and A. V. Kirsanov, *Zh. Obshch. Khim.*, **29**, 1474 (1959).
62. E. Leininger and T. Chulski, *J. Am. Chem. Soc.*, **71**, 2385 (1949); *Inorg. Syn.* **4**, 68 (1953).
63. J. Probst, *Z. Anorg. Allgem. Chem.*, **179**, 155 (1929).
64. M. Speter, *Rec. Trav. Chim.*, **46**, 588 (1927).
65. F. Vogel, *Angew. Chem.*, **42**, 263 (1929).
66. T. Milobedzki, J. H. Kolitowska, and Z. Berkan, *Roczniki Chem.*, **17**, 620 (1937); *Chem. Abstr.*, **32**, 3717 (1938).
67. R. G. van Name and W. J. Huff, *Am. J. Sci.* (*4*), **46**, 587 (1918).
68. W. Jung, *Anales Asoc. Quim Arg.*, **30**, 99 (1942).
69. A. Rosenheim and J. Pinsker, *Z. Anorg. Allgem. Chem.*, **64**, 327 (1909).
70. H. Schumann, H. Köpf, and M. Schmidt, *Chem. Ber.*, **97**, 1458 (1964).
71. M. M. Rauhut and A. M. Semsel, *J. Org. Chem.*, **28**, 473 (1963).
72. M. M. Rauhut and A. M. Semsel, U.S. Pat. 3, 3,099,690, 1963; *Chem. Abstr.*, **60**, 556a (1964).
73. A. H. Cowley and R. P. Pinnell, *Inorg. Chem.*, **5**, 1463 (1966).
74. W. H. Watson, *Texas J. Sci.*, **11**, 471 (1959); *Chem. Abstr.*, **54**, 13928 c (1960).
75. C. Dörken, *Ber. Dtsch. Chem. Ges.*, **21**, 1505 (1888).
76. W. Kuchen and H. Buchwald, *Chem. Ber.*, **91**, 2871 (1958).
77. A. B. Burg, *J. Am. Chem. Soc.*, **83**, 2226 (1961).
78. L. R. Grant and A. B. Burg, *J. Am. Chem. Soc.*, **84**, 1834 (1962).
79. K. Issleib and K. Krech, *Chem. Ber.*, **98**, 1093 (1965).
80. E. C. Evers and E. H. Street, *J. Am. Chem. Soc.*, **78**, 5726 (1956).
81. K. Issleib and A. Tzschach, *Chem. Ber.*, **93**, 1852 (1960).
82. K. Issleib and W. Seidel, *Chem. Ber.*, **92**, 2681 (1959).
83. H. Niebergall and B. Langenfeld, *Chem. Ber.*, **95**, 64 (1962).
84. L. Maier, *J. Inorg. Nucl. Chem.*, **24**, 275 (1962).
85. K. Issleib and E. Priebe, *Chem. Ber.*, **92**, 3183 (1959).
86. K. Issleib and G. Thomas, *Chem. Ber.*, **93**, 803 (1960).
87. F. W. Bennett, H. J. Emeléus, and R. N. Haszeldine, *J. Chem. Soc.*, **1953**, 1565.
88. H. Nöth and H.-J. Vetter, *Chem. Ber.*, **94**, 1505 (1961).
89. L. Maier, in *Progress in Inorganic Chemistry*, Vol. 5, F. A. Cotton, Ed., New York, 1962, p. 27.
90. J. W. B. Reesor and G. F. Wright, *J. Org. Chem.*, **22**, 385 (1957).
91. W. Seidel and K. Issleib, *Z. Anorg. Allgem. Chem.*, **325**, 113 (1963).
92. K. Issleib and F. Krech, *Chem. Ber.*, **94**, 2656 (1961).

93. R. Schmutzler, *Inorg. Chem.*, **3**, 421 (1964).
94. A. B. Burg and K. K. Joshi, *J. Am. Chem. Soc.*, **86**, 353 (1964).
95. H. Köhler and A. Michaelis, *Chem. Ber.*, **10**, 807 (1877).
96. M. Baudler and K. Hammerström, *Z. Naturforsch.*, **20b**, 810 (1965).
97. K. Issleib and B. Mitcherling, *Z. Naturforsch.*, **15b**, 267 (1957).
98. Wm. A. Henderson, M. Epstein, and F. S. Seichter, *J. Am. Chem. Soc.*, **85**, 2462 (1963).
99. K. Issleib and W. Seidel, *Z. Anorg. Allgem. Chem.*, **303**, 155 (1960).
100. Wm. A. Henderson, S. A. Buckler, and M. Epstein, U.S. Pat. 3,032,591, 1962; *Chem. Abstr.*, **57**, 11239f (1962).
101. W. Kuchen and H. Buchwald, *Angew. Chem.*, **68**, 791 (1956).
102. W. Kuchen and W. Grünewald, *Chem. Ber.*, **98**, 480 (1965).
103. F. Pass and H. Schindlbauer, *Monatsh. Chem.*, **90**, 148 (1959).
104. L. Maier, *Helv. Chim. Acta*, 49, 1119 (1966).
105. L. Maier, *Nature*, **208**, 383 (1965).
106. Th. Weil, B. Prijs, and H. Erlenmeyer, *Helv. Chim. Acta*, **35**, 616 (1952).
107. L. Maier and J. J. Daly, see ref. 37.
108. M. Fild, I. Hollenberg, and O. Glemser, *Naturwiss.*, **54**, 89 (1967).
109. Wm. A. Henderson, U.S. Pat. 3,029,289, 1962; *Chem. Abstr.*, **57**, 8618e (1962).
110. V. N. Kulakova, Y. M. Zinov'ev, and L. Z. Soborovskii, *Zh. Obshch. Khim.*, **29**, 3957 (1959); **54**, 20846e (1960).
111. F. Seel, K. Rudolph, and R. Budenz, *Z. Anorg. Allgem. Chem.*, **341**, 196 (1965).
112. A. B. Burg and W. Mahler, U.S.Pat. 2,923,742, 1960; *Chem. Abstr.*, **54**, 9765b (1960).
113. A. B. Burg and J. F. Nixon, *J. Am. Chem. Soc.*, **86**, 356 (1964).
114. A. B. Burg and W. Mahler, U.S. Pat. 2,923,741, 1960; *Chem. Abstr.*, **54**, 9767b (1960).
115. A. B. Burg and L. K. Peterson, *Inorg. Chem.*, **5**, 943 (1966).
116. A. H. Cowley and R. P. Pinnell, *Inorg. Chem.*, **5**, 1459 (1966).
117. K. Issleib and M. Hoffman, *Chem. Ber.*, **99**, 1320 (1966).
118. L. Horner, H. Hoffmann, and P. Beck, *Chem. Ber.*, **91**, 1583 (1958).
119. W. Kuchen and H. Buchwald, *Chem. Ber.*, **91**, 2296 (1958).
120. H. Schindlbauer, see ref. 37.
121. A. H. Cowley and R. P. Pinnell, *J. Am. Chem. Soc.*, **88**, 4533 (1966).
122. W. Hewertson and H. R. Watson, *J. Chem. Soc.*, **1962**, 1490.
123. W. Kuchen and H. Buchwald, *Angew. Chem.*, **69**, 307 (1957).
124. H. Niebergall and B. Langenfeld, U.S. Pat. 2,959,621, 1960; *Chem. Abstr.*, **55**, 7289 (1961).
125. H. Nöth and H.-J. Vetter, *Naturwiss.*, **47**, 204 (1960).
126. H. G. Ang and J. M. Miller, *Chem. Ind. (London)*, **1966**, 944.
127. A. B. Burg and W. Mahler, *J. Am. Chem. Soc.*, **79**, 4242 (1957).
128. A. B. Burg and J. E. Griffiths, *J. Am. Chem. Soc.*, **82**, 3514 (1960).
129. P. R. Bloomfield and K. Parvin, *Chem. Ind. (London)*, **1959**, 541.
130. K. Issleib and M. Hoffmann, *Chem. Ber.*, **99**, 1320 (1966).
131. W. Kuchen and W. Grünewald, *Angew. Chem.*, **75**, 576 (1963).
132. W. Kuchen and W. Grünewald, *Chem. Ber.*, **98**, 480 (1965).
133. U. Schmidt and Ch. Osterroht, *Angew. Chem.*, **77**, 455 (1965).
134. A. B. Burg, "Proceedings of the Robert A. Welch Foundation Conferences on Chemical Research," *Topics in Modern Inorganic Chemistry*, Vol. 6, Houston, Texas, 1962, p. 133.

135. E. Fluck and K. Issleib, *Z. Naturforsch.*, **21b**, 736 (1966).
136. A. Besson, *Compt. Rend.*, **122**, 140,814,1200 (1896); **124**, 763, 1346 (1897).
137. M. Baudler, *Z. Naturforsch.*, **9b**, 447 (1954).
138. M. Baudler, *Z. Anorg. Allgem. Chem.*, **228**, 171 (1956).
139. W. Kuchen and H. G. Beckers, *Angew. Chem.*, **71**, 163 (1959).
140. H. Spandau and A. Beyer, *Naturwiss.*, **46**, 400 (1959).
141. M. A. A. Beg and H. C. Clark, *Can. J. Chem.*, **39**, 564 (1961).
142. W. Strecker and Ch. Grossmann, *Ber. Dtsch. Chem. Ges.*, **49**, 63 (1916).
143. M. I. Kabachnik and E. S. Shepeleva, *Izvest. Akad. Nauk SSSR, Otdel. Khim. Nauk*, **1949**, 56; *Chem. Abstr.*, **43**, 5739 (1949).
144. K. Issleib and A. Tzschach, *Chem. Ber.* **92**, 704 (1959).
145. W. Kuchen and H. Buchwald, *Angew. Chem.*, **71**, 162 (1959).
146. P. J. Christen, L. M. van der Linde, and F. N. Hooge, *Rec. Trav. Chim.*, **78**, 161 (1959).
147. L. Maier, *Chem. Ber.*, **94**, 3043 (1961).
148. K. A. Pollart and H. J. Harwood, *J. Org. Chem.*, **27**, 4444 (1962).
149. A. B. Burg, *J. Inorg. Nucl. Chem.*, **11**, 258 (1959).
150. A. B. Burg, U.S. Pat. 3,242,216, 1966; *Chem. Abstr.*, **64**, 15922f (1966).
151. K. Issleib and D.-W. Müller, *Chem. Ber.*, **92**, 3175 (1959).
152. K. Issleib and G. Döll, *Chem. Ber.*, **94**, 2664 (1961).
153. K. Issleib and G. Döll, *Z. Anorg. Allgem. Chem.*, **324**, 259 (1963).
154. K. Issleib and K. Krech, *Chem. Ber.*, **99**, 1310 (1966).
155. H.-J. Vetter and H. Nöth, *Chem. Ber.*, **96**, 1816 (1963).
156. M. J. Gallagher and I. D. Jenkins, *Chem. Commun. (London)*, **1965**, 587; *J. Chem. Soc. (London)*, **1966**, 2176.
157. N. G. Feshchenko and A. V. Kirsanov, *Zh. Obshch. Khim.*, **30**, 3041 (1960); *Chem. Abstr.*, **55**, 14145 (1961).
158. N. G. Feshchenko and A. V. Kirsanov, *Zh. Obshch. Khim.*, **31**, 1399 (1961); *Chem. Abstr.*, **55**, 27,169 (1961).
159. K. Issleib and A. Tzschach, *Chem. Ber.*, **92**, 1397 (1959).
160. K. Issleib and W. Wenschuh, *Chem. Ber.*, **97**, 715 (1964).
161. H. Fritzsche, U. Hasserodt, and F. Korte, *Angew. Chem.*, **75**, 1205 (1963).
162. A. P. Lane and D. S. Payne, *Proc. Chem. Soc.*, **1964**, 403.
163. G. B. Postnikova and I. F. Lutsenko, *Zh. Obshch. Khim.*, **33**, 4028 (1963); *Chem. Abstr.*, **60**, 9309 (1964).
164. F. Seel, K. Rudolph, and W. Gombler, *Angew. Chem.*, **79**, 686 (1967).
165. K. Issleib and K. Krech, *Chem. Ber.*, **98**, 2545 (1965).
166. A. Finch, *Can. J. Chem.*, **37**, 1793 (1959).
167. K. Moedritzer, L. Maier, and L. C. D. Groenweghe, *J. Chem. Eng. Data*, **7**, 307 (1962).
168. E. Fluck and K. Issleib, *Chem. Ber.*, **98**, 2674 (1965).
169. F. A. Hart and F. G. Mann, *J. Chem. Soc. (London)*, **1957**, 3939.
170. L. Maier, *Helv. Chim. Acta*, **45**, 2381 (1962).
171. E. Fluck and K. Issleib, *Z. Anorg. Allgem. Chem.*, **339**, 274 (1965).
172. E. Fluck and K. Issleib, *Chem. Ber.*, **98**, 2674 (1965).
173. M. L. Nielsen, J. V. Pustinger, and J. Strobel, *J. Chem. Eng. Data*, **9**, 167 (1964).
174. E. Fluck, H. Bürger, and U. Goetze, *Z. Naturforsch.*, **22b**, 912 (1967).
175. J. R. Van Wazer, C. F. Callis, J. N. Shoolery, and R. C. Jones, *J. Am. Chem. Soc.*, **78**, 5715 (1956).

176. H. O. Fröhlich, Ph.D. thesis, University of Jena, Germany, 1962.
177. P. J. Wheatley, *J. Chem. Soc. (London)*, **1960**, 523.
178. J. Goubeau, H. Reinhardt, and D. Bianchi, *Z. Physik. Chem. NF*, **12**, 387 (1957).
179. K. A. Pollart and H. J. Harwood, Abstr. of Papers, 136th Meeting of the Am. Chem. Soc., Atlantic City, 1959, p. 102-P.
180. G. M. Bogolyubov and A. A. Petrov, *Zh. Obshch. Khim.*, **36**, 1505 (1966).
181. E. Lindner and H. Schless, *Chem. Ber.*, **99**, 3331 (1966).
182. A. A. Sandoval and H. C. Moser, *Inorg. Chem.*, **2**, 27 (1963).
183. A. A. Sandoval and H. C. Moser, Abstr. of Papers, 141st National Meeting of the American Chemical Society, Washington, D.C., March 21–29, 1962, p. 21-M.
184. A. A. Sandoval, H. C. Moser, and R. W. Kiser, *J. Phys. Chem.*, **67**, 124 (1963).
185. E. Fluck and H. Binder, unpublished work.
186. M. Baudler and L. Schmidt, *Naturwiss.*, **46**, 577 (1959).
187. M. Baudler, H. Ständeke, M. Borgardt, and H. Strabel, *Naturwiss.*, **52**, 345 (1965).
188. M. Baudler, H. Ständeke, M. Borgardt, H. Strabel, and J. Dobbers, *Naturwiss.*, **53**, 106 (1966).
189. W. Seidel, *Z. Anorg. Allgem. Chem.*, **330**, 141 (1964).
190. E. Fluck and H. Binder, *Z. Anorg. Allgem. Chem.*, in press.
191. G. J. Palenik and J. Donohue, *Acta Cryst.*, **15**, 564, 1962.
192. C. J. Spencer and W. N. Lipscomb, *Acta Cryst.*, **14**, 250 (1961).
193. E. Fluck and V. Novobilský, unpublished work.
194. R. L. Carroll and R. P. Carter, *Inorg. Chem.*, **6**, 401 (1967).
195. S. E. Frazier, R. P. Nielsen, and H. H. Sisler, *Inorg. Chem.*, **3**, 292 (1964).
196. E. Fluck, *Die kernmagnetische Resonanzspektroskopie und ihre Anwendung in der Anorganischen Chemie*, Springer-Verlag, Berlin, 1963.
197. M. M. Crutchfield, C. H. Dungan, J. H. Letcher, V. Mark, and J. R. Van Wazer, "P^{31} Nuclear Magnetic Resonance," in *Topics in Phosphorus Chemistry*, Vol. 5, M. Grayson, and E. J. Griffith, Eds. Interscience, New York, 1967.
198. R. W. Rudolf, R. C. Talyor, and R. W. Parry, *J. Am. Chem. Soc.*, **88**, 3729 (1966).
199. E. R. Andrew and V. T. Wynn, *Proc. Roy. Soc. (London)*, **291A**, 257 (1966).
200. K. Irgolic, R. A. Zingaro, and M. Kudchadker, *Inorg. Chem.*, **4**, 1421 (1965).
201. R. M. Lynden-Bell, *Trans. Faraday Soc.*, **57**, 888 (1961).
202. S. L. Manatt, G. L. Juvinall, and D. D. Elleman, *J. Am. Chem. Soc.*, **85**, 2664 (1963).
203. K. Issleib and E. Fluck, *Angew. Chem., Intern. Ed. Engl.*, **5**, 587 (1966).
204. W. Mahler, *J. Am. Chem. Soc.*, **86**, 2306 (1964).
205. L. Maier, reference in 197.
206. K. Issleib and M. Hoffman, *Chem. Ber.*, **99**, 1320 (1966).
207. C. F. Callis, J. R. Van Wazer, J. N. Shoolery, and W. A. Anderson, *J. Am. Chem. Soc.*, **79**, 2719 (1957).
208. A. H. Cowley and S. T. Cohen, *Inorg. Chem.*, **3**, 780 (1964).
209. R. Schmutzler, *Inorg. Chem.*, **3**, 421 (1964).

Condensed Phosphoric Acids and Condensed Phosphates

C. Y. SHEN and D. R. DYROFF

Inorganic Chemical Division, Monsanto Company
St. Louis, Missouri

CONTENTS

I. Introduction	158
II. Scope and Limitations	158
III. Methods of Preparation	160
A. Thermal Dehydration of Less-Condensed Acid Phosphates	160
B. Dehydration of Acid Phosphates at Relatively Low Temperature	162
C. Dehydration of Phosphates Containing No Hydrogen	162
D. Reaction with Formation of a Volatile By-Product Other Than Water	163
E. Reactions Involving the Breakage of P—O—P Linkages	164
1. Reactions at High Temperatures	164
2. Reactions in Solution	164
F. Precipitation from Solution in a Solvent	164
G. Crystallization of Equilibrium Products in the Absence of a Solvent	166
H. Separation of Mixtures	167
I. Other Preparation Methods	168
IV. Methods of Characterization	168
A. Phase Equilibrium Approach	169
B. Characterization of Individual Products	169
V. Pyrophosphoric Acid and its Salts	171
A. Pyrophosphoric Acid	171
B. Alkali Metal Pyrophosphates	171
1. Ammonium Pyrophosphates	171
2. Lithium Pyrophosphates	172
3. Sodium and Potassium Pyrophosphates	173
4. Cesium and Rubidium Pyrophosphates	173
C. Alkaline Earth Pyrophosphates	173
D. Pyrophosphates Containing an Alkali Metal and One or More Other Metals	175
E. Pyrophosphates Containing No Alkali Metal	182
VI. Tripolyphosphoric Acid and its Derivatives	189
A. Tripolyphosphoric Acid	189
B. Alkali Metal Tripolyphosphates	189
C. Other Tripolyphosphates	192

VII. Tetra and Higher Polyphosphates 196
    A. Condensed Phosphoric Acids 196
    B. Tetra and Higher Polyphosphates of a Single Chain Length 196
    C. Phosphate Glasses 202
VIII. Metaphosphoric Acids and their Salts 202
    A. Metaphosphoric Acids 202
    B. Trimetaphosphates 203
    C. Tetrametaphosphates 205
    D. Higher Metaphosphates 207
IX. Branched Phosphates 208
X. Condensed Phosphates of Uncertain Anion Type 208
References 212

I. INTRODUCTION

This article on the preparation of condensed phosphoric acids and condensed phosphates is a sequel to an earlier article on orthophosphoric acid and its salts.[1] In other recent reviews on condensed phosphates,[2-5] the emphasis has been on structure and classification rather than on preparation methods.

The large number of condensed phosphate anion types and the frequent occurrence of several crystal structures (polymorphs) for the same combination of cations and anions has caused a nomenclature problem, with the term metaphosphate being used particularly loosely. This article will conform to Van Wazer's suggestion[5] to use the term metaphosphate only for phosphate rings and the term polyphosphate for unbranched chains. For simplicity's sake, formulas of the type $(MPO_3)_x$ will be used to represent the long-chain polyphosphates even though this composition is only approached as a limit in the case of polyphosphates. Branched phosphates can be present as mixtures in preparations with lower cation content than the metaphosphates,[1] but because of the antibranching rule,[5] compounds containing a single species of this type anion are rare. Because of their rarity, all such compounds will be grouped together. The numbering of polymorphs has been chaotic and seems too inconsistent and well-entrenched to systematize. In each case, the presently prevailing designation(s) will be used here.

II. SCOPE AND LIMITATIONS

By definition, phosphates are compounds in the anions of which each phosphorus atom is surrounded approximately tetrahedrally by four oxygen atoms. A phosphate is condensed when some oxygen atoms (except for impurities) are shared between phosphorus atoms. In this discussion, the following types of phosphate species will be excluded except for isolated cases:

1. Complexes which are observed only in solution
2. Species which are partly organic
3. Materials of indefinite composition (including some poorly characterized products precipitated from solution)
4. Species which contain anions other than phosphate, oxide, or hydroxyl ions
5. Substituted phosphates such as fluorophosphates or amidophosphates (which are outside the above definition of phosphates)
6. Peroxyphosphates

A material is not considered to be of indefinite composition if only the amount of hydrate water is variable. Where polymorphism occurs, all of the forms which have been characterized will be discussed. In some cases there are conflicting reports concerning the nature of a compound or system, and there is little basis for determining who is right. Unless one of the reports is less reliable because it is much older and therefore based upon less powerful techniques, both sides will be presented.

In classifying the various phosphates, the emphasis will be on the anion structure rather than upon the M_2O/P_2O_5 ratio,[5] since the latter concept has a number of pitfalls which can lead to confusion. For instance, within a single system, one would normally expect the degree of anion condensation to vary inversely with the M_2O/P_2O_5 ratio. Yet in the system UO_2–P_2O_5,[6] the compound $2UO_2 \cdot P_2O_5$ is more condensed than the compound $3UO_2 \cdot 2P_2O_5$, the more descriptive formulas being $(UO)_2P_2O_7$ and $U_3(PO_4)_4$, respectively. It is also interesting to try to predict whether the compound $UO_2 \cdot P_2O_5$ has the structure UP_2O_7 or $[UO(PO_3)_2]_x$. Thus, to correctly apply the double oxide concept, one must know not only the valence of the metal but also the oxygen content of the cation. Also, if M_2O excludes the H_2O content, compounds of different degrees of condensation can have the same M_2O/P_2O_5 ratio, for example Na_2HPO_4 and $Na_4P_2O_7$. If H_2O is included, one must carefully distinguish between molecularly bound water and hydrate water. Finally, the M_2O/P_2O_5 ratio does not distinguish between a straight chain and a branched chain containing the same number of phosphorus atoms, and it could also be misleading in the case of a double salt of a metal oxide and a condensed phosphate.

III. METHODS OF PREPARATION

A. Thermal Dehydration of Less-Condensed Acid Phosphates

This common method, known for over a century,[7,8] is illustrated by eq. 1. Usually, when this method is used, the system is easily kept far from

$$2NaH_2PO_4 \rightleftharpoons Na_2H_2P_2O_7 + H_2O(g) \tag{1}$$

equilibrium so that the reaction proceeds mainly to the right. The importance of the potential reverse reaction and of the other phases present along with the crystalline starting material and product have been recognized only recently.[9,10]

The water vapor pressure can affect both the final product composition and the reaction rate.[10] The vapor pressure of the cation is usually negligible but is occasionally critical as in the case of preparing ammonium polyphosphates.[297] At relatively high temperatures, the vapor pressure of P_2O_5 may also require control in order to obtain the desired product.[23]

One saving feature of such thermal dehydration reactions is their large activation energy (40–60 kcal/mole[11,12] for several sodium phosphates) which results in a fairly narrow temperature range within which a given reaction proceeds at practical rates.

In order to convert a mixture of two crystalline nonvolatile salts to another crystalline salt of different composition at an appreciable rate, it seems reasonable that a molten intermediate phase must be present to facilitate diffusion of reactants to the growing crystals or nuclei.[14] Even in the simpler case of a single reactant, some diffusion will be necessary, and a molten phase may be required in very small amounts. In some systems a molten phase can easily be observed directly as in the case of dehydrating NaH_2PO_4 to $Na_2H_2P_2O_7$ in a sealed tube where a clear melt forms at 140° and the product crystallizes after the temperature is increased to 170°. A hot stage microscope can sometimes be used to directly observe the molten phase in systems where it is present in smaller amounts.[12]

In some systems the molten phase may be present in such small amounts that it must be observed indirectly if at all. Two methods used for this purpose are x-ray diffraction and paper chromatography of quenched samples. Electrical conductivity can also be used to follow the formation and disappearance of molten phases as shown in Figure 1. Pioneering work on the role of small amounts of molten phase in the absence of fluxing agents was carried out on the reaction by which $Na_5P_3O_{10}$ is prepared by dehydrating a mixture of sodium orthophosphates.[13]

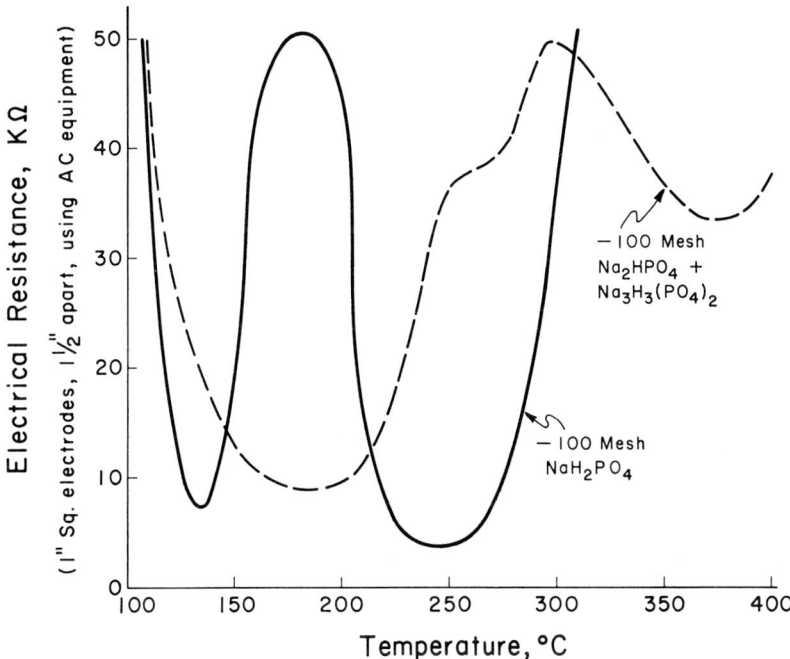

Fig. 1. Variation of electrical resistance during thermal dehydration of sodium orthophosphates.

Often the amount of molten phase can be profitably increased by adding a fluxing agent or a small amount of impurity which produces a lower-melting eutectic. The larger amount of molten phase may greatly enhance the formation of nuclei which is often the rate-controlling step in a solid state reaction.[14] Of course, seeding is another means of obtaining nuclei quickly and may sometimes even affect the course of the reaction as in the case of aluminum polyphosphates.[41]

A good flux should melt or decompose somewhat below the final reaction temperature without producing any undesirable product characteristics such as a color which is off. Among the common fluxes are nitrites, urea, and nitrates.[14a] Ammonium nitrate is used as a flux in the preparation of $Na_3P_3O_9$ of high purity.[15] Urea does double duty as a flux and dehydrating agent (eq. 2) and has been applied to the preparation of alkali metal polyphosphates.[14a,16,104]

$$2\;\overset{\mathrm{OH}}{\underset{|}{-\mathrm{P}-}}\; + (\mathrm{NH}_2)_2\mathrm{CO} \longrightarrow \;-\mathrm{P}-\mathrm{O}-\mathrm{P}-\; + \mathrm{CO}_2 + 2\mathrm{NH}_3 \qquad (2)$$

There are other factors which affect reaction rate and product purity. When two or more crystalline species are used as reactants, intimate mixing is required to produce a pure product, and intimate mixing requires a small particle size. Also, there is usually an optimum heating profile since the amount of molten phase present may be greatly diminished by removal of too much water. For instance, if too low a heating rate is used, the molten phase may have formed and dried up again before the temperature is high enough for rapid formation of the new crystalline phase. If the temperature is allowed to get too high, the molten phase may dry up faster than it can react to give crystals. Reduced pressure would be expected to aggravate such problems, while a controlled application of water vapor pressure (less than the equilibrium value) should minimize them.[10]

B. Dehydration of Acid Phosphates at Relatively Low Temperature

Equation 1 indicates that the condensation reaction can be forced to completion by lowering the activity of water. In fact, the equilibrium constant is such that about 0.16% pyrophosphate is already present at equilibrium in a concentrated aqueous solution of NaH_2PO_4 at 100°.[17] There are other ways to remove water besides simple application of heat, one of the most successful being the use of acetic anhydride.[18] The alkali metal dihydrogen orthophosphates have been dehydrated to condensed phosphates by refluxing with excess acetic anhydride, in some cases leading to products different from those produced by the corresponding thermal dehydration. Similar approaches with SO_3 and P_2O_5 usually give less pure products. The successful use of urea was mentioned earlier (eq. 2). In preparing condensed phosphoric acid, H_2O can be removed by electrolysis.[19]

C. Dehydration of Phosphates Containing No Hydrogen

The above dehydration methods require that hydrogen be present in the phosphate starting materials. However, as shown in eq. 3, it is possible to start from a hydrogen-free phosphate, add the hydrogen as a gas, and then remove it as water, taking with it the oxygen that one must remove in order to effect a condensation.[20] This novel approach is interesting chemi-

$$2FePO_4 + H_2 \xrightarrow{600°} Fe_2P_2O_7 + H_2O(g) \tag{3}$$

cally but is limited in scope because the metal ion in the starting material must have a lower positively charged valence state of considerable stability.

Otherwise, the free metal or a phosphide is produced. In all cases, the conditions must be carefully controlled to avoid side reactions and undesired reactions by the condensed phosphate product.

D. Reaction with Formation of a Volatile By-Product Other Than Water

The simplest reactions of this sort are decompositions of a single starting material as illustrated by eqs. 4[21] and 5.[22] Another example is the

$$4Na_2(C_2H_5)PO_4 \longrightarrow 2Na_4P_2O_7 + 3C_2H_4(g) + C_2H_5OH(g) + H_2O(g) \quad (4)$$

$$2M(NH_4)PO_4 \cdot xH_2O \longrightarrow M_2P_2O_7 + 2NH_3(g) + (1 + 2x)H_2O(g) \quad (5)$$

thermal decomposition at 1300° of $U(PO_3)_4$ with evolution of P_2O_5 to give UP_2O_7.[23] Equation 6 illustrates a double decomposition reaction with formation of a volatile by-product.[24]

$$3NaCl + 3NH_4H_2PO_4 \longrightarrow (NaPO_3)_3 + NH_4Cl(g) \quad (6)$$

Reaction of alkali metal halides with phosphoric acid (or P and O_2) to form polyphosphates is of commercial importance and has been extensively studied.[25-29] The production of shorter chain condensed phosphates from alkali halide and H_3PO_4 (eq. 7) is more difficult because of a

$$5NaCl + 3H_3PO_4 \longrightarrow Na_5P_3O_{10} + 5HCl(g) + 2H_2O(g) \quad (7)$$

greater solubility of HCl in the melt. The dependence of the HCl content of the melt upon the HCl vapor pressure in the gas phase was studied thermodynamically,[30] and countercurrent steam stripping of the melt has been found effective in reducing chloride contamination in the product.[27]

Using phosphorus and oxygen instead of H_3PO_4 leads to chlorine, as shown by eq. 8.[26,28]

$$P_4 + 6O_2 + 4KCl \longrightarrow 4/x(KPO_3)_x + 2Cl_2(g) \quad (8)$$

This approach is intriguing because the heat of combustion of phosphorus is used to form the condensed phosphate instead of being dissipated while producing orthophosphates and later replaced at added expense to drive the condensation. Unavailability of materials of construction to withstand the corrosive species involved has prevented commercial exploitation.

In isolated instances, reactions of $P_2O_5 \cdot xH_2O$ with metals or metal oxides have led to products which would require that H_2 or O_2 be evolved as a by-product. For instance, $P_2O_5 \cdot xH_2O$ has been reported to react with iron[31] to give $Fe_2P_4O_{12}$ (and H_2) and with U_3O_8 [23] to give $U(PO_3)_4$ (and O_2).

E. Reactions Involving the Breakage of P—O—P Linkages

1. Reactions at High Temperatures

Condensed phosphates can react at high temperatures with metal oxides, with metal salts which decompose to the oxides, or with less-condensed metal phosphates to produce phosphates with a lower degree of condensation than that of the starting phosphate (or more condensed starting phosphate). Such reactions, which take place in melts or in the solid state, are illustrated by eqs. 9–12.[32–34] In the solid state reactions of

$$[-\overset{\overset{\displaystyle O}{|}}{\underset{\underset{\displaystyle O}{|}}{P}}-O-\overset{\overset{\displaystyle O}{|}}{\underset{\underset{\displaystyle O}{|}}{P}}-]^{-n} + O^{-2} \longrightarrow [-\overset{\overset{\displaystyle O}{|}}{\underset{\underset{\displaystyle O}{|}}{P}}-O]^{-(n-m+2)} + [-\overset{\overset{\displaystyle O}{|}}{\underset{\underset{\displaystyle O}{|}}{P}}-O]^{-m} \quad (9)$$

$$(NaPO_3)_x + xNa_4P_2O_7 \longrightarrow xNa_5P_3O_{10} \quad (10)$$

$$Ca_3(PO_4)_2 + 2P_2O_5 \longrightarrow \frac{3}{x}[Ca(PO_3)_2]_x \quad (11)$$

$$Na_2CO_3 + P_2O_5 \longrightarrow \frac{2}{x}[NaPO_3]_x + CO_2(g) \quad (12)$$

this type, the effects of water vapor pressure, etc., are similar to those described in Section III-A.[14] The effect of water vapor is particularly dramatic in the case of eq. 12, where a drop of water added to a stable mixture of dry reactants at room temperature initiates a self-sustaining exothermic reaction which can be observed propagating through the reaction mass.[12]

2. Reactions in Solution

The most useful reactions of this type are the controlled hydrolysis of P_4O_{10} to produce tetrametaphosphates[35] and the ring cleavage of pure metaphosphates to produce the polyphosphates with the same number of phosphorus atoms. The latter method can be used to produce $Na_5P_3O_{10} \cdot 6H_2O$ which is more pure than that prepared in other ways.[36,37]

F. Precipitation from Solution in a Solvent

In most of these reactions, because of the low temperatures used, the same type of phosphate anion is present in the product and in the reactants. A large number of precipitated products have been obtained by mixing a solution of a soluble condensed phosphate with a solution of a soluble salt of a metal.[38,39] Many such products described in the older literature have been claimed by more recent authors to be mixtures, and in general, precipitated products can be arranged in a sort of hierarchy of scientific interest. If a precipitate can be shown to consist entirely of a single crystalline species, it immediately gains in stature because it can be expected to have a definite composition except possibly for variations in hydrate water content. Some amorphous precipitates also have a fairly constant composition over a wide range of conditions of precipitation, and if this is demonstrated to be the case (as described in Sec. IV-A), the compound is still of considerable interest. If, however, the composition varies continuously with the conditions or cannot be reproduced, the preparation is of little interest and will not be discussed here.

In cases where it is not possible to use metathesis reactions to generate relatively insoluble salts, one can often obtain a precipitate by evaporating water from or adding an agent such as alcohol to a solution of a soluble phosphate.

Frequently, a well-crystallized precipitate can be obtained only if the precipitation proceeds very slowly and uniformly, and several tricks have been discovered to accomplish this. It may be possible to slowly generate a precipitant or the product itself by a chemical reaction proceeding in the liquid phase at a slow rate. If the product is more soluble at lower pH, it may be possible to solubilize a slurry by adding SO_2 and then slowly recrystallize the product by removing the SO_2 at a controlled rate with a vacuum or an inert gas purge.[40] Evaporation of solvent or addition of precipitant can be done slowly too, but the concentration gradients involved in these methods are usually larger. Stirring of the solution is sometimes helpful and sometimes harmful (by generating new nuclei instead of allowing the growth to occur mainly on the crystals already present). Careful seeding is usually helpful.

Some precipitated products can be obtained only with seeding and were obtained for the first time by seeding with crystals of an isomorphous compound.[41]

The nature of the precipitated product depends mainly on the stoichiometry of the reactants, the nature of the solvent, the temperature, and the pH; and all of these variables should be considered independently. For instance, early workers had reported a series of hydrated acid pyrophosphates which later workers failed to reproduce. The existence of these salts was considered doubtful until a recent investigation[42] showed that one of them could be reproduced even at relatively high pH by using a ratio of metal ion to pyrophosphate corresponding to the acid salt. It was noted that earlier attempts to reproduce the preparation of the acid salts had failed because, although the pH was varied, the M/P_2O_7 ratio was held constant at a value corresponding to the *normal* salt rather than the acid salt. Apparently, in this case, the stoichiometry had a greater effect than the pH upon whether the normal or acid salt was obtained. Solvents other than water have been used, but only rarely.[43] The use of a nonaqueous solvent could conceivably lead to products impossible to precipitate from aqueous solution by avoiding hydrolysis of the cation or anion, by avoiding hydrate formation, by virtue of the acidity or basicity of the solvent, or by more subtle effects. The usual method of varying the nature of the solvent though, is via additives to an aqueous system.

Since ammonium phosphates are difficult to prepare thermally because of the volatility of the cation, they are often prepared by reaction of ammonium sulfide with the corresponding lead or copper salt generating a solution of the ammonium phosphate and an insoluble Pb or Cu sulfide by-product.[44] After separation of the sulfide, the ammonium phosphate can be precipitated out by one of the methods discussed above for water-soluble salts.

G. Crystallization of Equilibrium Products in the Absence of a Solvent

In this method, which overlaps several of the preceding methods, an appropriate mixture is tempered at a temperature within the region of stability of the desired equilibrium crystal phase and high enough that the phosphate species in the starting materials react with each other to give exclusively the species required by the growing crystals. The crystal growth provides the driving force for the phosphate rearrangements.[5]

What species is formed depends upon what species can be packed into a crystal structure stable under the prevailing conditions of stoichiometry, temperature, total pressure, and vapor pressure.[5] It has also been shown to depend upon the size and shape of the cation.[31] For instance, with the stoichiometry adjusted to give a metaphosphate, the use of one cation can lead to a tetrametaphosphate while the use of another cation under the same conditions can lead to a long-chain polyphosphate.[31]

The growth of an equilibrium crystal phase is generally the final step in the methods (Sec. III-A to III-E-1) above, but one may also simply use a different allotropic form of the desired product as the starting material.[201] In the latter case, the reaction could conceivably proceed without making or breaking any P—O—P bonds.[201]

Depending upon the vapor pressures involved, the system may be open or sealed during the tempering period. The reaction mixture may be heated above the melting point and then slowly cooled to the tempering temperature, thus allowing the product to crystallize from the melt; or it may be held at a temperature where there is little if any liquid present and the reaction proceeds via a solid state reaction. In the latter case, as discussed in Section III-A, a certain amount of molten intermediate phase may be present during the reaction period.[13]

Once the desired product is formed, it can be cooled at any rate if it is the stable phase at all temperatures between room temperature and the tempering temperature. Otherwise, it must be cooled fast enough to ensure that negligible amounts of undesired phases are formed while it is too cool to be within its region of stability but too hot for the phase transformation rate to be vanishingly small. Sometimes transitions are so fast that drastic quenching methods are required such as plunging the hot product into liquid nitrogen. Cooling too slowly can give a mixture of crystalline phases which can be mistaken for a single species. Occasionally, the phase transformation is fast even at room temperature, and the high-temperature phase cannot be obtained at room temperature.[201]

It is quite common to observe a series of allotropic forms of the same composition which can be obtained by tempering samples at various temperatures and rapidly cooling to room temperature. Recently, the preparation of a high *pressure* form of ZrP_2O_7 was accomplished by tempering a sample at 750–1000° and 55–100 kbars pressure and then rapidly cooling to room temperature before releasing the pressure.[45] The new phase did not revert to a low pressure form when the pressure was released. This approach has not been explored enough and could lead to many new phases.

H. Separation of Mixtures

Individual large-ring phosphates or individual polyphosphates of intermediate chain length such as those containing between six and twenty phosphorus atoms per anion have not been prepared by any method using other types of phosphates as starting material, but they are present in mixtures in easily prepared phosphate glasses and condensed phosphoric acids. So far, the only successful approaches to the preparation of the pure species have involved their isolation from mixtures.

The pentameta- and hexametaphosphates were isolated from Graham's salt by a combination of fractional crystallization techniques.[46] The penta- through octapolyphosphates have been isolated from a glass by anion exchange chromatography.[47,48] Polyphosphates have also been separated from condensed phosphoric acid using selective extraction with a long-chain alkylamine.[49,50]

Once a relatively pure anion species had been obtained in solution, it was possible to obtain crystalline salts by precipitation techniques. However, the number of cations which give crystalline precipitates with these anions is severely limited, and often a large organic cation has been used.[48,51-53]

I. Other Preparation Methods

Methods which can be described briefly and which are not very closely related to the methods described in the preceding sections are discussed here.

P—O—P linkages can be formed by reaction of silver phosphates with chlorophosphates, with AgCl forming as a by-product.[54] This method has some potential for use in positional synthesis but has not been well developed yet.

$Si(PO_3)_4 \cdot 2H_2O$ has been prepared by reaction of $(CH_3)_2Si(OC_2H_5)_2$ with P_2O_5,[55] and $H_2NP(O)(OH)_2$ has been reported to thermally rearrange to a product believed to be a long-chain ammonium polyphosphate.[56]

Ordinarily, condensed phosphoric acids consist of a mixture of species of different molecular weight with the exact equilibrium distribution depending upon the temperature and water content. However, by equilibrating salts (containing a single anion type) with the acid form of a cation exchanger at fairly low temperatures, it has been possible to prepare solutions containing essentially a single acid species. Such solutions have occasionally been prepared and promptly reacted with a base or metal salt solution before the acid had time to revert to the equilibrium mixture.[57]

In another sort of ion exchange process, solid $AlH_2P_3O_{10} \cdot xH_2O$ has been equilibrated with solutions of various metal salts to produce the corresponding insoluble double salts.[41]

IV. METHODS OF CHARACTERIZATION

For a reported preparation method to be of much interest at all, it is imperative that evidence be given that a pure compound was prepared rather than a simple mixture, and enough information on the composition and properties of the product must be given to allow future investigators to identify it in their preparations. In addition to this, in the case of phosphates, it is highly desirable that the type or types of phosphate anions present be determined experimentally.

A. Phase Equilibrium Approach

The essence of this type of approach to the identification of pure compounds is that one does not consider a single preparation by itself, but rather one measures one or more properties for a series of preparations covering a whole range of compositions. By studying the way in which the properties vary with changing overall composition, it is possible to decide which compositions correspond to pure compounds and which correspond to mixtures.

In the case of condensed phosphates which crystallize from the melt, the melting range is determined as a function of composition, and the compositions at which pure compounds are formed can be recognized from the appearance of a phase diagram based upon the data. There is not room here for a discussion of the interpretation of phase diagrams, but it should be pointed out that the appearance of the liquidus curve can only reveal the existence of the phases stable near the melting point, and it may even fail to reveal the high-temperature form if the effect of the phase on the liquidus curve is subtle and the data points are limited in number or in accuracy.

In the case of condensed phosphates precipitated from solution, phase diagrams are constructed on the basis of solubility data and again can be interpreted to reveal the existence of solid phases of definite composition.

Specific examples of phase diagrams of both types for phosphate systems can be found in Van Wazer's book on phosphorus chemistry,[5] and information on their interpretation can be found in any good physical chemistry text.

The power of the above methods is increased by studying other physical properties as a function of composition in addition to the melting point or solubility. Among the techniques used are x-ray diffraction, heterometry,[58] potentiometry,[59] and conductometry.[59] The last three methods are particularly useful in characterizing amorphous products precipitated from solution.

B. Characterization of Individual Products

Wet chemical methods[60] are still used to establish the chemical composition, but physical methods must be used to establish what type of covalently bonded groups are present and whether the product is a single compound or a mixture.

If single crystals large enough to handle individually are present, the most reliable way to establish that a single compound is present is to determine the unit cell by single-crystal x-ray diffraction. The powder pattern calculated from the unit cell should then be compared with the pattern observed for a bulk sample to prove that the single crystal was representative of the entire product. Also, the chemical formula and unit cell size should be used to calculate the density for comparison with an experimental value.

When crystals too small to handle individually are obtained, petrographic microscopy can often still be used to establish the number of species present.[61] More complete optical data can, of course, be obtained for large single crystals, but the data, while good for identification, cannot be used to check the proposed chemical formula as can the unit cell data obtainable from such crystals. The powder x-ray diffraction pattern of microcrystalline samples can sometimes be used to determine a unit cell but with less reliability than that provided by single crystals. A powder pattern which is not correlated with a set of unit cell constants can provide no more than an educated guess with regard to the number of phases present, but it is still very valuable as a fingerprint.[62,63]

A number of methods are available for establishing whether a product is a phosphate and if so, what type and size phosphate anions are present. Included among these methods are solid state infrared spectroscopy,[64-66] ^{31}P nuclear magnetic resonance spectroscopy of solutions,[67] end-group titration,[68] light scattering,[69,70] viscometry,[69,70] solubility fractionation,[71] dialysis,[72] and chromatography. NMR and/or IR can usually establish unambiguously whether or not the sample is indeed a phosphate, and they generally give good indications also, regarding the type of phosphate present.

For distinguishing one phosphate species from another, the chromatographic procedures have been the most useful. Paper chromatography can separate both metaphosphates and polyphosphates with up to eight phosphorus atoms per anion.[73-78] Formulas have been developed to predict the spot position for each anion type.[4,5,73] For large anions, a modified paper chromatographic procedure[79] or paper electrophoresis[80] can be used. Other chromatographic methods which have been used are ion exchange chromatography,[51,52,81-85] thin layer chromatography,[86-88] and gel permeation chromatography.[89]

Chromatography is applied to insoluble phosphates by treating the sample preliminarily to convert the phosphate to a soluble salt.[31] In doing this, one must be careful to avoid degradation reactions or selective conversions which solubilize only one of several types of anions present in the original sample.

V. PYROPHOSPHORIC ACID AND ITS SALTS

A. Pyrophosphoric Acid

Pyrophosphoric acid, $H_4P_2O_7$, is the only condensed phosphoric acid which can be prepared in crystalline form, and the crystals, in contrast to other condensed phosphoric acids, contain a single type of phosphate.[127] The first preparation described in detail utilized slow crystallization of a phosphoric acid containing 79.8% P_2O_5.[90] Recently, pyrophosphoric acid was found to exist in two crystalline forms.[91] Form I, which is the common form and usually crystallizes spontaneously from the liquid, melts at 54.3°, considerably below the earlier reported value of 61°.[92]

Form II, melting at 71.5°, is easily obtained by heating form I crystals in a sealed tube at about 50° for several hours. Form II is the stable form at room temperature because when mixtures of form I, form II, and 79.8% liquid phosphoric acid are mixed, the final product is form II. Both forms are orthorhombic with unit cell dimensions $a = 13.69$ and 11.05, $b = 20.08$ and 19.21, and $c = 6.49$ and 10.40 Å for forms I and II, respectively.[93]

Studies on the rate of crystallization of pyrophosphoric acid show that nucleation and diffusion are the rate-controlling steps.[93] The induction time can be reduced to a few minutes with $\geq 50\%$ by weight of seeds. The diffusion resistance is lowered by operating about 10° below the melting point. The rate law is first order, and at 50° the rate constants are 0.098 and 0.068 min^{-1} for forms I and II, respectively. Finely divided, $99+\%$ pure crystalline pyrophosphoric acids have been prepared by addition of 79.8% liquid phosphoric acid into a stirred bed of solid acid with indirect cooling to remove the heat of crystallization [-3.7 (I) and -4.8 (II) kcal/mole].[93]

At acid concentrations higher or lower than the theoretical composition of $H_4P_2O_7$, the crystallization products will contain either long chain or orthophosphoric acids.[94] These impurities are difficult to remove. When pyrophosphoric acid melts or dissolves in an organic solvent such as acetone, it reorganizes into a mixture of ortho, pyro, and long-chain phosphoric acids. If it is dissolved carefully in ice-cold water, the pyrophosphoric acid is relatively stable. The half-life at 0° is over 300 hr,[95–97] and pyrophosphoric acid solution is a convenient raw material for preparation of some pyrophosphates such as $(NH_4)_4P_2O_7$.[98]

B. Alkali Metal Pyrophosphates

1. Ammonium Pyrophosphates

In laboratory preparations,[99–102] ammonium pyrophosphates are generally made by ammoniation of a low-temperature pyrophosphoric acid solution to a given pH followed by precipitation with the addition of ethanol. The pyrophosphoric acid is obtained either by simple dissolution of a high-purity crystalline pyrophosphoric acid or by passing an alkali metal pyrophosphate through a cation exchanger. Simple heating of an ammonium orthophosphate yields a mixture of phosphates,[103] but it is possible to produce ammonium pyrophosphate by heating an orthophosphate in the presence of a dehydration agent.[104,105]

a. $(NH_4)_2H_2P_2O_7$: Three polymorphic forms (α, β, γ) were reported[100]; however, only the α form (orthorhombic) and γ form (monoclinic) were confirmed by other investigators.[101,102,104] The monoclinic form is stable at or below room temperature while the orthorhombic form is stable at 50–125°. The monoclinic form is crystallized by addition of ethanol to a room temperature solution of pH = 2.4–4.0. The orthorhombic form is obtained by addition of solid triammonium pyrophosphate to warm glacial acetic acid.[100] The orthorhombic form can also be obtained along with some glassy material by melting a mixture of P_2O_5, $2(NH_4)_2HPO_4$, and $2(NH_4)H_2PO_4$ in a covered dish at temperatures below 200°, seeding with the orthorhombic form, and tempering at 125° for about six hours to complete crystallization.[104]

b. $(NH_4)_3HP_2O_7 \cdot H_2O$: Triclinic crystals are crystallized from a solution of pH 6 at temperatures below 55° by addition of methanol.[101,102]

c. $(NH_4)_3HP_2O_7$: It is generally obtained by dehydration of the monohydrate but can also be obtained by heating a mixture of $NH_4H_2PO_4$, H_3PO_4, and urea[104,105] to 140–150° (eq. 13).

$$NH_4H_2PO_4 + H_3PO_4 + H_2NCONH_2 \longrightarrow (NH_4)_3HP_2O_7 + CO_2 \qquad (13)$$

d. $(NH_4)_4P_2O_7$: Addition of ethanol to an ammoniated pyrophosphoric acid solution at pH ≥ 6.5 will yield $(NH_4)_4P_2O_7 \cdot H_2O$ at 0° and the anhydrous salt at 25° or above.[98,102] Both salts lose ammonia on exposure to the atmosphere.

2. Lithium Pyrophosphates

a. $Li_2H_2P_2O_7$: Dehydration of LiH_2PO_4 at 190–230° yields a mixture consisting mainly of $Li_2H_2P_2O_7$ and $(LiPO_3)_x$.[103,106] Isolation and full characterization of $Li_2H_2P_2O_7$ have not been reported.

b. $Li_4P_2O_7$: Tetralithium pyrophosphate can be prepared by heating mixtures of $2LiH_2PO_4 + Li_2CO_3$, $2LiPO_3 + Li_2CO_3$, or $Li_2CO_3 + (NH_4)_2HPO_4$.[107] Dehydration of Li_2HPO_4 is not used as a preparation method because for some unknown reason Li_2HPO_4 cannot be crystallized from a lithium orthophosphate solution.[1]

3. Sodium and Potassium Pyrophosphates

The preparation of various sodium and potassium pyrophosphates has been under intensive investigation during the past century. Phase diagrams, such as $NaPO_3-Na_4P_2O_7$,[108] $H_2O-NaPO_3$,[109] $KPO_3-K_4P_2O_7$,[110] $Li_4P_2O_7-K_4P_2O_7$,[111] $Li_4P_2O_7-Na_4P_2O_7$,[112] $Na_4P_2O_7-K_4P_2O_7$,[113] and $Li_4P_2O_7-Na_4P_2O_7-K_4P_2O_7$ [113] are completely defined. There is a series of quenchable solid solutions of $xNa_4P_2O_7 \cdot yK_4P_2O_7$ which forms above 800°, but otherwise no double salts were formed in the $Li_4-K_4-Na_4P_2O_7$ system. The eutectic point of the ternary system is 652° (28% Li_{4-}, 36% K_{4-}, and 36% $Na_4P_2O_7$).[113]

The production of $M_4P_2O_7$ is straightforward by dehydration of M_2HPO_4. On the other hand, dehydration of MH_2PO_4 often involves the formation of a number of intermediates.[103] Preparation of $K_2H_2P_2O_7$ is easier via acidification of $K_4P_2O_7$ than via dehydration of KH_2PO_4, although under water vapor $K_2H_2P_2O_7$ is claimed to be formed from KH_2PO_4 at about 200°.[114]

$Na_2H_2P_2O_7$ is the only alkali dihydrogen pyrophosphate which can be prepared in pure form by heating of the corresponding dihydrogen orthophosphate. KH_2PO_4 has an unusually stable hydrogen bonded crystal structure[115] which causes its vapor pressure to be about equal to that of $K_2H_2P_2O_7$.[9] Thus, it is very difficult to complete the dehydration of KH_2PO_4 without further dehydration of the product $K_2H_2P_2O_7$.

The hydrates of pyrophosphates are usually obtained by crystallization from an aqueous solution. The best methods for preparation of various sodium and potassium pyrophosphates are summarized in Table I.

4. Cesium and Rubidium Pyrophosphates

Reported work on cesium and rubidium pyrophosphates is limited. $Cs_2H_2P_2O_7$ appeared to be formed by heating CsH_2PO_4 to 250–270°,[106] but further tempering at this same temperature range resulted in further dehydration. $Cs_2H_2P_2O_7$ was prepared pure by adding ethanol to a solution of $Cs_4P_2O_7$ in acetic acid.[106] Conversion of Cs_2HPO_4 to $Cs_4P_2O_7$ took place at 339°, and a reversible crystallographic transition was noted at 238°.[106] $Rb_4P_2O_7$ was made by dehydration of Rb_2HPO_4.[107] $Rb_2H_2P_2O_7$ has not been reported.

C. Alkaline Earth Pyrophosphates

Table II summarizes the preparation of alkaline earth pyrophosphates. The recent interest in phosphors, which usually contain alkaline earth

TABLE I. Methods of Preparation of Sodium and Potassium Pyrophosphates

Compounds	Preparation methods	Refs.
$Na_2H_2P_2O_7 \cdot 6H_2O$	Cooling a 15% $Na_2H_2P_2O_7$ solution from 30–10°.	5,116
α-$Na_2H_2P_2O_7$	Heating NaH_2PO_4 to 210–250°.	103,117 118
β-$Na_2H_2P_2O_7$	Addition of ethanol to an aqueous solution at pH 4.2. The resulting oil was vacuum dried to produce crystals.	118
$Na_3HP_2O_7 \cdot 9H_2O$	Cooling a hot saturated equimolar solution of $Na_2H_2P_2O_7$ and NaOH to 20°.	116,119
$Na_3HP_2O_7 \cdot H_2O$	Drying $Na_3HP_2O_7 \cdot 9H_2O$ at room temperature over P_2O_5.	119,120
$Na_3HP_2O_7$	Heating $Na_3HP_2O_7 \cdot 9H_2O$ at 150° to constant weight.	119,120
$Na_4P_2O_7 \cdot 10H_2O$	Cooling a filtered saturated $Na_4P_2O_7$ solution from 70–25°. Filter and air-dry.	5,116
$Na_4P_2O_7$ (form III)	Heating Na_2HPO_4 to 500° for 4 hr. Other forms revert to this form at room temperature.	108
$Na_4P_2O_7$ (form II)	Stable form between 530–980°.	108
$Na_4P_2O_7$ (form I)	Stable form between 985° and the melting point.	108
$KH_3P_2O_7$	Heating $KH_2PO_4 \cdot H_3PO_4$ to 100–104° under vacuum to constant weight.	121
$K_2H_2P_2O_7 \cdot \frac{1}{2}H_2O$ ($K_2H_2P_2O_7 \cdot \frac{1}{3}H_2O$)	Adding ethanol to a solution of $K_4P_2O_7$ acidified with HAc gives an oil which slowly crystallizes.	116[a] (122)[a]
$K_2H_2P_2O_7$	Heating KH_2PO_4 at 200° under steam.	114
$K_3HP_2O_7 \cdot \frac{1}{2}H_2O$	Adding ethanol to a solution containing one mole of HAc per mole $K_4P_2O_7$.	122
$K_3HP_2O_7$	a. Heating $K_3HP_2O_7 \cdot \frac{1}{2}H_2O$ at 220°.	122
	b. $K_2HPO_4 + KH_2PO_4 \rightarrow K_3HP_2O_7 + H_2O$ at 245°.	123
	c. An isomorphic form was claimed to be prepared by vacuum drying of an oil obtained by adding ethanol to a solution of $K_2H_2P_2O_7$ and KOH.	122
$K_4P_2O_7 \cdot 3H_2O$ ($K_4P_2O_7 \cdot 3.5H_2O$)	Slow addition of acetone to a 40% $K_4P_2O_7$ solution at room temperature.	124[a] (125)[a]
$K_4P_2O_7 \cdot H_2O$ ($K_4P_2O_7 \cdot \frac{1}{2}H_2O$)	Drying $K_4P_2O_7 \cdot 3H_2O$ at about 15% relative humidity at 27°.	124[a] (125)[a]
$K_4P_2O_7$	$2K_2HPO_4 \rightarrow K_4P_2O_7 + H_2O$ at 400–600°.	123,124
$(xK_4P_2O_7 \cdot yNa_4P_2O_7) \cdot 3(x+y)H_2O$	Equilibrium solid phase of solution region between $K_4P_2O_7$–H_2O and the invariant point (13.2% $Na_4P_2O_7$, 45% $K_4P_2O_7$, and 41.8% H_2O) at 25°.	126

[a] There is only one compound, but the amount of hydrate water has not been completely resolved.

pyrophosphates and activators,[128-129] has led to considerable study of alkaline earth pyrophosphates.[130] Calcium pyrophosphates are also important as plant nutrients, and many new compounds have been discovered by investigating the hydrolysis of calcium polyphosphates[131,132] and formation of double salts between the hydrolysis products and ammonium or potassium ions.[133,134] Judging from the large number of interesting calcium pyrophosphates obtained from solution, it is suspected that there are additional hydrated strontium pyrophosphates which remain to be discovered. For example, it has been shown by heterometry that $Sr_2P_2O_7 \cdot xH_2O$ can be precipitated from aqueous solution, but this salt has not yet been isolated or analyzed for water content.[135]

The chemistry of dehydration of alkaline earth dihydrogen orthophosphates is more complicated than the direct conversion to the acid pyrophosphates shown in Table II. Extensive x-ray data[136,137] and paper chromatographic analysis[31] demonstrated the existence of intermediates which often are polyphosphates of unspecified composition.[31]

In addition to the methods given in Table II, $Ba_2P_2O_7$ has also been prepared by reaction of BaO with molten $Na_3P_3O_9$ followed by leaching of the cooled reaction products with water.[138] Only the chemical analysis was reported, and therefore it cannot be established whether this product is the α or β form or perhaps a third form.

D. Pyrophosphates Containing an Alkali Metal and One or More Other Metals

Table III gives preparation methods for a large number of these compounds, and most of the preparations listed were fairly recent. However, numerous pyrophosphates were prepared as early as 1848,[39] and because the analytical methods were often crude, many of the compositions described in the older literature are doubtful. No attempt was made to cover the very early literature.

The following compounds were described[152] in 1926 and are not included in the table: $Na_2UO_2P_2O_7$, $Li_2UO_2P_2O_7$, $Na_8Fe_4(P_2O_7)_5 \cdot 28H_2O$, $Na_2Fe(OH)P_2O_7 \cdot xH_2O$, $Na_3Fe(OH)_2(P_2O_7) \cdot 4H_2O$, $Na_2CdP_2O_7 \cdot 4H_2O$, $K_2CdP_2O_7 \cdot 3H_2O$, $(NH_4)_2CdP_2O_7 \cdot H_2O$, $Na_2MnP_2O_7 \cdot 4H_2O$, $Na_2FeP_2O_7 \cdot 2.5H_2O$, $(NH_4)_2FeP_2O_7 \cdot 2H_2O$, $K_2PbP_2O_7 \cdot 5H_2O$, and $Na_4Pb_4(P_2O_7)_3 \cdot 6H_2O$. In 1936, a number of double salts of $Na_4P_2O_7$ and $M_2P_2O_7$ were described[140] where M = Mg, Mn, Co, Ni, Cu, Zn, and Cd; and some of the formulas given were rather bizarre because the author was trying to fit the compounds to Werner's theory of coordination. These compounds are

TABLE II
Methods of Preparation of Alkaline Earth Pyrophosphates

Compounds	Preparation methods	Refs.
$MgH_2P_2O_7 \cdot 11H_2O$	Reacted Mg(II) with pyrophosphoric acid solution.	139
$Mg_2P_2O_7 \cdot xH_2O$ $x = 3–7$	Reacted $MgCl_2$ solution with a solution of $Na_4P_2O_7$ and HAc.	140,141
α-$Mg_2P_2O_7$ (monoclinic)	Calcined $MgNH_4PO_4 \cdot 6H_2O$ at 1100° and cooled to room temperature.	142
β-$Mg_2P_2O_7$	Heated- $\alpha Mg_2P_2O_7$ above 63°.	143
$Ca_{1.37}H_{1.26}P_2O_7 \cdot H_2O$	Sirupy $H_4P_2O_7$ and $CaCO_3$ at 0° and at $CaO/P_2O_5 = 1$ were crushed in a mortar. The slurry was filtered, and alcohol was added to the filtrate. The precipitate was claimed to be a double salt of the indicated composition.	144
$CaH_2P_2O_7$	a. Heated $Ca(H_2PO_4)_2 \cdot H_2O$ to 270–280° under a steam atmosphere. b. Slowly added $Ca(H_2PO_4)_2 \cdot H_2O$ to agitated phosphoric acid at 210° to crystallize $CaH_2P_2O_7$.	31,136, 145 133,146
$Ca_3H_2(P_2O_7)_2$	Heated $Ca(H_2PO_4)_2$ to 240° and washed the product with water.	118
$Ca_3H_2(P_2O_7)_2 \cdot 4H_2O$	Treated one part $CaH_2P_2O_7$ with less than 19 parts by wt. of water at room temperature for several hours producing triclinic, long, bladelike crystals.	133
$Ca_3H_2(P_2O_7)_2 \cdot H_2O$	Treated one part $CaH_2P_2O_7$ with less than 10 parts by wt. of water at 65–75° for about 24 hr producing orthorhombic, beveled, rectangular plates.	133
$Ca_2P_2O_7 \cdot 4H_2O$ (orthorhombic)	Added 1.1 g $CaH_2P_2O_7$ to 200 ml 0.05M NH_4OH and agitated at 25° for 24 hr. Stable in dilute solution.	133
$Ca_2P_2O_7 \cdot 4H_2O$ (monoclinic)	Mixed 35 ml 2.2M $Ca(Ac)_2$ with a solution of 17 g $CaH_2P_2O_7$ in 425 ml water at room temperature. Precipitate was then aged.	133
$Ca_2P_2O_7 \cdot 2H_2O$	Crystallized from slowly hydrolyzing calcium polyphosphate glass slurry.	131
$Ca_2P_2O_7 \cdot H_2O$ (triclinic)	Allowed a solution of 1.1 g $CaH_2P_2O_7$ in 100 ml water to stand at room temperature for 3–7 days.	133
$Ca_2P_2O_7 \cdot H_2O$ (monoclinic)	Added a saturated solution of $CaCl_2$ to a solution of 1 g $K_4P_2O_7$ and 10 g KCl in 60 ml water at pH = 6–7. Diluted the mixture with 100 ml water and allowed to stand at 45° for 1 week.	133

(continued)

TABLE II (continued)

Compounds	Preparation methods	Refs.
α-$Ca_2P_2O_7$	Heated $CaHPO_4$ or $CaHPO_4 \cdot 2H_2O$ to 320–430°.	136,147
β-$Ca_2P_2O_7$	Heated $CaHPO_4$ or $CaHPO_4 \cdot 2H_2O$ to 700–850°.	116,136, 147
γ-$Ca_2P_2O_7$	Heated $CaHPO_4$ or $CaHPO_4 \cdot 2H_2O$ to 1140–1220°.	136,147
$SrH_2P_2O_7$	$Sr(H_2PO_4)_2 \xrightarrow{190-210°} SrH_2P_2O_7 + H_2O$	31,137
$Sr_2P_2O_7 \cdot \tfrac{1}{2}H_2O$	$2(\beta\text{-}SrHPO_4) \xrightarrow{300°} Sr_2P_2O_7 \cdot \tfrac{1}{2}H_2O + \tfrac{1}{2}H_2O$	137
β-$Sr_2P_2O_7$	$2(\alpha\text{-}SrHPO_4) \xrightarrow{400°} \beta\text{-}Sr_2P_2O_7 + H_2O$	137
	$Sr_2P_2O_7 \cdot \tfrac{1}{2}H_2O \xrightarrow{575°} \beta\text{-}Sr_2P_2O_7 + \tfrac{1}{2}H_2O$	137
α-$Sr_2P_2O_7$	$\beta\text{-}Sr_2P_2O_7 \xrightarrow{760-800°} \alpha\text{-}Sr_2P_2O_7$	137,148
$Ba_2P_2O_7 \cdot xH_2O$	Crystallized from aqueous solution.	135,148
$Ba_7H_2(P_2O_7)_4 \cdot 6H_2O$	Crystallized from aqueous solution.	148
α-$Ba_2P_2O_7$	$2BaHPO_4 \xrightarrow{385°} \alpha\text{-}Ba_2P_2O_7 + H_2O$	149,150
β-$Ba_2P_2O_7$ (also called δ-$Ba_2P_2O_7$)	Heated $BaHPO_4$ to 790°.	149,151

also excluded from Table III. A few precipitated double salts from the early literature which were later described by Van Wazer[5] as crystalline are included in Table III.

Physical properties of $Na_4U(P_2O_7)_2 \cdot 8H_2O$[153] and $Nb_2O_5 \cdot 2P_2O_5$[154] were given without mentioning the method of preparation or giving a reference to a preparation method. Evidence was given that the latter compound contains the pyrophosphate ion. The following compounds were mentioned in various recent discussions, and references were given to unpublished data or literature which was either very old or very obscure. The following references refer to the more recent discussions which usually do not themselves describe the preparation methods: $LaNaP_2O_7$,[155] KYP_2O_7,[155] $NaYP_2O_7$,[155] $MnNa_2P_2O_7$,[156] $FeNaP_2O_7$,[156] $CuNa_2P_2O_7 \cdot xH_2O$,[156,157] $CrNaP_2O_7$,[156] $CeNaP_2O_7$,[155] $CeKP_2O_7$,[158] $CdNa_2P_2O_7$,[159] and $(Na_xM_yP_2O_7)_7 \cdot Na_2O$ where M = Fe, Co, Ni, Cu, or Ti.[160] $Ag_3Na \cdot P_2O_7$ was mentioned in an early paper without any reference to its preparation being given.

The composition of the following preparations was questioned by later authors: $Na_5(TiO)_5P_7O_{25}$,[156] $Fe_4Na_3P_7O_{25}$,[156] $Fe_2Na_6(P_2O_7)_3 \cdot 9H_2O$,[59] $Na_7Ni_4P_7O_{25}$.[156] Conductometric titrations indicated that $CeKP_2O_7 \cdot xH_2O$,[161] $CeNa_4(P_2O_7)_2 \cdot xH_2O$,[161] and $Ce_2K_6(P_2O_7)_3 \cdot xH_2O$[161] can be

TABLE III
Preparation of Pyrophosphates Containing an Alkali Metal and One or More Other Metals

Compounds	Preparation methods	Refs.
$NH_4Ca_2H_3(P_2O_7)_2 \cdot 3H_2O$	Allowed 2 g $CaH_2P_2O_7$ to stand in 25 ml 1.5M NH_4Cl at 25–45°.	133,134
$NH_4Ca_2H_3(P_2O_7)_2 \cdot H_2O$	Allowed 10 g $CaH_2P_2O_7$ to stand in 20 ml 4.7M NH_4Cl in a stoppered container at room temperature for 6 months.	133
$(NH_4)_2CaH_4(P_2O_7)_2$	Allowed 5 g $CaH_2P_2O_7$ to stand at room temperature in 20 ml saturated solution of NH_4Cl. Crystals were filtered and washed with ethanol to remove NH_4Cl.	133
$(NH_4)CaHP_2O_7$	Digested 1 g $CaH_2P_2O_7$ in 75 ml 0.6M NH_4Cl at 65–70° for 1 hr. It is the stable phase of the system $CaH_2P_2O_7$–NH_4OH–H_2O at pH = 2.5–5 with less than 95% H_2O.	133,134
$(NH_4)_4Ca_3H_6(P_2O_7)_4 \cdot 3H_2O$	1 g $CaH_2P_2O_7$ was mixed with 50 ml 6.5M NH_4Cl at room temperature for 1 hr. Filtered and washed with ethanol.	133
$(NH_4)_2Ca_5(P_2O_7)_3 \cdot 6H_2O$	1 g monoclinic $Ca_2P_2O_7 \cdot 4H_2O$ was mixed with 50 ml 5M NH_4Cl at room temperature.	133
$(NH_4)_2Ca_3(P_2O_7)_2 \cdot 6H_2O$	1 g $CaH_2P_2O_7$ was mixed with 15 ml 0.6M NH_4OH at room temperature for 1–2 days. This is the stable phase in the $CaH_2P_2O_7$–NH_4OH–H_2O system at pH = 5–8 with 80–90% H_2O.	133,134
$(NH_4)_2CaP_2O_7 \cdot H_2O$	Mixed $CaH_2P_2O_7$ with NH_4OH at pH > 7 with 95+% water at room temperature.	133,134
$NH_4CrP_2O_7 \cdot 6H_2O$	Crystallized from aqueous solution.	5
$(NH_4)_2NiP_2O_7 \cdot 2H_2O$	$NiSO_4$ + $(NH_3)_3PO_4$ digested at 80° and pH 10. Characterized by IR.	164
$LiLaP_2O_7 \cdot 4H_2O$	$La(NO_3)_3$ solution + $Li_4P_2O_7$ solution, semi-amorphous.	165
$LiSmP_2O_7 \cdot 4H_2O$	$Sm(NO_3)_3$ solution + $Li_4P_2O_7$ solution, amorphous.	155
$LiSmP_2O_7$	Heated tetrahydrate to 500°, XRD pattern.	155
$NaAlP_2O_7$	Heated a mixture of Na_3PO_4, $AlPO_4$, and H_3PO_4 to 700–800°, XRD pattern.	166
$Na_2CaP_2O_7$	Prepared from a melt of $Na_4P_2O_7$ and $Ca_2P_2O_7$.	113
$Na_2CaP_2O_7 \cdot 4H_2O$	$CaCl_2$ solution + $Na_4P_4O_7$ solution, room temperature.	62
$Na_4Cd_4(P_2O_7)_3$	Prepared from a melt of $Cd_2P_2O_7$ and $Na_4P_4O_7$, XRD pattern.	159

(continued)

TABLE III (continued)

Compounds	Preparation methods	Refs.
$Na_6Ce_2(P_2O_7)_3$	$CeO_2 + NaPO_3 \xrightarrow{1100°} Na_6Ce_2(P_2O_7)_3 + O_2(g)$. Composition is not certain because of possible loss of P_2O_5, but some new XRD lines were observed.	160
$NaCeP_2O_7 \cdot \sim 4.5H_2O$	Crystallized from aqueous solution, XRD pattern.	171
$Na_2Co_3(P_2O_7)_2 \cdot 4H_2O$	Prepared from a melt of CoO and $Na_3P_3O_9$, cooled and leached with H_2O. Other amounts of hydrate water have been reported.	156
$Na_2Co_3(P_2O_7)_2$	Heated tetrahydrate to 800° or heated a mixture of $Na_4P_2O_7$ and $Co_2P_2O_7$ to 1000°. Products had identical XRD patterns.	156
$Na_2CoP_2O_7 \cdot 2H_2O$	Added alcohol to a mixture of $CoCl_2$ solution and $Na_4P_2O_7$ solution. Precipitate was washed and dried at 110° to constant weight.	167
$Na_2CoP_2O_7$	Prepared from a melt of CoO and $Na_3P_3O_9$.	156
$NaCo(NH_3)_6P_2O_7 \cdot xH_2O$	Precipitated from aqueous solution.	162
$NaCrP_2O_7 \cdot 5H_2O$	Crystallized from aqueous solution.	5
$NaCrP_2O_7 \cdot 8H_2O$	Crystallized from aqueous solution.	5
$Na_4Cu_8(P_2O_7)_5 \cdot 17H_2O$	Crystallized from aqueous solution.	5
$Na_{32}Cu_{14}(P_2O_7)_{15} \cdot 13H_2O$	Crystallized from aqueous solution.	5
$Na_2Cu_3(P_2O_7)_2 \cdot xH_2O$	Crystallized from aqueous solution.	157
$Na_6Cu(P_2O_7)_2 \cdot xH_2O$	Crystallized from aqueous solution.	5,168
$NaFeP_2O_7 \cdot 4H_2O$	$Na_4P_2O_7$ solution + $Fe(NO_3)_3$ solution, amorphous, analysis not exact.	59
$Na_8In_4(P_2O_7)_5$	Melted a mixture of $In_4(P_2O_7)_3$ and $Na_4P_2O_7$ and cooled slowly, crystalline.	169
$Na_5In(P_2O_7)_2 \cdot 7H_2O$	Dissolved freshly prepared $In_4(P_2O_7)_3 \cdot xH_2O$ in $Na_4P_2O_7$ solution with heating and allowed product to crystallize.	169
$NaLaP_2O_7 \cdot 2-4H_2O$	Crystallized from aqueous solution.	155,158, 165
$Na_2MgP_2O_7$	Melted 2 moles $Na_3P_3O_9$ and 3 moles MgO together. The melt was slowly cooled; the cake was powdered and washed.	138
$NaMnP_2O_7 \cdot 5H_2O$	Crystallized from aqueous solution.	5
$Na_2Mn_3(P_2O_7)_2 \cdot 4H_2O$	Melted a mixture of $MnCO_3$ and $Na_3P_3O_9$ cooled, and leached with H_2O	156
$Na_6Nd_2(P_2O_7)_3$	Prepared from a melt of $NaPO_3$ and Nd_2O_3, composition is not too well defined but new XRD lines were observed.	160

(continued)

TABLE III (*continued*)

Compounds	Preparation methods	Refs.
$Na_2NiP_2O_7$	Melted a mixture of NiO and $Na_3P_3O_9$, cooled, and leached with H_2O, XRD pattern.	156
$Na_2NiP_2O_7 \cdot 6H_2O$	Added alcohol to a solution of $NiCl_2$ and $Na_4P_2O_7$.	167
$Na_2NiP_2O_7 \cdot 2H_2O$	Heated hexahydrate to 250°, no evidence was given that it was crystalline.	164
$Na_2PbP_2O_7$	Reacted PbO with molten $Na_3P_3O_9$, XRD pattern.	159
$Na_2(Pb,Sr)P_2O_7$	Reacted SrO and PbO with molten $Na_3P_3O_9$ giving mixed crystals in which the Sr/Pb ratio can be varied.	156
$Na_6Sm_2(P_2O_7)_3$	Prepared from a melt of $NaPO_3$ and Sm_2O_3. Composition is not very well defined, but new XRD lines were observed.	160
$NaSmP_2O_7 \cdot 4H_2O$	$Sm(NO_3)_3$ solution + $Na_4P_2O_7$ solution, amorphous. Dihydrate was obtained by aging.	155
$NaSmP_2O_7$	Heated tetrahydrate to 545°, XRD pattern.	155
$Na_4Th(P_2O_7)_2 \cdot 6H_2O$	Crystallized from an aqueous solution prepared from ThP_2O_7 and $Na_4P_2O_7$.	231
$Na_2(TiO)P_2O_7$	Reacted TiO_2 with excess molten $Na_3P_3O_9$ at 1200°, cooled slowly, and leached with H_2O, crystalline.	156
$Na_4(UO_2)_8(P_2O_7)_5 \cdot xH_2O$	$UO_2(NO_3)_2$ solution + $Na_4P_2O_7$ solution.	170
$Na_5Yb(P_2O_7)_2 \cdot 4H_2O$	$YbCl_3$ solution + $Na_4P_2O_7$ solution.	158
$Na_2ZnP_2O_7$	Reacted ZnO with molten $Na_3P_3O_9$, cooled, and leached with H_2O.	138
$KAlP_2O_7$	a. Isolated from residue formed by heating $K_2AlP_3O_{10}$, XRD pattern.	41
	b. Heated $Al_4(P_4O_{12})_3$ with excess KCl at 900°, cooled, and leached with H_2O, XRD pattern.	173a
$KCa_3H(P_2O_7)_2 \cdot 4H_2O$	Mixed 4.4 g $K_4P_2O_7$ with 6.1 g $CaH_2P_2O_7$ in 500 ml H_2O at room temperature for 4 days.	133,134
$KCa_2H_3(P_2O_7)_2 \cdot 3H_2O$	Mixed 1 g $CaH_2P_2O_7$ with 10 ml 2.35M KCl at room temperature.	133,134
$KCaHP_2O_7 \cdot 2H_2O$	Mixed 1 g $CaH_2P_2O_7$ with 50 ml 2M KCl and added $K_4P_2O_7$ or KOH to pH = 4. Rod-like crystals formed in 1 hr.	133,134

(*continued*)

TABLE III (continued)

Compounds	Preparation methods	Refs.
$K_2CaH_4(P_2O_7)_2$	Mixed 5 g $CaH_2P_2O_7$ with 20 ml saturated KCl solution and 10 g KCl at room temperature for 12–24 hr. Filtered and washed KCl away with ethanol.	133,134
$K_4CaH_2(P_2O_7)_2$	Mixed 6 g $K_4P_2O_7$ and 4 g $CaH_2P_2O_7$ with 100 ml saturated KCl solution at room temperature. Crystals formed after about 24 hr.	133,134
$K_2Ca_5(P_2O_7)_3 \cdot 6H_2O$	Mixed $Ca_2P_2O_7 \cdot 4H_2O$ in $5M$ KCl at 70–75° and allowed the mixture to stand at room temperature for about a week. Excess KCl was washed away from product with ethanol.	133,134
$K_2Ca_3(P_2O_7)_2 \cdot 2H_2O$	Mixed 6.8 g $K_4P_2O_7$, 4.27 g $CaH_2P_2O_7$, and 100 ml H_2O at room temperature for 4–7 days.	133,134
$KCaHP_2O_7$	Mixed 1 g $CaH_2P_2O_7$ in 100 ml saturated KCl solution at about 50° for 2–3 hr.	133,134
$K_2CaP_2O_7 \cdot 4H_2O$	Mixed 15 ml $0.5M$ $Ca(Ac)_2$ solution and 5 g $K_4P_2O_7$ in 50 ml H_2O at room temperature.	133,134
$K_2CaP_2O_7$	Mixed $CaH_2P_2O_7$ with $K_4P_2O_7$ solution at pH = 6–8 with less than 95% H_2O.	133,134
$KCrP_2O_7 \cdot 5H_2O$	Crystallized from aqueous solution.	5
$KFeP_2O_7 \cdot 3.5H_2O$	$K_4P_2O_7$ solution + $Fe(NO_3)_3$ solution, amorphous, composition was not exact.	59
$KGdP_2O_7 \cdot 3H_2O$	$GdCl_3$ solution + $K_4P_2O_7$ solution, XRD pattern.	172
α-$KGdP_2O_7$	Heated trihydrate to 150°, XRD pattern.	173
β-$KGdP_2O_7$	Heated trihydrate to 620°, XRD pattern.	173
γ-$KGdP_2O_7$	Heated trihydrate to 1000°, XRD pattern, partially reverted to β form when quenched.	173
$KLaP_2O_7 \cdot 4H_2O$	$La(NO_3)_3$ solution + $K_4P_2O_7$ solution, semi-amorphous.	165
$KMnP_2O_7 \cdot 3H_2O$	Crystallized from aqueous solution.	5
$KMnP_2O_7 \cdot 5H_2O$	Crystallized from aqueous solution.	5
$KSmP_2O_7 \cdot 4H_2O$	$Sm(NO_3)_3$ solution + $K_4P_2O_7$ solution, amorphous.	155
$KSmP_2O_7$	Heated the tetrahydrate to 600°, XRD pattern.	155
$CsGdP_2O_7 \cdot 4.5H_2O$	$GdCl_3$ solution + $Cs_4(P_2O_7)$ solution.	183
α-$CsGdP_2O_7$	Heated the 4.5 hydrate to 710°, XRD pattern.	173
β-$CsGdP_2O_7$	Heated the 4.5 hydrate to 1000°, XRD pattern.	173

precipitated from aqueous solution, but the precipitates have not been isolated and analyzed. The existence of $Be_2Na_4(P_2O_7)_3 \cdot 5H_2O$ was said to be confirmed,[163] but the preparation method was not given.

E. Pyrophosphates Containing No Alkali Metal

The simple alkaline earth pyrophosphates were covered separately in Section V-C, but the alkaline earth pyrophosphate double salts such as $BaMgP_2O_7$ are included here. Table IV lists all the preparation methods found except for a few cases discussed below in which the description of the compound or preparation method was incomplete or doubtful. Also discussed below are a few points of special interest regarding the compounds in Table IV.

TABLE IV
Preparation of Pyrophosphates Containing no Alkali Metal
(Except for simple alkaline earth pyrophosphates which are discussed in Section V-C)

Compounds	Preparation methods	Refs.
$Ag_4P_2O_7$	$Na_4P_2O_7$ solution + $AgNO_3$ solution, IR.	178, 184
$AgMnP_2O_7 \cdot 3H_2O$	Crystallized from aqueous solution.	5
$Ag_3TlP_2O_7$	Added $Na_4P_2O_7$ solution to a solution of $AgNO_3$ and $TlNO_3$.	185
$Al_4(P_2O_7)_3 \cdot 18H_2O$	Precipitated from a solution of $NH_4Al(SO_4)_2$ and ammonium pyrophosphate at pH < 2.5.	186
$Al_8H_{12}(P_2O_7)_9$	Heated Al_2O_3 in phosphoric acid, XRD pattern.	41
$BaMgP_2O_7$	Equilibrium phase in $BaO-MgO-P_2O_5$ system, XRD pattern.	187
$(Ba,Mg)_2P_2O_7$	Heated mixtures of $BaHPO_4$ and $MgNH_4 \cdot PO_4 \cdot H_2O$. Three different equilibrium phases exist, each over a certain range of Ba/Mg and temperature. Small amounts of Sn were present but probably didn't affect the equilibria significantly. XRD patterns.	188
$(Ba,Sr)_2P_2O_7$	Heated mixtures of $BaHPO_4$ and $SrHPO_4$ to 1100°. Compositions low in Ba had crystal structures related to α-$Sr_2P_2O_7$. Compositions with more Ba had structures related to β-$Ba_2P_2O_7$ (also known as δ-$Ba_2P_2O_7$).	189
$BaZnP_2O_7$	Equilibrium phase in $BaO-ZnO-P_2O_5$ system, XRD pattern.	187

(continued)

TABLE IV (continued)

Compounds	Preparation methods	Refs.
$Be_2P_2O_7$	Heated $BeHPO_4 \cdot xH_2O$ or $BeNH_4PO_4 \cdot H_2O$. At least two forms exist. Form I is formed at $\geq 600°$; form II, at $> 900°$. XRD patterns.	190
$Bi_4(P_2O_7)_3$	Bi salt solution + solution of a pyrophosphate of a weak base.	191
$(BiO)_2H_2P_2O_7 \cdot 2H_2O$	Bi salt solution + solution of a pyrophosphate of a strong base.	191
$Cd_2P_2O_7$	Heated $CdNH_4PO_4 \cdot xH_2O$ to 980°, XRD pattern.	181
$CdH_2P_2O_7$	Heated $Cd(H_2PO_4)_2 \cdot 2H_2O$ to 210° or 180°, XRD pattern.	31,182
$Cd_5(PO_4)_2P_2O_7$	Heated $Cd_5H_2(PO_4)_4 \cdot 4H_2O$ to 340 or 700°, XRD pattern.	181,182
$(Cd,Zn)_2P_2O_7$	Heated mixtures of $Cd_2P_2O_7$ and $Zn_2P_2O_7$. Solid solutions with several different crystal structures can be prepared, each existing over a range of Cd/Zn and temperature. XRD patterns. β-$Zn_2P_2O_7$ containing >7 mole % $Cd_2P_2O_7$ can be quenched to room temperature whereas pure β-$Zn_2P_2O_7$ cannot.	192
$CeP_2O_7 \cdot xH_2O$	$Ce(SO_4)_2$ solution + $Na_4P_2O_7$ solution, amorphous.	193
CeP_2O_7	Heated $CeP_2O_7 \cdot xH_2O$ to 900°, XRD pattern.	193
$Ce_4(P_2O_7)_3 \cdot xH_2O$	$CeCl_3$ solution + $Na_4P_2O_7$ solution or $(NH_4)_4P_2O_7$; $Ce_2(SO_4)_3$ solution + $K_4P_2O_7$ solution.	161,194, 195
$Co_2P_2O_7 \cdot 6H_2O$	An amorphous salt was precipitated from aqueous solution which could be dried to the 6 hydrate at 20° or the 2 hydrate at 110°. Crystalline hexahydrate was prepared by recrystallization using SO_2.	40
$Co_2P_2O_7$	Heated $CoNH_4PO_4 \cdot xH_2O$ to $\geq 500°$. Well crystallized. There are two forms with the transition temperature near 304°.	196,197
$Co(NH_3)_6HP_2O_7$	Pyrophosphate solution + $Co(NH_3)_6Cl_3$ solution. Dried precipitate at 110°.	198
$Cr_4(P_2O_7)_3$	$4Cr(PO_3)_3 \xrightarrow{1350°} Cr_4(P_2O_7)_3 + 3P_2O_5(g)$, XRD pattern.	199
$Cu_2P_2O_7 \cdot \frac{4}{3}H_2O$	$Cu(NO_3)_2$ solution + $Na_4P_2O_7$ solution.	200
α-$Cu_2P_2O_7$	Crystallized from a melt and cooled to room temperature. Complete x-ray structure determination.	201

(continued)

TABLE IV (continued)

Compounds	Preparation methods	Refs.
β-$Cu_2P_2O_7$	α-$Cu_2P_2O_7$ $\underset{}{\overset{70-100+°}{\rightleftarrows}}$ β-$Cu_2P_2O_7$. Complete x-ray structure determination. Tends to revert to α form when cooled.	201
$Dy_4(P_2O_7)_3 \cdot 36H_2O$	$DyCl_3$ solution + $(NH_4)_4P_2O_7$ solution at pH 4.	195
$Er_4(P_2O_7)_3 \cdot 36H_2O$	$ErCl_3$ solution + $(NH_4)_4P_2O_7$ solution at pH 4.	195
$Fe_2P_2O_7$	Prepared by reduction of $FePO_4$ with H_2 or CO at elevated temperatures with elimination of water.	20, 202, 203
$Fe_4(P_2O_7)_3 \cdot 17H_2O$	$Na_4P_2O_7$ solution + $Fe(NO_3)_3$ solution at pH < 2. Amount of hydrate water is probably variable.	59, 204
$Fe_4(P_2O_7)_3$	Heated 17 hydrate to 700°, or heated a mixture of $FePO_4$ and $Fe(PO_3)_3$ to 940°. XRD pattern.	41, 204
$FeHP_2O_7$	Heated Fe_2O_3 in phosphoric acid. XRD pattern.	41
$Fe_3(OH)(P_2O_7)_2 \cdot 12H_2O$	$Na_4P_2O_7$ solution + $Fe(NO_3)_3$ solution, pH < 2.	59, 204
$Fe_3(OH)(P_2O_7)_2$	Heated 12 hydrate to 700°.	204
$Gd_4(P_2O_7)_3 \cdot xH_2O$	$GdCl_3$ solution + Li, Na, K, or NH_4 pyrophosphate solution, amorphous; x can be 3, 13, or 24.	172, 173, 183, 195
$Gd_4(P_2O_7)_3$	Heated a hydrate. XRD shows two crystalline forms.	173
$Gd_4(P_2O_7)_3 \cdot 0.5(Gd_2O_3)$	Heated $Gd_5(OH)_3(P_2O_7)_3 \cdot 9H_2O$ above 285°. May be more than one allotropic form.	173
$Gd_5(OH)_3(P_2O_7)_3 \cdot 9H_2O$	$GdCl_3$ + $Cs_4P_2O_7$ solution, XRD pattern.	173, 183
GeP_2O_7, cubic	Reacted GeO_2 with pyrophosphoric acid at 1300° or heated $Ge(HPO_4)_2$ to 1250°, XRD pattern.	193, 205
GeP_2O_7, monoclinic	Reacted GeO_2 with pyrophosphoric acid at > 500°. Difficult to avoid contamination with pseudohexagonal form. XRD pattern.	205
GeP_2O_7, pseudohexagonal	Reacted GeO_2 with pyrophosphoric acid at > 500°. Difficult to avoid contamination with monoclinic form. XRD pattern.	205
α- and β-GeP_2O_7	α was prepared by heating $Ge(HPO_4)_2 \cdot H_2O$ to 700°, and β was prepared by heating a mixture of GeO_2 and acid ammonium phosphate to 700°. It appears that both	206

(continued)

TABLE IV (continued)

Compounds	Preparation methods	Refs.
	preparations are mixtures with α containing the pseudohexagonal form and β containing the monoclinic form. XRD patterns and IR spectra are given.	
HfP_2O_7	Dissolved freshly precipitated Hf hydroxide in hot concentrated phosphoric acid and heated to 500°, cubic unit cell given.	193
$Hg_2P_2O_7$	$2Hg_3(PO_4)_2 \xrightarrow{630°} 2Hg_2P_2O_7 + 2Hg(g) + O_2(g)$	207
$In_4(P_2O_7)_3 \cdot xH_2O$	$In_2(SO_4)_3$ solution + $Na_4P_2O_7$ solution.	42,169
$InHP_2O_7 \cdot 5H_2O$	$In_2(SO_4)_3$ solution + $Na_4P_2O_7$ solution.	42
$InHP_2O_7$	Heated pentahydrate to 600–700°.	42
$La_4(P_2O_7)_3 \cdot xH_2O$	$Na_4P_2O_7$ solution + $La(NO_3)_3$ solution, or $(NH_4)_4P_2O_7$ solution + $LaCl_3$ solution at pH 4, amorphous.	158,165, 195
$La_4(P_2O_7)_3 \cdot 8H_2O$	$Na_4P_2O_7$ solution + $LaCl_3$ solution, XRD pattern.	208
$La_4(P_2O_7)_3$	Heated a hydrate to ≥214°, XRD pattern.	165
$LaHP_2O_7 \cdot 2H_2O$	$LaCl_3$ solution + $Na_4P_2O_7$ solution $\xrightarrow{HCl}$ $LaHP_2O_7 \cdot 2H_2O$.	158
$La_5(OH)_3(P_2O_7)_3 \cdot 12H_2O$	$La(NO_3)_3$ solution + $Cs_4P_2O_7$ solution, amorphous.	209
$Lu_4(P_2O_7)_3 \cdot 36H_2O$	$LuCl_3$ solution + $(NH_4)_4P_2O_7$ solution at pH 4.	195
$Mn_2P_2O_7$	Heated $MnNH_4PO_4 \cdot H_2O$ to ~500°; apparently only one allotrope exists; crystal structure was determined.	22,210, 211
$Mn_2P_2O_7 \cdot 3H_2O$	Conductivity studies indicated that the normal salt can be precipitated from aqueous solution, and paramagnetic resonance studies were reported for the trihydrate. Preparation wasn't described in the abstract.	212,213
$(MoO_2)_2P_2O_7$	Heated $Mo(OH)_3PO_4$ or MoO_2HPO_4; two forms exist, one of them being unstable; idealized crystal structure was done on the stable form.	214,215, 223
$Nd_4(P_2O_7)_3 \cdot xH_2O$	Precipitated from aqueous solution at pH 4–4.5.	195,216
$NiH_2P_2O_7 \cdot 2H_2O$	Heated $Ni(NH_4)_2P_2O_7 \cdot 2H_2O$ to 160°.	217
$Ni_2P_2O_7 \cdot xH_2O$	Precipitated from aqueous solution. H_2O content depends on drying temperature.	164,200, 217

(continued)

TABLE IV (*continued*)

Compounds	Preparation methods	Refs.
$Ni_2P_2O_7$	Heated $NiHPO_4$ or $NiNH_4PO_4$. There are two forms with the transition temperature near 575°.	164,197
$Pb_2P_2O_7$	Heated a mixture of $PbCO_3$ and $(NH_4)_2HPO_4$, XRD pattern.	218
PbP_2O_7	Heated PbO_2 in concentrated phosphoric acid to 350° with decanting; cubic unit cell given.	193
$PbH_2P_2O_7$	Formed as an unstable intermediate in the thermal dehydration of $Pb(H_2PO_4)_2$, XRD pattern.	219
$Pr_4(P_2O_7)_3 \cdot 24H_2O$	$PrCl_3$ solution + $(NH_4)_4P_2O_7$ solution at pH 4.	195
PuP_2O_7	Heated $PuH_2(PO_4)_2 \cdot xH_2C_2O_4$ to 950°, cubic unit cell is given.	220
$Sc_4(P_2O_7)_3 \cdot xH_2O$	Precipitated from aqueous solution, amorphous.	221
$Sc_4(P_2O_7)_3$	Heated hydrate to 800°, unit cell given.	221
SiP_2O_7	Reacted SiO_2 or the hydroxide with excess phosphoric acid at various temperatures. There are at least four allotropic forms. XRD patterns are given.	193,205, 222, 224, 225
$Sm_2P_2O_7$	$SmCl_2$ solution + pyrophosphate solution, decomposes if taken out of H_2O.	226
$Sm_4(P_2O_7)_3 \cdot xH_2O$	Precipitated from various aqueous solutions, amorphous.	155,195
$Sm_4(P_2O_7)_3$	Heated hydrate at 630°, XRD pattern.	155
$Sn_3(HP_2O_7)_2 \cdot xH_2O$	Precipitated from aqueous solution below pH 5, amorphous.	227
SnP_2O_7	Heated fresh Sn hydroxide in concentrated phosphoric acid to 500°, cubic unit cell given.	193
$Sn_2P_2O_7$	$SnCl_2$ solution + excess $Na_2H_2P_2O_7$ solution, precipitate vacuum dried over P_2O_5.	228
$(ThO)_2P_2O_7$	Equilibrium phase in $ThO_2-P_2O_5$ system at 1350°.	6
β-ThP_2O_7	Stable phase between 810 and 1290°, orthorhombic unit cell described.	6,229, 230
α-ThP_2O_7	Stable above and below interval of stability for β, but P_2O_5 is lost at 1300°, cubic unit cell given.	6,229, 230
$(Th,U)P_2O_7$	Solid solutions of varying U/Th ratio can be formed.	230

(*continued*)

TABLE IV (continued)

Compounds	Preparation methods	Refs.
$ThP_2O_7 \cdot xH_2O$	$Th(NO_3)_4$ solution + $Na_4P_4O_7$ solution, amorphous.	231
TiP_2O_7	Reacted TiO_2 or the hydroxide (fresh) with concentrated phosphoric acid at 500–1400°, cubic unit cell given. There may be more than one cubic or pseudocubic form.	193,232
$TlH_3P_2O_7$	Equilibrium phase in system P_2O_5–Tl_2O–H_2O, XRD pattern.	233
$U_2O_3P_2O_7$	Heated $NH_4UO_2PO_4 \cdot 3H_2O$ to 1000°, or heated a mixture of $UO_2(NO_3)_2 \cdot 6H_2O$ and $(NH_4)_2HPO_4$ to 1200°, XRD pattern.	234,235
$(UO_2)_2P_2O_7$	Heated $NH_4UO_2PO_4 \cdot 3H_2O$ to ~600° with strict control of temperature and calcining time.	234
$(UO)_2P_2O_7$	Equilibrium phase in the system UO_2–P_2O_5, XRD pattern.	6,235
$(UO_2)_2P_2O_7 \cdot xH_2O$	$UO_2(NO_3)_2$ solution + $Na_4P_2O_7$ solution.	170
α-UP_2O_7	Heated a mixture of $(NH_4)_2HPO_4$ and $UO_2(NO_3)_2 \cdot 6H_2O$ to 1100°. Product is a stable phase. Cubic unit cell given. May be as many as 3 cubic forms.	6,193, 230, 235
β-UP_2O_7	Stable phase in the system UO_2–P_2O_5, low temperature form. Orthorhombic unit cell given.	6,236
$(VO)_2P_2O_7$	ESR data indicate that the equilibrium phase $V_2O_4 \cdot P_2O_5$ is a pyrophosphate.	237
WOP_2O_7	Heated a mixture of WO_3 and P_2O_5 to 550°. Crystal structure determined.	238
$Y_4(P_2O_7)_3 \cdot 18H_2O$	YCl_3 solution + $(NH_4)_4P_2O_7$ solution at pH 4.	195
$Yb_4(P_2O_7)_3$	$YbCl_3$ solution + $Na_4P_2O_7$ solution.	158
α-$Zn_2P_2O_7$	Melted $ZnNH_4PO_4$ and cooled, unit cell given.	239
β- and γ-$Zn_2P_2O_7$	These forms exist above 132° but revert to the α form upon cooling. Structure of β form was determined, and ESR data indicated that the γ form exists above 155°.	239,210, 201
$Zn_2P_2O_7 \cdot 4H_2O$	Crystallized from aqueous solution. Unit cell given.	116
$Zn_2P_2O_7 \cdot 3H_2O$ and $Zn_2P_2O_7 \cdot 5H_2O$	Crystallized from aqueous solution. XRD patterns were different and differed from the ASTM pattern available at the time of the investigation.	240

TABLE IV (continued)

Compounds	Preparation methods	Refs.
$ZnH_2P_2O_7$	Heated $Zn(H_2PO_4)_2 \cdot xH_2O$ to 110–225°, crystalline.	31
ZrP_2O_7	Heated Zr hydroxide or oxychloride in concentrated phosphoric acid. There are two cubic forms with a transition at 300°.	193,241, 242
$(ZrO)_2P_2O_7 \cdot 5H_2O$	Precipitated from aqueous solution, amorphous.	243
$(ZrO)_2P_2O_7$	$2ZrP_2O_7 \xrightarrow{1550°} (ZrO)_2P_2O_7 + P_2O_5(g)$. Prepared at lower temperatures by heating an orthophosphate or a mixture of ZrO_2 and ZrP_2O_7. XRD pattern.	241,244
ZrP_2O_7 (high pressure)	Applied 55–100 kbars pressure to cubic ZrP_2O_7 at 750–1000° and cooled before releasing the pressure. Product is metastable at room temperature and has a different XRD pattern from the other forms.	45

A series of compounds $MHP_2O_7 \cdot xH_2O$ where M = La, Ce, Nd, Sc, Sm, In, or Y had been described in the early literature, but later attempts to confirm their existence had failed in most cases.[158,172] Still later it was shown that one of these compounds, $InHP_2O_7 \cdot xH_2O$, could indeed be prepared, but that the most critical reaction condition was the stoichiometry rather than the pH.[42] Most members of the series are omitted from Table IV because their existence has not yet been experimentally confirmed, but it is suspected that they can all be prepared by a method similar to that used for the In salt.

$CdSO_4$, $TiCl_4$, $Pb(NO_3)_2$, and Nb solutions all gave precipitates with pyrophosphate solution, but the precipitates were not analyzed.[174–176] Heterometry, potentiometry, and conductometry indicate that the Pb and Cd salts have the composition $M_2P_2O_7 \cdot xH_2O$.[176] There has been disagreement over whether the Cd salt can be obtained in crystalline form.[177] The existence of $Be_2P_2O_7 \cdot 9H_2O$ was confirmed, but the preparation method wasn't given.[163] The infrared spectrum of $Tl_4P_2O_7$ was reported, but no preparation method or reference was given.[178] Van Wazer[5] mentioned $BaMnP_2O_7 \cdot 3H_2O$, described in the early literature as crystalline, but consultation of the original literature indicates that the correct formula is $Ba(MnP_2O_7)_2 \cdot 5H_2O$. Formation of Al pyrophosphate by heating an acid orthophosphate was claimed but no analytical data were given,[179] and in similar studies later on, this compound wasn't mentioned.[180]

$Cd_5(PO_4)_2(P_2O_7)$ is particularly interesting because it contains both the orthophosphate and pyrophosphate anions.[181,182] Strong chromatographic and infrared spectral evidence was presented to show that both orthophosphate and pyrophosphate were present. However, the XRD pattern has a lot of peaks in common with $Cd_2P_2O_7$ or $Cd_3(PO_4)_2$ and even the few unique peaks could be due to only a portion of the sample with some other composition. In other words, the preparation could possibly be a mixture of two or more simpler phases. A single-crystal x-ray structure study on this compound would be of great interest.

It is also noteworthy that although the precipitation of Gd(III) with $Cs_4P_2O_7$ leads to a product free of alkali metal, as does precipitation with the other alkali metal pyrophosphates, the composition of the precipitate is quite different.[183] The Cs pyrophosphate precipitates $Gd_5(OH)_3(P_2O_7)_3 \cdot 9H_2O$ while the other alkali metals precipitate $Gd_4(P_2O_7)_3 \cdot xH_2O$.

VI. TRIPOLYPHOSPHORIC ACID AND ITS DERIVATIVES

A. Tripolyphosphoric Acid

Tripolyphosphoric acid has not been prepared in pure crystalline form, and a condensed phosphoric acid with a composition equivalent to $H_5P_3O_{10}$ contains only 18% tripolyphosphoric acid,[245] the rest of the product consisting of acids of other chain lengths. Pure tripolyphosphoric acid solution can be prepared by passing a solution of recrystallized $Na_5P_3O_{10} \cdot 6H_2O$ through a cation-exchange column at 0°.[246] It tends to degrade to ortho- and pyrophosphoric acids even at low temperatures, and therefore should be prepared fresh for each use.

B. Alkali Metal Tripolyphosphates

The preparation methods are summarized in Table V, and selected areas are discussed further in this section. The system $NH_3-H_5P_3O_{10}-H_2O$ has been studied at 0 and 25°,[247] and the morphological and optical properties and x-ray diffraction patterns of the different crystals were described.[100,101]

Conversion of other sodium phosphates to $Na_5P_3O_{10}$ is mechanistically complicated and has been fully studied.[10,13,33,53,118,248-250] The most common method is to heat an intimate mixture of Na_2HPO_4 and NaH_2PO_4 in a 2:1 mole ratio under a steam atmosphere. The steam

TABLE V
Tripolyphosphates with Alkali Metal Cations Only

Compounds	Preparation methods	Refs.
$(NH_4)_3H_2P_3O_{10}$	(1) Crystallized at room temperature from an oil produced by adding alcohol to an aqueous ammonium tripolyphosphate solution at pH < 3.	101
	(2) Crystallized at 0° and pH 2.5–3.1 from an ammonium tripolyphosphate solution prepared from $H_5P_3O_{10}$ solution and NH_3.	247
$(NH_4)_3H_2P_3O_{10} \cdot H_2O$	Crystallized at < 10° from an oil produced by adding alcohol to an aqueous solution with pH < 3. Probably a metastable phase.	101
$(NH_4)_4HP_3O_{10}$	(1) Crystallized at 0° by evaporation of an ammonium tripolyphosphate solution with pH = 4–4.9.	101,247
	(2) Crystallized from a mixture of 1 g $(NH_4)_5$-P_3O_{10} with 25 ml 33.3% HAc by volume.	100
$(NH_4)_9H(P_3O_{10})_2 \cdot 2H_2O$	(1) Crystallized at 0° by evaporating an ammonium tripolyphosphate solution with pH = 4.95–5.19.	247
	(2) Treated $(NH_4)_5P_3O_{10}$ solution with HAc to pH = 5. Added ethanol to separate an oil phase which crystallized at room temperature.	101
$(NH_4)_5P_3O_{10} \cdot H_2O$	Added ethanol to an ammonium tripolyphosphate solution at pH = 6 at 25–50°.	101
$(NH_4)_5P_3O_{10} \cdot 2H_2O$	Crystallized from ammonium tripolyphosphate solution at 0° with pH $\geq$ 6.	247
$Na_5P_3O_{10}$-II	(1) Dehydrated a finely milled mixture of $NaH_2PO_4 + 2Na_2HPO_4$ at 400° in a covered container.	10
	(2) Dehydrated $Na_5P_3O_{10} \cdot 6H_2O$ at 350–400°.	62,261
	(3) Tempered a melt of $Na_4P_2O_7$ and $NaPO_3$ at 400°.	108,253
$Na_5P_3O_{10}$-I	(1) Heated $Na_5P_3O_{10}$-II at 550° under a steam atmosphere for 24 hr.	261
	(2) Tempered a melt of $Na_4P_2O_7$ and $NaPO_3$ at 550°.	108
$Na_5P_3O_{10} \cdot 6H_2O$	(1) Added acetone or methanol to a filtered solution of $Na_5P_3O_{10}$.	53
	(2) Crystallized from aqueous solution with simultaneous generation of $Na_5P_3O_{10}$ by hydrolytic cleavage of $Na_3P_3O_9$.	62,36

(continued)

TABLE V (continued)

Compounds	Preparation methods	Refs.
$Na_4HP_3O_{10} \cdot H_2O$	Partially neutralized a solution of $Na_5P_3O_{10}$ with acetic acid and then added alcohol. Seeding is helpful.	262,263
$Na_3H_2P_3O_{10} \cdot 1.5H_2O$	Adjusted $Na_5P_3O_{10}$ solution with $HClO_4$ to pH = 7 and added ethanol.	62,262
$Na(NH_4)_4P_3O_{10} \cdot 2H_2O$	Mixed NH_4OH and $NaH_4P_3O_{10}$ solution (from cation exchanger and $Na_5P_3O_{10}$ solution) to produce a solution with pH ≥ 8, at 4°. Added ethanol.	264
$Na(NH_4)_4P_3O_{10} \cdot 4H_2O$	Recrystallized $Na(NH_4)_4P_3O_{10}$ from water at 50°.	264
$Na_3(NH_4)_2P_3O_{10}$	Reacted $Na_3NiP_3O_{10}$ with $(NH_4)_2S$, or reacted $Na(NH_4)_4P_3O_{10}$ with NaOH.	120
$Na_5P_3O_{10} \cdot K_5P_3O_{10}$	Cooled a melt of the stoichiometric composition.	12,257
$K_5P_3O_{10}$	$KH_2PO_4 + 2K_2HPO_4 \xrightarrow{325-400°} K_5P_3O_{10} + 2H_2O(g)$	123
α-$K_5P_3O_{10} \cdot 2H_2O$	Slowly evaporated a solution of $K_5P_3O_{10}$ under vacuum, or let $K_5P_3O_{10}$ absorb water by keeping it in a humid atmosphere.	118,265
β-$K_5P_3O_{10} \cdot 2H_2O$	Precipitated by adding ethanol to a $K_5P_3O_{10}$ solution. The product often contains some α form.	118,265
$K_3H_2P_3O_{10} \cdot H_2O$	(1) Treated $Ag_5P_3O_{10}$ [131] with cold KCl and HCl solution.	118
	(2) Added ethanol to a mixture of $K_5P_3O_{10}$ and HAc.	266
$K_3H_2P_3O_{10}$	Heated $K_3H_2P_3O_{10} \cdot H_2O$ under vacuum at 125°. Amorphous.	118

enhances the formation of a molten intermediate, prior to crystallization of $Na_5P_3O_{10}$.[10] Sodium tripolyphosphate occurs in two anhydrous crystalline forms and a hexahydrate, and the crystal structures have been determined for all three.[36,251,253] $Na_5P_3O_{10}$-II forms at 400° and can be converted to $Na_5P_3O_{10}$-I at 500–550°. It is difficult to convert I to II (in the absence of a molten phase) by tempering at temperatures below the phase transition temperature of 417°.[254] $Na_5P_3O_{10}$-I can be converted to form II by applying a pressure greater than 3000 psi.[248] It is difficult to prepare $Na_5P_3O_{10}$ by dehydration of the hexahydrate because of the tendency to degrade to ortho- and pyrophosphates.[53,255,256] Potassium tripolyphosphate, $K_5P_3O_{10}$, can be produced in much the same way as sodium tripolyphosphate.[110] The system $NaPO_3$–$Na_4P_2O_7$–$K_4P_2O_7$–KPO_3 was studied by Morey.[257] The region $Na_5P_3O_{10}$–$K_5P_3O_{10}$ was rechecked by DTA,[12] and the results support Morey in that the composition of the sodium–potassium double salt is $Na_{2.5}K_{2.5}P_3O_{10}$ rather than $Na_3K_2P_3O_{10}$. In the thermal dehydration of KH_2PO_4, compounds such as $K_2H_2P_2O_7$,[127,258] $K_3H_2P_3O_{10}$,[118,258] $2KH_2PO_4 \cdot K_2H_2P_2O_7$,[258,259] and $KH_2PO_4 \cdot K_2H_2P_2O_7$[259] have been claimed as intermediates. Work by Thilo and Grunze[103] with paper chromatographic identification, however, showed that in an open atmosphere, the dehydration products consisted almost exclusively of $(KPO_3)_x$ and starting material. In some cases, a mixture of lower molecular weight potassium polyphosphates could be obtained after a short heating period at 230°, but the proportion of each polyphosphate fraction was so small that this procedure could not be considered for preparation of potassium acid tripoly- or acid pyrophosphates. Under a steam atmosphere (in a sealed tube) the formation of Kurrol salt was inhibited as reported earlier,[114,260] but the compound, $K_3H_2P_3O_{10}$, claimed by Boulle et al.[258] was not found.

C. Other Tripolyphosphates

The preparation methods are summarized in Table VI, but some additional information not suitable for the table is given in the text.

No tripolyphosphate was found in the systems CaO–P_2O_5 [146,267–269] or SrO–P_2O_5.[270] Acid tripolyphosphate intermediates were observed in the thermal dehydration of $Ca(H_2PO_4)_2 \cdot 2H_2O$[31,118,271–272] but not in the thermal dehydration of $Sr(H_2PO_4)_2 \cdot H_2O$ or $Ba(H_2PO_4)_2$.[130]

A series of compounds $FeM_2P_3O_{10}$ and $AlM_2P_3O_{10}$ (where M = Ag, K, Li, Na, NH_3, or $\frac{1}{2}Ca$) was prepared by reacting a solution of a salt of M with solid $FeH_2P_3O_{10} \cdot xH_2O$ or solid $AlH_2P_3O_{10} \cdot xH_2O$.[41] The reactions were ion exchange processes and led to hydrated products which were characterized by their x-ray diffraction patterns. It was claimed that the hydrated salts could be thermally dehydrated to give the corresponding anhydrous tripolyphosphates which were also crystalline.

Potentiometry, conductometry, heterometry, viscometry, or hydrogen-ion displacement studies were used to show that precipitates of the following compositions can be obtained from aqueous solutions, but the products were not isolated and analyzed: $Co_5(P_3O_{10})_2 \cdot xH_2O$,[212] $Co_2NaP_3O_{10} \cdot xH_2O$,[212] $Cu_2NaP_3O_{10} \cdot xH_2O$,[212] $DyNa_2P_3O_{10} \cdot xH_2O$,[273] $GdNa_2P_3O_{10} \cdot xH_2O$,[273] $Mn_2NaP_3O_{10} \cdot xH_2O$,[212] $Na_8Ni(P_3O_{10})_2 \cdot xH_2O$,[274] $Na_8Zn(P_3O_{10})_2 \cdot xH_2O$,[274] $Na_2SmP_3O_{10} \cdot xH_2O$,[273] $Nd_5(P_3O_{10})_3 \cdot xH_2O$.[216,275]

TABLE VI
Preparation of Tripolyphosphates other than the Alkali Metal Salts

Compounds	Preparation methods	Refs.
$AlH_2P_3O_{10}$-I	Heated $Al(H_2PO_4)_3$ to 240° with washing by anhydrous ether, or heated Al_2O_3 in concentrated phosphoric acid. XRD pattern.	180,41
$AlH_2P_3O_{10} \cdot 2$–$3H_2O$	Obtained by the reversible hydration of $AlH_2P_3O_{10}$-I in contact with moist air or wash water. XRD pattern.	41
$AlH_2P_3O_{10}$-II	Heated Al_2O_3 in concentrated phosphoric acid and separated the form II from the form I which was produced simultaneously. XRD pattern.	41
$Ba_5(P_3O_{10})_2 \cdot xH_2O$	a. $Na_3P_3O_9$ solution + $BaCl_2$ solution $\xrightarrow{NaOH}$ $Ba_5(P_3O_{10})_2$(s).	280
	b. Crystallized from aqueous solution.	255
$Be_5(P_3O_{10})_2 \cdot xH_2O$	Microcrystalline precipitate with variable amount of hydrate water was obtained from aqueous solution.	255
$Be_2NaP_3O_{10} \cdot 5H_2O$	Crystallized by evaporating an aqueous solution.	255
$Ca_5(P_3O_{10})_2 \cdot 8H_2O$	Precipitated from aqueous solution.	281
$Ca_5(P_3O_{10})_2$	Dehydrated the hydrate.	282
$Ca_2HP_3O_{10}$	Heated $Ca(H_2PO_4)_2$ at 300–330° under steam.	271,272, 283
$Ca_2NaP_3O_{10} \cdot 4H_2O$	Precipitated from aqueous solution.	255
$CdNa_3P_3O_{10} \cdot 12H_2O$	$Na_5P_3O_{10}$ solution + $CdCl_2$ solution, crystalline.	255
$Cd_2NaP_3O_{10} \cdot 7H_2O$	Crystallized from aqueous solution.	255,284
$Co_2KP_3O_{10} \cdot xH_2O$	Crystallized from aqueous solution.	255
$CoNa_3P_3O_{10} \cdot 12H_2O$	Crystallized from aqueous solution.	285
$[Co(NH_3)_6]_3Na(P_3O_{10})_2$	Precipitated from aqueous solution and dried at 110°.	162
$CrH_2P_3O_{10}$-I$\cdot 2$–$3H_2O$	Heated $CrCl_3$ in concentrated phosphoric acid with seeding at 200–300°, cooled, and washed the crystals in H_2O.	199
$CrH_2P_3O_{10}$-I	Dehydrated the hydrate.	199
$CrH_2P_3O_{10}$-II	Heated $CrCl_3$ in concentrated phosphoric acid at 200–300° with seeding; cooled, and washed the crystals in H_2O.	199
$CrH_2P_3O_{10}$-III$\cdot 1$–$2H_2O$	Heated $CrCl_3$ in concentrated phosphoric acid at 200–300° with seeding; cooled, and washed the crystals in H_2O.	199
$CrH_2P_3O_{10}$-III	Dehydrated the hydrate.	199
$CrNa_2P_3O_{10} \cdot 6H_2O$	$Cr_2(SO_4)_3$ solution + $Na_5P_3O_{10}$ solution, crystalline.	255

(*continued*)

TABLE VI (continued)

Compounds	Preparation methods	Refs.
$Cu_5(P_3O_{10})_2 \cdot xH_2O$	Precipitated from aqueous solution.	255, 286
$CuNa_3P_3O_{10} \cdot 12H_2O$	Crystallized from aqueous solution.	255
$Dy_5(P_3O_{10})_3 \cdot xH_2O$	$DyCl_3$ solution + $Na_5P_3O_{10}$ solution.	273
$FeH_2P_3O_{10}\text{-I} \cdot 2\text{-}3H_2O$	Heated Fe_2O_3 in concentrated phosphoric acid at 220–260° with seeding; cooled, and washed with H_2O. Form II crystals often form at the same time and can be separated by a decantation method.	41
$FeH_2P_3O_{10}\text{-I}$	Dehydrated the hydrate. XRD pattern is related to but different from that of the hydrate.	41
$FeH_2P_3O_{10}\text{-II}$	Heated Fe_2O_3 in concentrated phosphoric acid at 220–260°; cooled, and washed with H_2O. Form II must be separated from form I which is synthesized simultaneously.	41
$FeH_2P_3O_{10}\text{-III} \cdot 1\text{-}2H_2O$	Heated Fe_2O_3 in concentrated phosphoric acid with seeding by the isomorphous Cr salt.	199
$FeNa_3P_3O_{10} \cdot 11.5H_2O$	Crystallized from aqueous solution.	255
$Gd_5(P_3O_{10})_3 \cdot xH_2O$	$GdCl_3$ solution + $Na_5P_3O_{10}$ solution.	273
$GeHP_3O_{10}$	Heated GeO_2 in excess phosphoric acid at 200–220°. XRD and paper chromatographic characterization.	205
$(HfO)_5(P_3O_{10})_2 \cdot 12H_2O$	$HfOCl_2$ solution + $Na_5P_3O_{10}$ solution, amorphous.	287
$(HfO)_2NaP_3O_{10} \cdot 4H_2O$	$HfOCl_2$ solution + $Na_5P_3O_{10}$ solution, amorphous.	287
$(HfO)_5Na_{10}(P_3O_{10})_4 \cdot 12H_2O$	$Hf(SO_4)_2$ solution + $Na_5P_3O_{10}$ solution, amorphous.	287
$La_5(P_3O_{10})_3 \cdot 16H_2O$	$LaCl_3$ solution + $Na_5P_3O_{10}$ solution, XRD pattern.	208
$Mn_5(P_3O_{10})_2$	Precipitated from aqueous solution.	255
$MnNa_3P_3O_{10} \cdot 12H_2O$	Crystallized from aqueous solution.	255
$Na_3NiP_3O_{10} \cdot 12H_2O$	$Na_5P_3O_{10}$ solution + $NiSO_4$ solution, crystalline.	164, 276
$Na_8Pb(P_3O_{10})_2 \cdot 14H_2O$	Crystallized from aqueous solution.	255
$NaPrHP_3O_{10} \cdot 3H_2O$	Precipitated from aqueous solution at pH 3.5.	288
$NaSr_2P_3O_{10} \cdot 7H_2O$	Crystallized from aqueous solution.	255
$Na_3(UO_2)P_3O_{10} \cdot 6\text{-}8H_2O$	Precipitated from aqueous solution.	289, 290
$Na_5(UO_2)_5(P_3O_{10})_3 \cdot 19H_2O$	Precipitated from aqueous solution, amorphous.	289
$Na_3ZnP_3O_{10} \cdot 11.5\text{-}12.5H_2O$	Crystallized from aqueous solution.	255, 291

(continued)

TABLE VI (*continued*)

Compounds	Preparation methods	Refs.
$NaZn_2P_3O_{10} \cdot 9-9.5H_2O$	Crystallized from aqueous solution.	240,255, 291
$Na(ZrO)_2P_3O_{10} \cdot 6H_2O$	Precipitated from aqueous solution, amorphous.	243
$Na_{10}(ZrO)_5(P_3O_{10})_4 \cdot 11.5H_2O$	Precipitated from aqueous solution, amorphous.	243
$Pb_5(P_3O_{10})_2 \cdot xH_2O$	Precipitated from aqueous solution.	176,255, 277
$SiHP_3O_{10}$	Heated SiO_2 in excess phosphoric acid at 200–220°. XRD and paper chromatography.	205
$Sm_5(P_3O_{10})_3 \cdot xH_2O$	$SmCl_3$ solution + $Na_5P_3O_{10}$ solution.	273
$Sr_5(P_3O_{10})_2 \cdot xH_2O$	Crystallized from aqueous solution.	255
$Th_5(P_3O_{10})_4 \cdot 14.5H_2O$	Precipitated from aqueous solution, amorphous.	289
$(UO_2)_5(P_3O_{10})_2 \cdot 12H_2O$	$Na_3P_3O_9 + UO_2(NO_3)_2$ in solution (ring cleavage), XRD, and paper chromatography.	292
$(ZrO)_5(P_3O_{10})_2 \cdot 5.5H_2O$	Precipitated from aqueous solution, amorphous.	243

Precipitates formed by reaction of Zn or Cd sulfate solution with $Na_5P_3O_{10}$ solution were shown by heterometry and conductometry to have approximately the composition $M_5(P_3O_{10})_2 \cdot xH_2O$.[176] However, later work on the reaction of $ZnSO_4$ and $Na_5P_3O_{10}$ solutions failed to confirm the existence of $Zn_5(P_3O_{10})_2 \cdot xH_2O$ and showed instead that $NaZn_2P_3O_{10} \cdot 9H_2O$ was formed.[276] Similarly, it seems likely that the true composition of the Cd salt is $NaCd_2P_3O_{10} \cdot xH_2O$. In the case of Ni, heterometry,[212] and amperometric titration[277] indicate formation of $NaNi_2P_3O_{10} \cdot xH_2O$, and the isolation of such a compound has been reported.[120] However, in the latest paper found on the subject, the preparation of $NaNi_2P_3O_{10} \cdot xH_2O$ could not be confirmed, and $Ni_5(P_3O_{10})_2 \cdot 17-18H_2O$ was formed instead.[276] Hydrogen ion displacement studies[275] showed that $Y_5(P_3O_{10})_3 \cdot xH_2O$ can be precipitated from solution. The isolation of this compound was reported; but the details of its preparation were not given in the abstract, and the original article was not available.[278] In another abstract for which the original article was unavailable, a series of compounds was described which probably have the formula $Ln_3Na(P_3O_{10})_2 \cdot xH_2O$ where Ln = Pr, Tb, Ho, Er, Tm, Yb, and Lu.[279] Potentiometric and conductometric titrations showed formation of insoluble compounds with $Ln^{3+}:P_3O_{10}^{5-} = 3:2$, and compounds were said to have been isolated with formula "$Ln_3(P_3O_{10})\cdot H_2O$." This formula is obviously wrong and is perhaps given correctly in the original paper. Physical properties of the following compounds were given without giving the preparation methods: $Na_3NiP_3O_{10}$,[64] $Ag_5P_3O_{10}$,[178] TaP_3O_{10},[154] and $NaZn_2P_3O_{10}$.[178]

VII. TETRA AND HIGHER POLYPHOSPHATES

A. Condensed Phosphoric Acids

As mentioned in Section V-A, all condensed phosphoric acids, except for crystalline pyrophosphoric acids, are mixtures of various chain lengths at equilibrium. Since the composition is continuously variable depending upon the temperature of preparation, these materials are outside the scope of this chapter. It has been claimed that $H_6P_4O_{13}$ can be crystallized from a condensed phosphoric acid,[293] but this could not be confirmed. On standing, a waxy solid formed which was shown by XRD to be a mixture of crystalline $H_4P_2O_7$-I and orthophosphoric acids. $(HPO_3)_x$ has never been crystallized although it was once believed that such an acid existed.[90,294]

On the other hand it is possible to prepare aqueous solutions of various polyphosphoric acids which contain predominantly a single-chain length and which are stable enough to be used as reagents in preparing their salts. Tetrapolyphosphoric, hexapolyphosphoric,[131,295] and long-chain polyphosphoric acid solutions[57] have been prepared from crystalline salts by means of ion exchangers. Other chain lengths have been obtained by separation of mixtures as discussed in Section III-H. Solvolysis of P_2O_5 in nonaqueous or mixed solvents can lead to solutions containing mostly pyrophosphoric acid or under other conditions to a solution rich in tripolosphyphoric acid (along with trimetaphosphoric acid).[296]

B. Tetra and Higher Polyphosphates of a Single Chain Length

The compounds given in Table VII are fairly well characterized with regard to anion type. Since the metaphosphates and the long-chain polyphosphates of a given cation have essentially the same chemical composition, it is necessary to use special analytical techniques such as paper chromatography to distinguish them. It is probable that some of the compounds with the metaphosphate composition which are discussed in

TABLE VII
Preparation of Tetra and Higher Polyphosphates

Compounds	Preparation methods	Refs.
$(AgPO_3)_x$	Crystallized from a melt prepared from $Ag_3P_3O_9$. Complete XRD structure determination.	302
$Ag_6P_4O_{13} \cdot xH_2O$	Precipitated from aqueous solution, amorphous.	4,303
$[Al(PO_3)_3]_x$-B	Heated $Al(H_2PO_4)_3$ to 800°, or thermally dehydrated $AlH_2P_3O_{10}$-I.	41,180
$[Al(PO_3)_3]_x$-C	Heated Al_2O_3 in phosphoric acid. The isomorphous Fe salt can be used for seeding, and the product can be separated from tetrametaphosphate impurities.	41,173a
$[Al(PO_3)_3]_x$-D	Thermally decomposed $Al(NH_4)_2P_3O_{10} \cdot H_2O$.	41
$[Al(PO_3)_3]_x$-E	Crystallized from a hot solution of Al_2O_3 in phosphoric acid seeded with the isomorphous Fe salt.	41
$Ba_3P_4O_{13}$	Heated a mixture of $(NH_4)_2HPO_4$ and $BaHPO_4$. Two forms related by a reversible inversion at 870° were described which could be obtained at room temperature by quenching. Both XRD patterns are different from that of the $Ba_3P_4O_{13}$ prepared by Langguth et al. at lower temperature. Thus it appears that there are at least three allotropic forms.	31,151, 219, 304
$[Ba(PO_3)_2]_x$	Heated $Ba(H_2PO_4)_2$ to various temperatures high enough to drive off the H_2O and found only one form. However there was one preparation later with a different XRD pattern which may indicate a second allotrope. The reported formation of the tetrametaphosphate could not be confirmed.	31,187, 219, 304
$[Ba(PO_3)_2 \cdot nH_2O]_x$	$Ba(OH)_2 + (HPO_3)_x$, amorphous, can be converted to crystalline $[Ba(PO_3)_2]_x$ by heating to 500°.	57
$Ba_2MgP_4O_{13}$	Equilibrium phase in the $BaO-MgO-P_2O_5$ system, XRD pattern.	187
$[Be(PO_3)_2]_x$	Heated $Be(H_2PO_4)_2$ to $\geq 550°$. It seems there are several crystalline forms.	305,306
$Bi_2P_4O_{13}$	Heated a mixture of Bi_2O_3 and H_3PO_4 to 700°, XRD pattern.	307
$[Bi(PO_3)_3]_x$	Heated Bi_2O_3 in excess phosphoric acid. XRD pattern.	31

(continued)

TABLE VII (*continued*)

Compounds	Preparation methods	Refs.
$Ca_4P_6O_{19}$	Equilibrium phase in $CaO-P_2O_5$ system.	267,271, 295, 308
α-$[Ca(PO_3)_2]_x$	Heated $Ca(H_2PO_4)_2$ to 963–985°.	145
β-$[Ca(PO_3)_2]_x$	Heated $Ca(H_2PO_4)_2$ to 600°, or tempered γ-$[Ca(PO_3)_2]_x$ at 450–500°.	31,136, 271, 308, 309
γ-$[Ca(PO_3)_2]_x$	Heated $CaH_2P_2O_7$ to 340–360°.	136,271, 308
δ-$[Ca(PO_3)_2]_x$	Heated a mixture of $Ca(H_2PO_4)_2$ and $CaHPO_4$ at 400°.	136,271, 308
$[Cd(PO_3)_2]_x$	Heated CdO with a slight excess of H_3PO_4 to 300–400°. XRD pattern resembles that reported for α-$Cd(PO_3)_2$.	31,182, 310
$[Co(NH_3)_6]_2P_4O_{13} \cdot 6H_2O$	Crystallized from an oil formed by mixing solutions of tetrapolyphosphate and $Co(NH_3)_6Cl_3$.	295
$[Co(NH_3)_6]_8(P_6O_{19})_3 \cdot 20H_2O$	$Ca_4P_6O_{19}$ was dissolved in H_2O by means of EDTA, and $Co(NH_3)_6Cl_3$ solution was added giving an oil which slowly crystallized.	295
$[Co(NH_3)_6(PO_3)_3 \cdot nH_2O]_x$	$Co(NH_3)_6(OH)_3$ solution + $(HPO_3)_x$ solution. XRD showed that the precipitate was poorly crystallized.	57
$[Cr(PO_3)_3]_x$	Forms B, C, and E were prepared by heating solutions of $CrCl_3$ in H_3PO_4 to 400–800°.	199
$(CsPO_3 \cdot nH_2O)_x$	Small crystals were precipitated from aqueous solution.	4
$(CsPO_3)_x$	Heated CsH_2PO_4 or a mixture of Cs_2CO_3 and $NH_4H_2PO_4$. There are at least two forms with the transition temperature at 467–480°. The Kurrol's salt was crystallized from a melt.	31,106, 311, 312
$[CsLi(PO_3)_2]_x$	Equilibrium phase in $LiPO_3$–$CsPO_3$ system.	312
$[Cu(NH_3)_4(PO_3)_2 \cdot nH_2O]_x$	Added acetone to a mixture of $Cu(NH_3)_4(OH)_2$ and $(HPO_3)_x$ in aqueous solution. XRD pattern.	57
$[Fe(PO_3)_3]_x$-B	Heated Fe_2O_3 in phosphoric acid and seeded with the isomorphous Al salt. Forms A and C which are produced simultaneously can be removed by a decantation procedure.	41

(*continued*)

TABLE VII (continued)

Compounds	Preparation methods	Refs.
$[Fe(PO_3)_3]_x$-C	Heated $Fe(H_2PO_4)_3$ to $\geq 850°$ or other forms of $Fe(PO_3)_3$ to 800–900°.	41,180
$[Fe(PO_3)_3]_x$-D	Prepared by thermal decomposition of $FeH_2P_3O_{10}$-I.	41
$[Fe(PO_3)_3]_x$-E	Heated Fe_2O_3 in phosphoric acid. Seeding is helpful.	41
$[Fe_2(OH)_3(PO_3)_3]_x$	$FeCl_3$ solution + $(NaPO_3)_x$ solution.	313
$[Hg(PO_3)_2]_x$	Heated a mixture of $HgCl_2$ and H_3PO_4 to 400°, XRD pattern.	31
$(KPO_3)_x$	Heated KH_2PO_4 to $>200°$, or heated a mixture of KCl and H_3PO_4 to $\sim 500°$. Crystal structure was done.	9,29, 103, 123, 311, 314, 315
$[KLi(PO_3)_2]_x$	Equilibrium phase in $LiPO_3$–KPO_3 system.	312
$La_2P_4O_{13} \cdot xH_2O$	Precipitated from aqueous solution.	165
$[La(PO_3)_3]_x$	Prepared by thermal dehydration of an acid orthophosphate.	305
$(LiPO_3)_x$-I	Heated LiH_2PO_4 to 210°, monoclinic.	103, 316, 317
$(LiPO_3)_x$-II	Heated form I to 250–630°.	103
$[Li_nNa_2-n(PO_3)_2]_x$	Solid solutions formed in the system $LiPO_3$–$NaPO_3$.	312
$[Mn(PO_3)_2]_x$	Crystallized from a melt prepared from $MnCO_3$ and $Na_3(PO_3)_3$. By-products removed by leaching.	156
$[MoO_2(PO_3)_2]_x$	Heated MoO_3 in excess phosphoric acid to 450°, cooled, and leached; or heated a mixture of MoO_3 and P_2O_5 to 500°. Crystal structure was done.	318,319
$(NaPO_3)_x$-II	Heated $NaH_2PO_4 \cdot H_2O$ to 300° in a closed container. High temperature form of Maddrell's salt.	103,109, 320, 321
$(NaPO_3)_x$-III	Heated $NaH_2PO_4 \cdot H_2O$ to 265–275°, cooled, and leached. Low temperature form of Maddrell's salt.	103,109, 321, 322
$(NaPO_3)_x$-IVA	At 580–590°, seeded a supercooled melt and tempered for 70–90 min, cooled rapidly. Na Kurrol's salt A.	311, 323, 324
$(NaPO_3)_x$-IVB	Seeded a supercooled melt at 570–590° with $<1\%$ Al_2O_3, tempered for an hour at 560–580°, and cooled. Na Kurrol's salt B. Also	311,321, 324, 327

(continued)

TABLE VII (continued)

Compounds	Preparation methods	Refs.
	form A can be converted to form B by pressure or moisture.	
$[Na_2H(PO_3)_3]_x$	Melted a mixture of $Na_2H_2P_4O_{12}$ and $Na_3P_3O_9$, quenched to a glass, and tempered several days at 200°. Crystal structure was done.	326,328
$[Na_3H(PO_3)_4]_x$	Heated a mixture of $NaH_2PO_4 \cdot H_2O$ and H_3PO_4 at 600° for an hour and then at 350° for 12 hr.	328
$(NH_4)_6P_4O_{13} \cdot 6H_2O$	$Pb_3P_4O_{13}$ + $(NH_4)_2S$ solution.	44
$(NH_4PO_3)_x$-I	$(NH_4)_2HPO_4$ + $(NH_2)_2CO \rightarrow NH_4PO_3$ + $3NH_3(g)$ + $CO_2(g)$.	16,104, 105
$(NH_4PO_3)_x$-II	Heated $(NH_4PO_3)_x$-I at 300° in a sealed container.	297
$[Ni(NH_3)_6(PO_3)_2 \cdot nH_2O]_x$	Added acetone to a solution of $Ni(OH)_2$, NH_4OH, and $(HPO_3)_x$, and dried the resulting oil over P_2O_5, amorphous.	57
$[Pb(PO_3)_2]_x$	Calcined $Pb(H_2PO_4)_2$ or a mixture of $PbCO_3$ and $(NH_4)_2HPO_4$. Crystal structure done. The reported formation of the tetrametaphosphate could not be confirmed.	31,218, 219, 329
$Pb_3P_4O_{13}$	Equilibrium phase in $PbO-P_2O_5$ system.	31,151, 218, 219, 307
$(RbPO_3 \cdot nH_2O)_x$	Very small crystals precipitated from aqueous solution.	4
$(RbPO_3)_x$	Heated RbH_2PO_4 to 440° or to a temperature somewhat above the melting point followed by slow cooling. Crystal structure done.	330,331
$[Sr(PO_3)_2]_x$	Heated $Sr(H_2PO_4)_2$ to 850°, α-$Sr(PO_3)_2$.	31,137, 182
$[Ti(PO_3)_3]_x$	Heated Ti in molten $H_2NH_4PO_4$ with oxygen excluded, monoclinic.	299
$[Zn(PO_3)_2]_x$	This form known as β-$Zn(PO_3)_2$ is the stable one between ~700° and the melting point of ~860°.	31,332, 333
$Zn_3P_4O_{13} \cdot xH_2O$	Precipitated from aqueous solution, amorphous.	4

Section X are long-chain polyphosphates, but they are not included here because insufficient evidence was offered regarding the type of anion.

The thermally prepared compounds $[Ba(PO_3)_2]_x$ and $[Pb(PO_3)_2]_x$ are considered to be polyphosphates although it has been claimed that they are tetrametaphosphates.[219] Later work indicated that only the long-chain polyphosphates can be prepared thermally, and it was suggested that small amounts of tetrametaphosphate intermediate had been extracted for analysis from the earlier products and mistakenly assumed to be representative of the entire preparations.[31]

The type of phosphate ion present in products with the metaphosphate composition which are crystallized at high temperatures seems to be a function of the size and charge of the cation. Some cations always give long-chain polyphosphates; some always give tetrametaphosphates; and others give any of several types depending upon the conditions.[3,31,41]

Crystalline long-chain ammonium polyphosphate has been prepared and identified only recently[101,104,105,297] and exists in several forms. The most commonly prepared form[101,104,105] seems to be metastable (form I) and can be converted to a more stable form (form II) by tempering at 300° under an ammonia atmosphere. Form II is orthorhombic with $a = 4.256$, $b = 6.475$, and $c = 12.04 \pm 0.01$ Å, and the probable space group is $P2_12_12_1$.[298] The density calculated for $4NH_4PO_3$ per cell is 1.94 g/cm^3 compared to a measured value of 1.90 g/cm^3.

The long-chain polyphosphates of Al, Cr, and Fe are interesting because of the large number of allotropic forms and because of the isomorphism exhibited.[41,199] Some of the forms were prepared for the first time by seeding with an isomorphous salt which is more easily prepared. This technique could probably lead to additional forms of long-chain polyphosphates such as $[Ti(PO_3)_3]_x$ for which only a single form has been described which is isomorphous with the monoclinic long-chain polyphosphates of Fe and Cr.[299]

Thermal preparation of polyphosphates containing a single type of anion with over three phosphorus atoms is rare. The long-chain polyphosphates contain a distribution of chain lengths, but the chains are all so long and similar in their properties and composition that these preparations are for all practical purposes pure compounds rather than mixtures. Several thermally prepared tetrapolyphosphates are included in Table VII, but the only other polyphosphate of intermediate chain length which has been prepared thermally is Troemelite which was recently shown to be a hexapolyphosphate rather than a pentapolyphosphate as had been previously believed.[295]

A number of long-chain polyphosphates have been prepared thermally which were characterized as highly polymeric but which were not categorized as either glasses, crystalline compounds, or mixtures.[300] Reaction of Graham's salt with various metal salts in solution has given a variety of amorphous precipitates containing large and variable amounts of water.[4,301] The relative amounts of Na and the other cation in the precipitate can vary also.

C. Phosphate Glasses

Phosphate glasses are noncrystalline solids of continuously variable composition and are thus outside the scope of this chapter. Nevertheless, because of the extensive work which has been done in this area, a brief discussion is included here. A more detailed discussion was given by Van Wazer.[5]

Phosphate glasses are usually prepared by rapid quenching of a melt between two heavy copper or aluminum blocks, which can in some cases provide cooling rates in excess of 300°/sec. The range of useful compositions depends upon the cations used; but, in general, the upper limit for P_2O_5 content is set by requirements for chemical resistance of the product, and the lower limit for P_2O_5 content is set by the tendency of compositions of high cation content to crystallize (devitrify) during or after cooling.

The effect of the cation on the distribution of the various phosphate species in phosphate glass has been studied for systems containing H, Li, Na, K, and Rb.[245,334-338] Duration and temperature of heating the melt and aging time at room temperature were found to have little effect on the phosphate distribution.[339] The water vapor pressure over the melt and the efficiency of removing volatile by-products can have great effects on the distribution.

Phosphate glasses containing various amounts of Li, Na, K, Mg, Zn, Ba, Pb, Al, and Ca were evaluated as fusible enamels.[340] Some other studies of phosphate glass formation dealt with the systems Ca–Zn–alkali metal phosphate,[339] $CaO-Al_2O_3-P_2O_5$,[269] $Na_2O-P_2O_5-Al_2O_3-B_2O_3$,[341] $GeO_2-P_2O_5-H_2O$,[205] $SiO_2-P_2O_5-H_2O$,[205] $Na_2O-V_2O_5-P_2O_5$,[342] $PbO-WO_3-P_2O_5$,[343] $NaPO_3-PbO$,[344] $NaPO_3-CdO$,[344] and $NaPO_3-MnO$.[344]

VIII. METAPHOSPHORIC ACIDS AND THEIR SALTS

A. Metaphosphoric Acids

The best method for preparing these acids is by reaction of a solution of a soluble salt at low temperature with an ion exchanger.[345] The acid solution must then be used quickly (for the preparation of new salts) in order to avoid hydrolysis to other phosphate species. Tetrametaphosphoric acid solution can also be prepared from P_4O_{10} in 70–80% yield by low temperature hydrolysis.[35]

B. Trimetaphosphates

The preparation of trimetaphosphates is summarized in Table VIII. As with the other types of phosphates, it was not attempted to cover the old literature because of the lack of reliable analytical data. A number of trimetaphosphates, crystallized from aqueous solution, were listed in a summary of the older literature by Van Wazer.[5] Infrared spectra were given for the following compounds without specifying the preparation methods: $Ba_3(P_3O_9)_2 \cdot 4H_2O$,[64] $Mn_3(P_3O_9)_2 \cdot 11H_2O$,[64] $Sr_3(P_3O_9)_2 \cdot 7H_2O$,[64] and $Tl_3P_3O_9$.[178]

TABLE VIII
Preparation of Trimetaphosphates

Compounds	Preparation methods	Refs.
$Ag_3P_3O_9 \cdot xH_2O$	Crystallized from aqueous solutions. Preparations with $x = 0$, 1, and 13–15 have been described. Unit cell was determined for the monohydrate.	2,7,347
$AgCdP_3O_9$	Calcined a mixture of $(NH_4)_2HPO_4$, $CdCO_3$, and AgH_2PO_4. Unit cell given.	348
$Ca_3(P_3O_9)_2 \cdot 10H_2O$	Added acetone to a solution prepared from trimetaphosphoric acid and $Ca(ClO_4)_2$.	31
$Ca_3(P_3O_9)_2$	Heated the hydrate to 110–170°, or heated $Ca(H_2PO_4)_2$ to about 300°.	31
$CaKP_3O_9$ (hexagonal)	Calcined a mixture of $(NH_4)_2HPO_4$, $CaCO_3$, and KH_2PO_4. Unit cell given.	349
$CaNa(P_3O_9) \cdot 3H_2O$	Added alcohol to a solution prepared from $Na_3P_3O_9$ and $Ca(ClO_4)_2$.	31
$CaNa_4(P_3O_9)_2$	Crystallized from a melt of $2Na_3P_3O_9$ and $Ca(PO_3)_2$.	350
$CdKP_3O_9$ (hexagonal)	Calcined a mixture of $(NH_4)_2HPO_4$, $CdCO_3$, and KH_2PO_4. Unit cell given.	349
$CdRbP_3O_9$	Calcined a mixture of $(NH_4)_2HPO_4$, $CdCO_3$, and RbH_2PO_4. Unit cell given.	348
$CdTlP_3O_9$	Calcined a mixture of $(NH_4)_2HPO_4$, $CdCO_3$, and TlH_2PO_4. Unit cell given.	348
$CoKP_3O_9$ (hexagonal)	Calcined a mixture of $(NH_4)_2HPO_4$, $CoCO_3$, and KH_2PO_4. Unit cell given.	349
$Cs_3P_3O_9$	Heated CsH_2PO_4 in the presence of HCl, NH_4NO_3, other alkali metal ions, or acetic anhydride.	4,18,33
$K_3P_3O_9$	Heated KH_2PO_4 in the presence of HCl, NH_4NO_3, other alkali metal ions, or acetic anhydride.	18,33

(continued)

TABLE VIII (*continued*)

Compounds	Preparation methods	Refs.
$K_2LiP_3O_9 \cdot H_2O$	Added ethanol to a solution prepared by reacting $Ag_3P_3O_9$ slurry with LiCl and KCl solutions and filtering.	351
$KMgP_3O_9$ (hexagonal)	Calcined a mixture of $(NH_4)_2HPO_4$, $MgCO_3$, and KH_2PO_4. Unit cell given.	349
$KMnP_3O_9$ (hexagonal)	Calcined a mixture of $(NH_4)_2HPO_4$, $MnCO_3$, and KH_2PO_4. Unit cell given.	349
$K_9Na_3(P_3O_9)_4$	Crystallized from a melt or by adding methanol to a solution of $Na_3P_3O_9$, KCl, and KOH.	352,110
$KZnP_3O_9$ (hexagonal)	Calcined a mixture of $(NH_4)_2HPO_4$, $ZnCO_3$, and KH_2PO_4. Unit cell given.	349
$Li_3P_3O_9 \cdot 3H_2O$	Crystallized from a solution prepared from LiOH and trimetaphosphoric acid.	353
$Na_3P_3O_9 \cdot 6H_2O$	Added alcohol to an aqueous solution at 0–20°.	354,355
$Na_3P_3O_9 \cdot 3H_2O$	Rapidly added an equal volume of ethanol to a 5–10% aqueous solution at 20° and immediately filtered. Product is efflorescent.	354
$Na_3P_3O_9 \cdot 1.5H_2O$	Slowly added ethanol to 10% aqueous $Na_3P_3O_9$ solution at 30° with agitation.	354
$Na_3P_3O_9 \cdot H_2O$	Evaporated an aqueous solution at 60°.	354–357
$Na_3P_3O_9$-I	Heated NaH_2PO_4 at 475–500°. Crystal structure done.	321,356
$Na_3P_3O_9$-I′	Cooled a melt slowly to 500–525° and then quenched.	321
$Na_3P_3O_9$-I″	Cooled a melt slowly to $\geq 375°$ and then quenched.	321
$Na_2HP_3O_9$	Cooled a melt to $\geq 300°$, tempered, and quenched.	328
$(NH_4)_3P_3O_9$	Crystallized from an aqueous solution of NH_3 and trimetaphosphoric acid.	353
$Ni_3(P_3O_9)_2$	$3Ni(NH_4)_2P_2O_7 \xrightarrow{660°} Ni_3(P_3O_9)_2 + 3H_2O(g) + 6NH_3(g)$.	164
$Ni_3(P_3O_9)_2 \cdot xH_2O$	$Na_3P_3O_9$ solution + $NiSO_4$ solution. Only evidence given for the identity of the product was the preparation method and the IR spectrum. It would seem that the product could also be a Ni–Na double salt.	164
$Rb_3P_3O_9$	Heated RbH_2PO_4 in the presence of HCl, NH_4NO_3, other alkali metal ions, or acetic anhydride.	4,18,33
$Sr_3(P_3O_9)_2$	Formed as a minor constituent (~10%) during heating of $Sr(H_2PO_4)_2$ to 400–500°.	31

In the thermal preparation of phosphates of the metaphosphate composition, the formation of the trimetaphosphate anion seems to be enhanced by acidic conditions such as the presence of HCl[33] or acetic anhydride.[18] Under such conditions the trimetaphosphates of K, Rb, and Cs are formed, although they are not formed by simple thermal dehydration of the corresponding acid orthophosphates.[18]

In the precipitation of insoluble trimetaphosphates from aqueous solutions there is a great tendency to form more soluble double salts with ions such as sodium, and this property has been used to separate trimetaphosphate from other phosphate species.[346]

C. Tetrametaphosphates

The preparation of tetrametaphosphates is summarized in Table IX.

TABLE IX
Preparation of Tetrametaphosphates

Compounds	Preparation methods	Refs.
$Al_4(P_4O_{12})_3$	Heated other forms of $Al(PO_3)_3$ to 900–1000°, or heated $Al(H_2PO_4)_3$.	31,41, 363, 364
$Ba_2P_4O_{12} \cdot 4H_2O$	Added acetone to an aqueous solution of $Ba(ClO_4)_2$ and tetrametaphosphoric acid.	31
$Cd_2P_4O_{12}$	Heated $Cd(H_2PO_4)_2 \cdot 2H_2O$ to 300–500°.	31,365
$Co_2P_4O_{12}$	Heated $Co(H_2PO_4)_2 \cdot 2H_2O$, $[Co(NH_3)_6(PO_3)_3 \cdot nH_2O]_x$, or a mixture of $CoCO_3$ or CoO with a slight excess of H_3PO_4 to 400–500°.	31,57, 365
$Cr_4(P_4O_{12})_3$	Crystallized from a solution of $CrCl_3$ in H_3PO_4 heated to 400–800°, or heated $Cr(H_2PO_4)_3$ to 400–500°.	199,299
$Cu_2P_4O_{12}$	Heated $Cu(H_2PO_4)_2$, $[Cu(NH_3)_4(PO_3)_2 \cdot nH_2O]_x$, or a mixture of $CuCO_3$ or CuO with a slight excess of H_3PO_4 to 400–540°.	31,57, 361, 365
$Fe_2P_4O_{12}$	Heated $Fe(H_2PO_4)_2$ or a mixture of Fe and H_3PO_4 (under N_2 or CO_2) to 400–500°.	31,41
$Fe_4(P_4O_{12})_3$	Heated $Fe(H_2PO_4)_3$ to 400–500° or heated Fe_2O_3 in phosphoric acid with seeding by the isomorphous Al salt.	41,199, 299
$K_4P_4O_{12} \cdot 2H_2O$	Added ethanol to a solution prepared by reacting K_2S solution with $Cu_2P_4O_{12}$ and filtering.	353,366
$K_4P_4O_{12}$	Heated the hydrate to ~100°.	353
$K_2SrP_4O_{12}$	Calcined a mixture of $SrCO_3$, K_2CO_3, and excess $(NH_4)_2HPO_4$. Unit cell given.	367

(continued)

TABLE IX (*continued*)

Compounds	Preparation methods	Refs.
$Li_4P_4O_{12} \cdot 4H_2O$	Added ethanol to an aqueous solution prepared by reacting Li_2S solution with $Cu_2P_4O_{12}$ and filtering.	353,366
$Li_4P_4O_{12} \cdot 2H_2O$	Heated $Li_4P_4O_{12} \cdot 4H_2O$ to 60–100°.	353
$Li_4P_4O_{12}$	Heated a hydrate to 100–180°. Some decomposition to Li polyphosphates occurs.	353
$Mg_2P_4O_{12}$	Heated $Mg(H_2PO_4)_2$ to 400–500°.	31,365
$Mn_2P_4O_{12}$	Heated $Mn(H_2PO_4)_2$, or a mixture of $MnCO_3$ or MnO with a slight excess of H_3PO_4, to 400–500°.	31,365
$Na_2H_2P_4O_{12}$	Crystallized from a melt prepared from equal molar amounts of NaH_2PO_4 and H_3PO_4. Crystal structure done.	18,328, 362
$Na_4P_4O_{12} \cdot 10H_2O$	Added alcohol to a solution prepared by reacting Na_2S solution with $Cu_2P_4O_{12}$ and filtering, or added NaCl to a solution prepared by low-temperature hydrolysis of P_4O_{10}. Elution chromatography on an ion exchange resin has been used for purification.	35,368
$Na_4P_4O_{12} \cdot 4H_2O$ (monoclinic)	Added NaCl to an aqueous solution at 40–75°.	35,358, 359
$Na_4P_4O_{12} \cdot 4H_2O$ (triclinic)	Added NaCl to an aqueous solution at 75–100°.	35,358, 360
$Na_4P_4O_{12}$	Heated a hydrate to 75–200°.	35
$(NH_4)_4P_4O_{12}$	Reacted $(NH_4)_2S$ solution with $Cu_2P_4O_{12}$.	353,359, 366
$(NH_4)_2SrP_4O_{12}$	Calcined a mixture of $SrCO_3$, $(NH_4)_2CO_3$, and excess $(NH_4)_2HPO_4$. Unit cell given.	367
$Ni_2P_4O_{12}$	Heated $Ni(H_2PO_4)_2 \cdot 2H_2O$, $[Ni(NH_3)_6(PO_3)_2 \cdot nH_2O]_x$, or a mixture of NiO or $NiCO_3$ with H_3PO_4 to 400–1000°.	31,57, 164, 365
$Pb_2P_4O_{12} \cdot 3H_2O$	$Na_4P_4O_{12}$ solution + $Pb(NO_3)_2$ solution, crystalline.	31
$SrRb_2P_4O_{12}$	Calcined a mixture of $SrCO_3$, Rb_2CO_3, and excess $(NH_4)_2HPO_4$, Unit cell given.	367
$SrTl_2P_4O_{12}$	Calcined a mixture of $SrCO_3$, Tl_2CO_3, and excess $(NH_4)_2HPO_4$. Unit cell given.	367
$Ti_4(P_4O_{12})_3$	Heated $Ti(H_2PO_4)_3$ to 400–500°, or heated a mixture of Ti and H_3PO_4 with exclusion of oxygen.	199,299
$Zn_2P_4O_{12}$	Heated $Zn(H_2PO_4)_2$ to 400–500°, or heated ZnO or $ZnCO_3$ with a slight excess of H_3PO_4.	31,365

Infrared spectra were given for the following compounds without giving the preparation methods: $Ag_4P_4O_{12}$,[178] $Ag_4P_4O_{12} \cdot 2H_2O$,[64] $Cu_2P_4O_{12} \cdot 8H_2O$,[64] $Mn_2P_4O_{12} \cdot 9H_2O$,[64] $Pb_2P_4O_{12} \cdot 4H_2O$,[64] and $Sr_2P_4O_{12} \cdot 5H_2O$.[64]

It has been postulated[5,358] that two different conformations of the tetrametaphosphate anion exist which are not easily interconverted and that in the case of certain dimorphs, one form contains one type of anion and the other form contains the other type of anion. X-ray structure determinations of the monoclinic and triclinic forms of $Na_4P_4O_{12} \cdot 4H_2O$[359,360] failed to support this theory, which could still apply in other cases, however.

$Pb_2P_4O_{12}$[219] and $Ba_2P_4O_{12}$[219,361] had been claimed to be formed thermally but later studies[31] showed them to be present in only minor amounts as already discussed in Section VII-B.

The rather recent thermal preparation of $Na_2H_2P_4O_{12}$[18,328,362] is particularly significant because it offers the most direct route to those tetrametaphosphates which are precipitated by mixing a solution of a soluble tetrametaphosphate with a solution of a soluble salt of a multiply charged or heavy metal. A number of such preparations were described in the older literature but are not discussed here because their identification and purity are often doubtful.[5] This area needs reinvestigation with modern tools and techniques.

D. Higher Metaphosphates

The presence of large ring metaphosphates in Graham's salt was noted by Van Wazer and Karl-Kroupa through careful two-dimensional paper chromatographic separation.[74] The successful isolation of sodium pentametaphosphate tetrahydrate and sodium hexametaphosphate hexahydrate from Graham's salt by fractional precipitation with acetone and hexammine cobalt(III) chloride was reported recently.[46] In addition to the derivatives listed in Table X, precipitates were obtained with $Co(NH_3)_6Cl_3$, but their analysis was not given.[46]

Hexametaphosphate was also discovered as a minor constituent (5–20%) in the product obtained by calcining a mixture of Li_2CO_3 and H_3PO_4 with a molar ratio of $Li/P = 7/5$ at 275–300° for 5–6 hr.[50] Since the solution of this product contains only pyro- and hexametaphosphate, preparation of $Na_6P_6O_{18} \cdot 6H_2O$ can easily be accomplished by using an ion-exchange resin to convert the Li salts to the acids, neutralizing to the Na salts, and employing fractional precipitation to isolate the sodium hexametaphosphate.

TABLE X
Preparation of Higher Metaphosphates

Compounds	Preparation methods	Refs.
$Ag_5P_5O_{15} \cdot 2.6H_2O$	$AgNO_3$ solution + $Na_5P_5O_{15}$ solution.	46
$Ag_6P_6O_{18} \cdot 2.7H_2O$	$AgNO_3$ solution + $Na_6P_6O_{18}$ solution.	46
$Ba_5(P_5O_{15})_2 \cdot 10H_2O$	$BaCl_2$ solution + $Na_5P_5O_{15}$ solution.	46
$Ba_6(P_6O_{18})_2 \cdot 9.1H_2O$	$BaCl_2$ solution + $Na_6P_6O_{18}$ solution.	46
$K_6P_6O_{18}$	Added alcohol to a solution prepared from K_2CO_3 and $H_6P_6O_{18}$.	50
$Li_6P_6O_{18} \cdot 6H_2O$	Li_2CO_3 + $H_6P_6O_{18}$ solution prepared from $Na_6P_6O_{18}$ by ion exchange.	50
$Li_6P_6O_{18}$	Heated the hydrate to 120–550°.	50
$Na_5P_5O_{15} \cdot 4H_2O$	Prepared by fractionation of Graham's salt.	46
$Na_5P_5O_{15} \cdot 2H_2O$	Dehydrated the tetrahydrate over P_4O_{10} at 30° under vacuum.	46
$Na_5P_5O_{15} \cdot 0.6H_2O$	Dehydrated the dihydrate over P_4O_{10} at 100° under vacuum.	46
$Na_6P_6O_{18} \cdot 6H_2O$	Prepared by fractionation of Graham's salt or a special Li phosphate mixture. Crystal structure done.	46,50, 369
$Na_6P_6O_{18}$	Heated the hydrate to 120°.	50

IX. BRANCHED PHOSPHATES

Branched phosphates, in which at least one PO_4 group contains three oxygen atoms shared with other PO_4 groups, are found mostly in amorphous acids and salts of high P_2O_5 content which are outside the scope of this chapter because of their indefinite composition. However, the following salts of definite composition have been prepared. CaP_4O_{11} and $Ca_2P_6O_{17}$ are equilibrium phases in the $CaO-P_2O_5$ system.[267] $PbSr_2(P_4O_{11})_3$ was prepared by reaction of SrO and PbO with molten sodium trimetaphosphate.[156]

X. CONDENSED PHOSPHATES OF UNCERTAIN ANION TYPE

The compounds listed in Table XI all have the metaphosphate composition, but it was not determined whether they are long-chain polyphosphates or metaphosphates. For the following compositions either no structural formula was given at all or the formula given was doubtful.

Aluminum pyrophosphate and $2P_2O_5 \cdot Al_2O_3$ (equivalent in composition to the tetrapolyphosphate) were claimed to be formed by dehydration

TABLE XI

Preparations with Metaphosphate Composition but Uncertain Anion Type

Compounds	Preparation methods	Refs.
$Al(PO_3)_3$-G	Heated Al_2O_3 in phosphoric acid at 400–600° and separated the product from tetrametaphosphate formed simultaneously. XRD pattern. IR indicates a ring structure.	173a
$CaK(PO_3)_3$ (rhombohedral)	Calcined a mixture of $(NH_4)_2HPO_4$, $CaCO_3$, and KH_2PO_4. Although the high-temperature form is a trimetaphosphate, the anion type in the low-temperature form (rhombohedral) was not specified.	349
$Cd(PO_3)_2$ (α and β)	The tetrametaphosphate which can be prepared at 300–410° converts to the α form at $\sim$550° which in turn converts to the β form at $\sim$810°. α is "probably" a long-chain polyphosphate. No information was given on the anion type in the β form.	31, 182, 192, 310
$CdK(PO_3)_3$ (rhombohedral)	Calcined a mixture of $(NH_4)_2HPO_4$, $CdCO_3$, and KH_2PO_4. Although the high-temperature form is a trimetaphosphate, the anion type in the low-temperature form (rhombohedral) was not specified.	349
$(Cd,Mg,Zn)(PO_3)_2$	There is a large region of solid solution in the ternary system $Cd(PO_3)_2$–$Mg(PO_3)_2$–$Zn(PO_3)_2$ in which the crystals have distorted $Mg(PO_3)_2$ structures.	310
$(Cd,Zn)(PO_3)_2$	In the system $Cd(PO_3)_2$–$Zn(PO_3)_2$, there are five different types of solid solution with structures similar to α- or β-$Cd(PO_3)_2$, α- or β-$Zn(PO_3)_2$, or $Mg(PO_3)_2$.	310
$CoK(PO_3)_3$ (rhombohedral)	Calcined a mixture of $(NH_4)_2HPO_4$, $CoCO_3$, and KH_2PO_4. Although the high-temperature form is a trimetaphosphate, the anion type in the low-temperature (rhombohedral) form was not specified.	349, 377
$CsLi(PO_3)_2$	There are two thermally prepared forms with a transition temperature near 410°. No information on anion type.	312
$In(PO_3)_3$	Heated a mixture of $InPO_4$ and H_3PO_4, cooled, and washed with H_2O. Did not specify whether the product was crystalline or what type of anion was present.	169

(*continued*)

TABLE XI (*continued*)

Compounds	Preparation methods	Refs.
$KMg(PO_3)_3$ (rhombohedral)	Calcined a mixture of $(NH_4)_2HPO_4$, $MgCO_3$, and KH_2PO_4. Although the high-temperature form is a trimetaphosphate, the anion type in the low-temperature (rhombohedral) form was not specified.	349, 377
$KMn(PO_3)_3$ (rhombohedral)	Calcined a mixture of $(NH_4)_2HPO_4$, $MnCO_3$, and KH_2PO_4. Although the high-temperature form is a trimetaphosphate, the anion type in the low-temperature (rhombohedral) form was not specified.	349
$KNi(PO_3)_3$	Crystallizes with a rhombohedral unit cell. Preparation method was not described in the abstract but is probably similar to that for $CoK(PO_3)_3$.	377
$KZn(PO_3)_3$ (rhombohedral)	Calcined a mixture of $(NH_4)_2HPO_4$, $ZnCO_3$, and KH_2PO_4. Although the high-temperature form is a trimetaphosphate, the anion type in the low-temperature (rhombohedral) form was not specified.	349, 377
$(Mg,Zn)(PO_3)_2$	Heated mixtures of ZnO, $MgCO_3$, and $(NH_4)_2HPO_4$. Solid solutions were found with structures similar to $Mg(PO_3)_2$, α-$Zn(PO_3)_2$, and β-$Zn(PO_3)_2$.	332
$Mo(PO_3)_3$	Reacted Mo with metaphosphoric acid at high-temperature and pressure. It is probably a long-chain polyphosphate since it is isotypic with $[Ti(PO_3)_3]_x$.	299, 378
$NH_4PO_3 \cdot 6H_2O$	Reacted $(NH_4)_2S$ solution with $Cu(PO_3)_2$. Anion type not given.	379
$Pu(PO_3)_4$	Orthorhombic crystals were grown from a solution of PuO_2 in metaphosphoric acid. No information on anion type.	380
$Si(PO_3)_4 \cdot 2H_2O$	$(CH_3)_2Si(OC_2H_5)_2 + P_2O_5$, crystalline, structural formula unknown.	55
$Sr_3(PO_3)_6$	The product crystallized from aqueous solution and was referred to as a hexametaphosphate but it seems doubtful that it really was the hexametaphosphate.	381
$Sr(PO_3)_2$ (β and γ)	Heated $Sr(H_2PO_4)_2$ to ~325° to get γ form which converts to β form at ~400–420°. At still higher temperatures, the α form is produced which is a long-chain polyphosphate. The evidence regarding the type of anions in the β and γ forms is inconclusive.	31, 137, 182, 382

(*continued*)

TABLE XI (continued)

Compounds	Preparation methods	Refs.
Th(PO$_3$)$_4$ and (Th,U)(PO$_3$)$_4$	Two forms, orthorhombic and monoclinic, occur in the ThO$_2$–P$_2$O$_5$ system with a transition temperature near 750°. Both phases form solid solutions with the isomorphous uranium compounds. Anion types were not discussed.	6,23
β-U(PO$_3$)$_4$	Reacted UO$_2$ with excess phosphoric acid at ≲500°. This phase is orthorhombic and is believed to be metastable. β → α transition is irreversible and occurs at ∼970°. Anion type was not discussed.	23,380
α-U(PO$_3$)$_4$	Reacted UO$_2$ with excess phosphoric acid at 1000°. This phase is monoclinic and not triclinic as had been reported earlier. Anion type was not discussed.	23
α-Zn(PO$_3$)$_2$	Said to be the stable form below ∼700° but cannot be prepared by tempering the β form below the inversion temperature or by devitrifying a glass. α form appears to be a mixture of the tetrametaphosphate and one or more other phases based upon the XRD patterns.	31,332, 333

of acid phosphates, but the amounts present and the analytical methods were not specified.[179] In later work on the same area, these compounds weren't mentioned.[180]

BaO·TiO$_2$·P$_2$O$_5$ was prepared by calcining a mixture of TiO$_2$, BaHPO$_4$, and (NH$_4$)$_2$HPO$_4$ and underwent a rapidly reversible phase transition at 967 ± 5°.[304] The structural formula could be either Ba(TiO)P$_2$O$_7$ or BaTi(PO$_4$)$_2$. 4KPO$_3$·V$_2$O$_5$ was prepared by heating a mixture of KPO$_3$ and V$_2$O$_5$ and was characterized by XRD.[370] The structural formula may be K$_4$(VO$_2$)$_2$P$_4$O$_{13}$. 2Ta$_2$O$_5$·P$_2$O$_5$ was prepared by roasting at 1000° a Ta phosphate precipitated from an acidic solution and was characterized by XRD.[371] The structural formula could be (TaO$_2$)$_4$·P$_2$O$_7$. 5TiO$_2$·2P$_2$O$_5$ [232] and 5PbO·2P$_2$O$_5$ [218] were also prepared thermally and characterized by XRD but are less likely to be pure compounds because simple structural formulas cannot be written.

XRD showed that a new compound was formed from a melt of Nb$_2$O$_5$ and NaPO$_3$, and no orthophosphate was found in the product. The new compound was not isolated and analyzed.[372] Nb$_2$O$_5$·P$_2$O$_5$ and

$2Nb_2O_5 \cdot P_2O_5$ were prepared thermally[373,374]; and the former, by analogy with the uranium salt, could be $Nb_2O_3P_2O_7$ while the latter could possibly be $(NbO_2)_4P_2O_7$.

The formation of $Fe_2P_4O_{13}$ in the system Fe_2O_3–H_2O–P_2O_5 at temperatures $\leq 80°$ was reported[375,376] but seems unlikely. Only chemical analyses were given, and in one case the analysis was just as consistent with $Fe_2H_6(PO_4)_4$.

References

1. C. Shen and C. Callis, "Orthophosphoric Acids and Orthophosphates," in *Preparative Inorganic Reactions*, Vol. 2, W. Jolly, Ed., Interscience, New York, 1965.
2. D. Corbridge, "The Structural Chemistry of Phosphorus Compounds," in *Topics in Phosphorus Chemistry*, Vol. 3, E. Griffith and M. Grayson, Eds., Interscience, New York, 1966.
3. E. Thilo, *Angew. Chem. Intern. Ed. Engl.*, **4**, 1065 (1965).
4. E. Thilo, in *Advan. Inorg. Chem. Radiochem.*, **4**, 1–66 (1962).
5. J. R. Van Wazer, *Phosphorus and Its Compounds*, Vol. 1, Interscience, New York, 1958.
6. A. Burdese and M. Borlera, *Ann. Chim. (Rome)*, **53**, 344 (1963).
7. T. Clark, *Edinburgh J. Sci.*, **7**, 298 (1827); *Ann. Chim. Phys.* (2), **41**, 276 (1829).
8. T. Graham, *Phil. Trans. Roy. Soc. London, Ser. A*, **123**, 253 (1833).
9. S. Kiehl and G. Wallace, *J. Am. Chem. Soc.*, **49**, 376 (1927).
10. J. McGilvery and A. Scott, *Can. J. Chem.*, **32**, 1100 (1954).
11. E. Kowalska, W. Kowalski, and A. Truszkowski, *Prezemysel. Chem.*, **42**, 212 (1963).
12. C. Shen, unpublished data.
13. J. Edwards and A. Herzog, *J. Am. Chem. Soc.*, **79**, 3647 (1957).
14. W. Garner, *Chemistry of the Solid State*, Butterworths, London, 1955.
14a. O. Pfrengle, U.S. Patent Reissue No. 24, 381, October 29, 1957.
15. R. Hisar, *Bull. Soc. Chim. France*, **1953**, 47–51.
16. S. Ueda, K. Oyama, and K. Koma, *Kogyo Kagaku Zasshi*, **66**, 586 (1963).
17. C. Schmulbach, J. Van Wazer, and R. Irani, *J. Am. Chem. Soc.*, **81**, 6347 (1959).
18. F. Kasparek, *Monatsh. Chem.*, **92**, 1023 (1961).
19. E. Griffith, U.S. Patent No. 2,839,408, June 17, 1958.
20. J. Korinth and P. Royen, *Z. Anorg. Allgem. Chem.*, **313**, 121 (1961).
21. E. Thilo and D. Schultze, *Chem. Ber.*, **93**, 2430 (1960).
22. J. Etienne, A. de Sallier Dupin, and A. Boulle, *Compt. Rend.*, **256**, 173 (1963).
23. Y. Baskin, *J. Inorg. Nucl. Chem.*, **29**, 383 (1967).
24. C. Shen, U.S. Patent No. 3,314,750, April 18, 1967.
25. F. Kerschbaum, U.S. Patent No. 2,142,944, January 3, 1939.
26. J. Alexander, U.S. Patent No. 2,792,284-285, May 14, 1957.
27. A. Lobdell, U.S. Patent Reissue No. 25,455, October 8, 1963.
28. G. Hartlapp, U.S. Patent No. 3,087,783, April 30, 1963.
29. B. Raistrick and J. Raitt, U.S. Patent No. 3,049,419, August 14, 1962.

30. W. Holmes, J. Jeffes, J. McCoubrey, and R. Tomlinson, *Trans. Faraday Soc.*, **60**, 1958 (1964).
31. E. Thilo and I. Grunze, *Z. Anorg. Allgem. Chem.*, **290**, 209, 223 (1957).
32. M. Markowitz, H. Stewart, and D. Boryta, *Inorg. Chem.*, **2**, 768 (1963).
33. M. Porthault and J. Merlin, *Compt. Rend.*, **250**, 1067 (1960).
34. N. Dombrovskii, *Ukr. Khim. Zh.*, **26**, 555 (1960).
35. R. Bell, L. Audrieth, and O. Hill, *Ind. Eng. Chem.*, **44**, 570 (1952).
36. D. Dyroff, Ph.D. thesis, California Institute of Technology, 1965.
37. C. Shen, *Ind. Eng. Chem. Prod. Res. Develop.*, **5**, 272 (1966).
38. T. Fleitmann and W. Henneberg, *Ann. Chem.*, **45**, 304, 387 (1845).
39. A. Schwarzenberg, *Ann. Chem.*, **65**, 133 (1848).
40. F. Filinov and V. Budanova, *Zh. Neorgan. Khim.*, **1**, 1915 (1956).
41. F. d'Yvoire, *Bull. Soc. Chim. France*, **1962**, 1224, 1237, and 1243.
42. R. Guzairov, V. Leitsin, and S. Grekov, *Zh. Neorgan. Khim.*, **9**, 20 (1964).
43. G. Fujioka and G. Cady, *J. Am. Chem. Soc.*, **79**, 2451 (1957).
44. E. Griffith, *J. Inorg. Nucl. Chem.*, **26**, 1381 (1964).
45. C. Sclar, L. Carrison, and C. Schwartz, *Nature*, **204** (1958), 573 (1964).
46. E. Thilo and U. Schuelke, *Z. Anorg. Allgem. Chem.*, **341**, 293 (1965).
47. H. Rothbart, H. Weymouth, and W. Rieman III, *Talanta*, **11**, 33 (1964).
48. M. Miura and Y. Moriguchi, *Bull. Chem. Soc. (Japan)*, **37**, 1522 (1964); **38**, 678 (1965).
49. C. Shen, *Ind. Eng. Chem. Process Design Develop.*, **6**, 414 (1967).
50. E. Griffith and R. Buxton, *J. Am. Chem. Soc.*, **89**, 2884 (1967).
51. W. Wieker, *Z. Elektrochem.*, **64**, 1047 (1960).
52. S. Otani, M. Miura, and T. Doi, *Kogyo Kagaku Zasshi*, **66**, 593 (1963).
53. O. Quimby, *J. Phys. Chem.*, **58**, 603 (1954).
54. A. Todd, *J. Chem. Soc.*, **1946**, 627; **1947**, 648.
55. A. Kreshkov and D. Karateev, *Zh. Prikl. Khim.*, **32**, 369 (1959).
56. M. Goehring and J. Sambeth, *Chem. Ber.*, **90**, 232 (1957).
57. R. Klement and R. Popp, *Chem. Ber.*, **93**, 156 (1960).
58. M. Bobtelsky, *Heterometry*, Van Nostrand, New York, 1960.
59. V. Leitsin and S. Grekov, *Russ. J. Inorg. Chem. (Engl. Transl)*, **11**, 1133 (1966).
60. R. Klement, in *Handbuch der Analytischen Chemie*, W. Fresenius and G. Jander, Eds., Springer, Berlin, 1953.
61. O. Quimby, *Chem. Rev.*, **40**, 141 (1947); see also refs. 101, 133, and 134.
62. D. Corbridge and F. Thomas, *Anal. Chem.*, **30**, 1101 (1958).
63. American Society for Testing and Materials, Philadelphia, X-ray powder data file, 1957.
64. D. Corbridge and E. Lowe, *J. Chem. Soc.*, **1954**, 493.
65. D. Corbridge, *J. Appl. Chem. (London)*, **6**, 456 (1956).
66. A. Chapman and L. Thirlwell, *Spectrochim. Acta*, **20**, 937 (1964).
67. M. Crutchfield, C. Dungan, J. Letcher, V. Mark, and J. R. Van Wazer, in *Topics in Phosphorus Chemistry*, Vol. 5, E. Griffith and M. Grayson, Eds., Interscience, New York, 1967.
68. J. Van Wazer, E. Griffith, and J. McCullough, *Anal. Chem.*, **26**, 1755 (1954); E. Griffith, *Anal. Chem.*, **28**, 525 (1956).
69. U. Strauss, E. Smith, and P. Wineman, *J. Am. Chem. Soc.*, **75**, 3935 (1953).
70. U. Strauss and T. Treitler, *J. Am. Chem. Soc.*, **77**, 1743 (1955); *J. Am. Chem. Soc.*, **78**, 3553 (1956).
71. J. Van Wazer, *J. Am. Chem. Soc.*, **72**, 647 (1950).

72. K. Karbe and G. Jander, *Kolloid-Beihefte*, **54**, 1 (1942); *Chem. Abstr.*, **36**, 5438 (1942).
73. J. Ebel, *Mikrochim. Acta*, **1954**, 679; *Compt. Rend.*, **233**, 415 (1951); *Compt. Rend.*, **234**, 621 (1952); *Bull. Soc. Chim. France*, **20**, 991 (1953).
74. E. Karl-Kroupa, *Anal. Chem.*, **28**, 1091 (1956); J. Van Wazer and E. Karl-Kroupa, *J. Am. Chem. Soc.*, **78**, 1772 (1956).
75. H. Grunze and E. Thilo, *Die Papierchromatographie der Kondensierten Phosphate*, Akademie-Verlag, Berlin, 1955.
76. G. Biberacher, *Z. Anorg. Allgem. Chem.*, **285**, 86 (1956).
77. O. Pfrengle, *Z. Anal. Chem.*, **158**, 81 (1957).
78. E. Thilo and W. Feldmann, *Z. Anorg. Allgem. Chem.*, **298**, 316 (1959).
79. S. Ohashi and J. Van Wazer, *Anal. Chem.*, **35**, 1984 (1963).
80. B. Sansoni and R. Klement, *Angew. Chem.*, **65**, 422 (1953); **66**, 598 (1954).
81. J. Beukenkamp, W. Rieman III, and S. Lindenbaum, *Anal. Chem.*, **26**, 505 (1954).
82. J. Grande and J. Beukenkamp, *Anal. Chem.*, **28**, 1497 (1956).
83. J. Ebel and N. Busch, *Compt. Rend.*, **242**, 647 (1956).
84. R. Kolloff, *ASTM (Am. Soc. Testing Materials) Bull.*, No. 237, 74 (1959).
85. D. Lundgren and N. Loeb, *Anal. Chem.*, **33**, 366 (1961).
86. T. Rossel, *Z. Anal. Chem.*, **196**, 6 (1963); **197**, 333 (1963).
87. M. Bundler and M. Mengel, *Z. Anal. Chem.*, **206**, 8 (1964).
88. M. Bundler and F. Stuhlmann, *Naturwiss.*, **51**, 57 (1964).
89. S. Ohashi, N. Yoza, and Y. Ueno, *J. Chromatog.*, **24**, 300 (1966).
90. H. Giran, *Compt. Rend.*, **134**, 1499 (1902); *Ann. Chim. Phys. Series 7*, **30**, 203 (1903).
91. Z. Wakefield and E. Egan, Jr., *J. Agr. Food Chem.*, **10**, 344 (1962).
92. J. Malowan, in *Inorganic Synthesis*, Vol. 3, L. Audrieth, Ed., McGraw-Hill, New York, 1950, pp. 96–98.
93. C. Shen, J. Lyons, and G. Nelson, unpublished data.
94. E. Thilo and R. Sauer, *J. Prakt. Chem.*, **4**, 325 (1957).
95. Monsanto Co., Technical Bulletin I-210, Phospholeum, St. Louis, Missouri.
96. C. Bunton and H. Chaimovich, *Inorg. Chem.*, **4**, 1763 (1965).
97. A. Nelson, *J. Chem. Eng. Data*, **9**, 357 (1964).
98. C. Swanson and F. McCollough, in *Inorg. Synthesis*, Vol. 7, J. Kleinberg, Ed., McGraw-Hill, New York, 1963, pp. 65–67.
99. T. Farr and J. Fleming, *J. Chem. Eng. Data*, **10**, 20 (1965).
100. R. Coates and G. Woodward, *J. Chem. Soc.*, **1964**, 1780.
101. A. Frazier, J. Smith, and J. Lehr, *J. Agr. Food Chem.*, **13**, 316 (1965).
102. Y. Arai and S. Nagai, *Kogyo Kagaku Zasshi*, **69**, 2077 (1966).
103. E. Thilo and H. Grunze, *Z. Anorg. Allgem. Chem.*, **281**, 262 (1955).
104. P. Sears and H. Vandersall, Belgian Patent No. 651,782, February 15, 1965, and unreported data.
105. E. Kobayashi, *Kogyo Kagaku Zasshi*, **69**, 2065 (1966).
106. R. Osterheld and M. Markowitz, *J. Phys. Chem.*, **60**, 863 (1956).
107. M. Markowitz, R. Harris, and W. Hawley, *J. Inorg. Nucl. Chem.*, **22**, 293 (1961).
108. E. Partridge, V. Hicks, and G. Smith, *J. Am. Chem. Soc.*, **63**, 454 (1941).
109. G. Morey, *J. Am. Chem. Soc.*, **75**, 5794 (1953).
110. G. Morey, *J. Am. Chem. Soc.*, **76**, 4724 (1954).
111. I. Belyaev and N. Sigida, *Russ. J. Inorg. Chem. (Engl. Transl.)*, **3**, 440 (1958).

112. V. Goryacheva, A. Bergman, and A. Kislova, *Russ. J. Inorg. Chem.* (*Engl. Transl.*), **4**, 1269 (1959).
113. A. Bergman and V. Goryacheva, *Russ. J. Inorg. Chem.* (*Engl. Transl.*), **7**, 1267 (1962).
114. C. McCullogh, U.S. Pat. No. 2,021,012, 1932.
115. J. West, *Z. Krist.*, **74**, 306 (1930).
116. D. Corbridge, *Acta Cryst.*, **10**, 85 (1957).
117. N. Dombrovskii, *Russ. J. Inorg. Chem.* (*Engl. Transl.*), **7**, 700 (1962).
118. C. Morin, *Bull. Soc. Chim. France*, **1961**, 1717–1740.
119. F. Hubbard, *Ind. Eng. Chem.*, **41**, 2908 (1948).
120. E. Thilo and R. Rätz, *Z. Anorg. Chem.*, **258**, 33 (1949).
121. P. Silber and A. Norbert, *Compt. Rend.*, **256**, 4023 (1963).
122. J. Verdier, C. Morin, and A. Boulle, *Compt. Rend.*, **257**, 917 (1963).
123. R. Osterheld and L. Audrieth, *J. Phys. Chem.*, **56**, 38 (1952).
124. C. Durgin and R. Foster, unpublished data, Monsanto Co., St. Louis, Missouri.
125. E. Kobayashi, *Tokyo Kogyo Shikensho Hokoku*, **59**, 26 (1964).
126. L. Watson and J. Metcalf, *J. Chem. Eng. Data*, **6**, 331 (1961).
127. D. Balareff, *Z. Anorg. Allgem. Chem.*, **118**, 123 (1921).
128. A. McKeag and E. Steward, *Brit. J. Appl. Phys.*, Suppl. No. 4, **6**, S26–S31 (1954).
129. H. Jenkins, A. McKeag, and P. Ranby, *J. Electrochem. Soc.*, **96**, 1 (1949).
130. R. Mooney and M. Aia, *Chem. Rev.*, **61**, 433 (1961).
131. E. Brown, J. Lehr, J. Smith, W. Brown, and A. Frazier, *J. Phys. Chem.*, **61**, 1669 (1957); **62**, 366, 625 (1958).
132. M. Siegel, J. Getsinger, and H. Mann, *J. Agr. Food Chem.*, **10**, 72 (1962).
133. E. Brown, J. Lehr, J. Smith, and A. Frazier, *J. Agr. Food Chem.*, **11**, 214 (1963).
134. E. Brown, J. Lehr, A. Frazier, and J. Smith, *J. Agr. Food Chem.*, **12**, 70 (1964).
135. M. Bobtelsky and S. Kertes, *J. Appl. Chem.* (*London*), **4**, 419 (1954).
136. A. McIntosh and W. Jablonski, *Anal. Chem.*, **28**, 1424 (1956).
137. R. Ropp, M. Aia, C. Hoffman, T. Veleker, and R. Mooney, *Anal. Chem.*, **31**, 1163 (1959); *J. Am. Chem. Soc.*, **81**, 826 (1959).
138. R. Klement, *Chem. Ber.*, **93**, 2314 (1960).
139. V. Osipov, *J. Gen. Chem. USSR* (*Engl. Transl.*), **6**, 933 (1936).
140. H. Basset, W. Bedwell, and J. Hutchinson, *J. Chem. Soc.*, **1936**, 1412.
141. H. Baer, *Ann. Physik.*, **75**, 152 (1848).
142. R. Roy, E. Middlesworth, and F. Hummel, *Am. Mineralogist*, **33**, 458 (1948).
143. C. Calvo, *Can. J. Chem.*, **43**, 1139 (1965).
144. M. Dubost, *Bull. Soc. Chim. France*, **1959**, 810.
145. W. Hill, D. Reynolds, S. Hendricks, and K. Jacob, *J. Assoc. Offic, Agr. Chemists*, **28**, 105 (1945).
146. H. Bassett, *J. Chem. Soc.*, **1958**, 2949.
147. R. Mesmer and R. Irani, *J. Chem. Eng. Data*, **8**, 530 (1963).
148. J. Fischer and G. Kraft, *Z. Anorg. Allgem. Chem.*, **334**, 15 (1964).
149. P. Ranby, D. Mash, and S. Henderson, *Brit. J. Appl. Phys.*, Suppl. No. 4, **6**, S18 (1954) (Pub. 1955).
150. T. Dupuis and C. Duval, *Anal. Chim. Acta*, **4**, 256 (1950).
151. R. Langguth, R. Osterheld, and E. Karl-Kroupa, *J. Phys. Chem.*, **60**, 1335 (1956).
152. A. Rosenheim, S. Frommer, and W. Handler, *Z. Anorg. Allgem. Chem.*, **153**, 126 (1926).

153. V. Aleksanyan, *Proc. Acad. Sci. USSR, Chem. Sect. (Engl. Transl.)*, **115**, 459 (1957).
154. S. Haider, *Anal. Chim. Acta*, **24**, 250 (1961).
155. I. Tananaev and G. Shevchenko, *Inorg. Materials USSR (Engl. Transl.)*, **1**, 340 (1965).
156. R. Klement and E. Petz, *Montash. Chem.*, **95**, 1403 (1964).
157. J. Watters and A. Aaron, *J. Am. Chem. Soc.*, **75**, 611 (1953).
158. A. Sheka and E. Sinyavskaya, *Russ. J. Inorg. Chem. (Engl. Transl.)*, **11**, 555 (1966).
159. R. Klement, *Naturwiss.*, **49**, 512 (1962).
160. S. Beryl and N. Voskresenskaya, *Russ. J. Inorg. Chem. (Engl. Transl.)*, **10**, 601 (1965).
161. L. Eliazyan and V. Tarayan, *Izv. Akad. Nauk Arm. SSR, Khim. Nauk*, **14**, 127 (1961); *Chem. Abstr.*, **56**, 9694d (1962); V. Serebrennikov, *Chem. Abstr.*, **49**, 3713h (1955).
162. H. McCune and G. Arquette, *Anal. Chem.*, **27**, 401 (1955).
163. B. Bleyer and B. Müller, *Z. Anorg. Chem.*, **79**, 273; *Chem. Abstr.*, **7**, 1454 (1913).
164. M. Viltange, *Mikrochim. Ichnoanal. Acta*, **1964**, 1.
165. I. Tananaev and V. Vasil'eva, *Russ. J. Inorg. Chem. (Engl. Transl.)*, **9**, 1141 (1964); R. Baxter, *Dissertation Abstr.*, **23**, 4511 (1963).
166. H. Huber, W. Dewald, and H. Schmidt, *Angew. Chem.*, **70**, 133 (1958).
167. R. Duval and C. Duval, *Compt. Rend.*, **207**, 994 (1938).
168. K. Menzi and J. Wegmann, German Patent No. 969,699, July 10, 1958; *Chem. Abstr.*, **54**, 13678f (1960).
169. F. Enslin, H. Dreyer, and O. Lessman, *Z. Anorg. Chem.*, **254**, 315 (1947).
170. C. Dragulescu, J. Julean, and N. Vilceanu, *Rev. Roumaine Chim.*, **10**, 809 (1955); *Chem. Abstr.*, **64**, 9221d (1966).
171. E. Giesbrecht and M. Perrier, *Anais Assoc. Brasil. Quim.*, **19**, 121 (1960); *Chem. Abstr.*, **56**, 13747d (1962).
172. I. Tananaev and S. Petushkova, *Russ. J. Inorg. Chem. (Engl. Transl.)*, **9**, 601 (1964).
173. I. Tananaev, V. Kuznetsov, and S. Petushkova, *Russ. J. Inorg. Chem. (Engl. Transl.)*, **11**, 157 (1966).
173a. Ya. Klyucharov and L. Skoblo, *Proc. Acad. Sci. USSR, Chem. Sect. (Engl. Transl.)*, **154**, 90 (1964).
174. V. Spitsyn and E. Ippolitova, *Zh. Analit. Khim.*, **6**, 5 (1951); *Chem. Abstr.*, **45**, 4592i (1951).
175. A. Sharova and A. Shtin, *Izv. Sibirsk. Otd. Akad. Nauk SSSR*, **1959**, 40; *Chem. Abstr.*, **54**, 6377C (1960).
176. M. Bobtelsky and S. Kertes, *J. Appl. Chem. (London)*, **5**, 125 (1955).
177. D. Campbell and M. Kilpatrick, *J. Am. Chem. Soc.*, **76**, 893 (1954).
178. J. Pustinger, W. Cave, and M. Nielson, *Spectrochim. Acta*, **1959**, 909.
179. H. Guérin and R. Martin, *Compt. Rend.*, **234**, 1777 (1952).
180. F. d'Yvoire, *Bull. Soc. Chim. France*, **1961**, 2277.
181. R. Ropp, R. Mooney, and C. Hoffman, *Anal. Chem.*, **33**, 1687 (1961).
182. R. Ropp and M. Aia, *Anal. Chem.*, **34**, 1288 (1962).
183. I. Tananaev and S. Petushkova, *Russ. J. Inorg. Chem. (Engl. Transl.)*, **11**, 154 (1966).
184. M. Kohn, *Anal. Chim. Acta*, **9**, 229 (1953).

185. C. Dragulescu and P. Tribunescu, *Acad. Rep. Populare Romine Baza Cercetari Stiint. Timisoara, Studii Certecari, Stiinte Chim.*, **4**, 9 (1959); *Chem. Abstr.*, **53**, 14822e (1959).
186. T. Kobayashi, *Yogyo Kyokai Shi*, **71**, 201 (1963); *Chem. Abstr.*, **63**, 1452c (1965).
187. M. Hoffman, *J. Electrochem. Soc.*, **110**, 1223 (1963).
188. R. Ropp, *J. Electrochem. Soc.*, **109**, 15 (1962).
189. R. Ropp and R. Mooney, *J. Electrochem. Soc.*, **107**, 15 (1960).
190. A. Boulle and A. Dupin, *Compt. Rend.*, **253**, 2948 (1961); **254**, 122 (1962).
191. C. Dragulescu, D. Ceausescu, and D. Lazar-Jucu, *Acad. Rep. Populare Romine Baza Cercetari Stiint. Timisoara, Studii Cercetari, Stiinte Chim.*, **5**, 11 (1958); *Chem. Abstr.*, **54**, 16126b (1960).
192. J. Brown and F. Hummel, *J. Electrochem. Soc.*, **111**, 1052 (1964).
193. H. Voellenkle, A. Wittmann, and H. Nowotny, *Monatsh. Chem.*, **94**, 956 (1963).
194. V. Tarayan and L. Eliazyan, *Izv. Akad. Nauk Arm. SSR, Khim. Nauk*, **11**, 243 (1958); *Chem. Abstr.*, **53**, 6855c (1959).
195. G. Vasil'ev and V. Serebrennikov, *Russ. J. Inorg. Chem.* (*Engl. Transl.*), **9**, 874 (1964).
196. J. Etienne and A. Boulle, *Compt. Rend.*, **260**, 3977 (1965).
197. J. Sarver, *Trans. Brit. Ceram. Soc.*, **65**, 191 (1966).
198. T. Matsuo, *Japan Analyst*, **4**, 108 (1955); *Chem. Abstr.*, **50**, 5465c (1956).
199. P. Remy and A. Boulle, *Compt. Rend.*, **258**, 927 (1964).
200. K. Yatsimirskii and V. Vasil'ev, *J. Anal. Chem. USSR* (*Engl. Transl.*), **11**, 573 (1956).
201. B. Robertson and C. Calvo, *Acta Cryst.*, **22**, 665 (1967).
202. P. Royen and J. Korinth, *Z. Anorg. Allgem. Chem.*, **291**, 227 (1957).
203. J. Hutter, *Ann. Chim.* (*Paris*), **8**, 450 (1953).
204. V. Leitsin and S. Grekov, *Russ. J. Inorg. Chem.* (*Engl. Transl.*), **11**, 152 (1966).
205. B. Lelong, *Ann. Chim. Paris*, **9**, 229 (1964).
206. K. Avdnevskaya and I. Tananaev, *Russ. J. Inorg. Chem.* (*Engl. Transl.*), **10**, 197 (1965).
207. S. Mehta and N. Patel, *J. Univ. Bombay, Science No. 30*, **1951**, 82; *Chem. Abstr.*, **47**, 992h (1953).
208. E. Giesbrecht, *Anais Assoc. Brasil. Quim.*, **21**, 13 (1964); *Chem. Abstr.*, **63**, 1453h (1965).
209. I. Tananaev and V. Vasil'eva, *Russ. J. Inorg. Chem.* (*Engl. Transl.*), **9**, 1237 (1964).
210. J. Chambers, W. Datars, and C. Calvo, *J. Chem. Phys.*, **41**, 806 (1964).
211. K. Lukaszewicz and R. Smajkiewicz, *Roczniki Chem.*, **35**, 741 (1961); *Chem. Abstr.*, **55**, 21743e (1961).
212. M. Bobtelsky and S. Kertes, *J. Appl. Chem.* (*London*), **5**, 675 (1955).
213. C. MacLean and G. Kor, *Appl. Sci. Res. Sect. B*, **4**, 425 (1955); *Chem. Abstr.*, **50**, 6186h (1956).
214. P. Kierkegaard, *Arkiv Kemi*, **19**, 1 (1962).
215. I. Schulz, *Z. Anorg. Allgem. Chem.*, **281**, 99 (1955).
216. A. Buyers, E. Giesbrecht, and L. Audrieth, *J. Inorg. Nucl. Chem.*, **5**, 133 (1957).
217. M. Jacquinot, *Compt. Rend.*, **256**, 2816 (1963).
218. I. Argyle and F. Hummel, *J. Am. Ceram. Soc.*, **43**, 452 (1960).
219. R. Osterheld and R. Langguth, *J. Phys. Chem.*, **59**, 76 (1955).
220. C. Bjorklund, *J. Am. Chem. Soc.*, **79**, 6347 (1957).
221. B. Hajek, *Z. Chem.*, **3**, 194 (1963).

222. T. Tien and F. Hummel, *J. Am. Ceram. Soc.*, **45**, 422 (1962).
223. P. Kierkegaard, *Arkiv Kemi*, **19**, 51 (1962).
224. H. Makart, *Helv. Chim. Acta*, **50**, 399 (1967).
225. L. Chang, E. Boichinova, and O. Setkina, *Russ. J. Inorg. Chem. (Engl. Transl.)*, **9**, 798 (1964).
226. A. Popov and W. Wendlandt, *Proc. Iowa Acad. Sci.*, **60**, 300 (1953); *Chem. Abstr.*, **48**, 7471b (1954).
227. R. Mesmer and R. Irani, *J. Inorg. Nucl. Chem.*, **28**, 493 (1966).
228. R. Klement and H. Haselbeck, *Chem. Ber.*, **96**, 1022 (1963).
229. A. Burdese and M. Borlera, *Ric. Sci.*, **30**, 1343 (1960); *Chem. Abstr.*, **55**, 7978h (1961).
230. A. Burdese and M. Borlera, *Ann. Chim. (Rome)*, **53**, 333 (1963); *Chem. Abstr.*, **59**, 7034e (1963).
231. F. Filinov, E. Tekster, A. Kolpakova, and E. Panteleeva, *Russ. J. Inorg. Chem. (Engl. Transl.)*, **5**, 552 (1960).
232. D. Harrison and F. Hummel, *J. Am. Ceram. Soc.*, **42**, 487 (1959).
233. L. LeDonche and A. de LaFouchardiere, *Compt. Rend.*, **258**, 164 (1964).
234. A. Klygin, D. Zavrazhnova, and N. Nikol'skaya, *J. Anal. Chem. USSR (Engl. Transl.)*, **16**, 297 (1961).
235. H. Kirchner, K. Merz, and W. Brown, *J. Am. Ceram. Soc.*, **46**, 137 (1963).
236. E. Staritzky and D. Cromer, *Anal. Chem.*, **28**, 1211 (1956).
237. B. Nador, *Acta Chim. Acad. Sci. Hung.*, **40**, 1 (1964); *Chem. Abstr.*, **61**, 9075f (1964).
238. P. Kierkegaard, *Acta Chem. Scand.*, **12**, 1715 (1958).
239. C. Calvo, *Can. J. Chem.*, **43**, 1147 (1965).
240. O. Quimby and H. McCune, *Anal. Chem.*, **29**, 248 (1957).
241. D. Harrison, H. McKinstry, and F. Hummel, *J. Am. Ceram. Soc.*, **37**, 277 (1954).
242. J. Lecomte, A. Boulle, C. Morin, and B. Lelong, *Compt. Rend.*, **258**, 131 (1964).
243. E. Giesbrecht and G. Vicentini, *Anais Assoc. Brasil. Quim.*, **18**, 63 (1959); *Chem. Abstr.*, **54**, 8390d (1960).
244. V. Vesely and V. Pekarek, *J. Inorg. Nucl. Chem.*, **25**, 697 (1963).
245. A. Huhti and P. Gartaganis, *Can. J. Chem.*, **34**, 785 (1956).
246. C. Monk, *J. Chem. Soc.*, **1949**, 443 and 429.
247. T. Farr, J. Fleming, and J. Hatfield, *J. Chem. Eng. Data*, **12**, 141 (1967).
248. N. Dombrovskii, *Russ. J. Inorg. Chem. (Engl. Transl.)*, **7**, 47 (1962).
249. J. McGilvery, *Detergent Age*, Dec. 1964.
250. A. Ichikawa and A. Kato, *Mem. Fac. Eng. Kyushu Univ.*, **24**, No. 2 (1965).
251. D. Davies and D. Corbridge, *Acta Cryst.*, **11**, 315 (1958).
252. D. Corbridge, *Acta Cryst.*, **13**, 262 (1960).
253. J. Dymon and A. King, *Acta Cryst.*, **4**, 378 (1951).
254. G. Morey, *J. Am. Chem. Soc.*, **80**, 775 (1958).
255. P. Bonneman-Bemia, *Compt. Rend.*, **204**, 433 (1937); *Compt. Rend.*, **206**, 1379 (1938); *Ann. Chim. (Paris)*, **16**, 395 (1941).
256. C. Shen, J. Metcalf, and E. O'Grady, *Ind. Eng. Chem.*, **51**, 717 (1959).
257. G. Morey, F. Boyd, J. England, and W. Chen, *J. Am. Chem. Soc.*, **77**, 5003 (1955).
258. A. Boulle, M. Domine-Berges, and C. Morin, *Compt. Rend.*, **241**, 1772 (1955).
259. N. Terem and S. Akalan, *Compt. Rend.*, **228**, 1374 (1949).
260. P. Askenasy and F. Nessler, *Z. Anorg. Allgem. Chem.*, **189**, 307 (1930).

261. C. Shen and J. Metcalf, *Ind. Eng. Chem. Prod. Res. Develop.*, **4**, 107 (1965).
262. M. Crutchfield, C. Callis, and E. Kaelble, *Inorg. Chem.*, **1**, 389 (1962).
263. H. Huber, *Z. Anorg. Allgem. Chem.*, **230**, 123 (1936).
264. R. Klement, *Z. Anorg. Chem.*, **260**, 18 (1949).
265. A. Boulle and C. Morin, *Compt. Rend.*, **250**, 1013 (1960).
266. W. Dewald, *Angew. Chem.*, **67**, 654 (1955).
267. W. Hill, G. Faust, and D. Reynolds, *Am. J. Sci.*, **242**, 457, 542 (1944).
268. G. Trömel, H. Harkort, and W. Hotop, *Z. Anorg. Chem.*, **256**, 253 (1948).
269. P. Stone, E. Egan, and J. Lehr, *J. Am. Ceram. Soc.*, **39**, 89 (1956).
270. E. Kreidler and F. Hummel, *Inorg. Chem.*, **6**, 884 (1967).
271. S. Ohashi and J. Van Wazer, *J. Am. Chem. Soc.*, **81**, 830 (1959).
272. W. Hill, S. Hendricks, E. Fox, and J. Cady, *Ind. Eng. Chem.*, **39**, 1667 (1947).
273. E. Giesbrecht and E. Melardi, *Anais Acad. Brasil. Cienc.*, **36**, 275 (1964); *Chem. Abstr.*, **63**, 2604f (1965).
274. L. Prodan and N. Ermolenko, *Dokl. Akad. Nauk Belorussk SSR*, **5**, 442 (1961); *Chem. Abstr.*, **58**, 12007c (1963).
275. E. Giesbrecht and L. Audrieth, *J. Inorg. Nucl. Chem.*, **6**, 308 (1958).
276. E. Prodan, *Russ. J. Inorg. Chem. (Engl. Transl.)*, **3**, 209 (1958).
277. M. Kobayashi, *J. Chem. Soc. Japan, Pure Chem. Sect.*, **76**, 793 (1955); *Chem. Abstr.*, **50**, 13649c (1956).
278. E. Giesbrecht and E. Melardi, *Anais. Acad. Brasil. Cienc.*, **35**, 527 (1963); *Chem. Abstr.*, **61**, 5172f (1964).
279. E. Melardi and E. Giesbrecht, *Anais Acad. Brasil. Cienc.*, **37**, 221 (1965); *Chem. Abstr.*, **64**, 18933g (1966).
280. A. Indelli, *Ann. Chim. (Rome)*, **46**, 717 (1956); *Chem. Abstr.*, **51**, 7925c (1957).
281. R. Bell, U.S. Patent No. 2,852,341, September 16, 1958.
282. F. Schwarz, *Z. Anorg. Chem.*, **9**, 249 (1895).
283. C. Morin, M. Dubost, and A. Boulle, *Compt. Rend.*, **249**, 1116 (1959).
284. N. Ermolenko and L. Prodan, *Vestsi Akad. Navuk Belarusk. SSR, Ser. Fiz-Techn. Navuk*, **1962**, 50; *Chem. Abstr.*, **58**, 8625a (1963).
285. M. Kobayashi, *Nippon Kagaku Zasshi*, **78**, 611 (1957); *Chem. Abstr.*, **52**, 10797i (1958).
286. J. Beuenkamp, U.S. Patent No. 3,004,824, Appl. March 18, 1957.
287. E. Giesbrecht, G. Vicentini, and M. Perrier, *J. Inorg. Nucl. Chem.*, **25**, 893 (1963).
288. S. Kundra, P. Subbaraman, and J. Gupta, *Indian J. Chem.*, **3**, 60 (1965); *Chem. Abstr.*, **63**, 233a (1965).
289. E. Giesbrecht and G. Vicentini, *Anais Assoc. Brasil. Quim.*, **19**, 61 (1960); *Chem. Abstr.*, **57**, 393h (1962).
290. B. Apte, P. Subbaraman, and J. Gupta, *Indian J. Chem.*, **3**, 454 (1965); *Chem. Abstr.*, **64**, 11837f (1966).
291. L. Prodan and E. Prodan, *Russ. J. Inorg. Chem. (Engl. Transl.)*, **4**, 749 (1959).
292. A. Indelli and G. Saglietto, *J. Inorg. Nucl. Chem.*, **25**, 1259 (1963).
293. M. Rakuzin and A. Arseniev, *Chemiker Zt.*, **47**, 195 (1923).
294. T. Farr, T. V. A. Chem. Eng. Report No. 8, 1950, U.S. Gov't. Printing Office.
295. W. Wieker, A. Grimmer, and E. Thilo, *Z. Anorg. Allgem. Chem.*, **330**, 78 (1964).
296. M. Shima, K. Hammamoto, and S. Utsumi, *Bull. Chem. Soc. (Japan)*, **33**, 1386 (1960).
297. C. Shen and E. Stahlheber, Belg. Patent No. 674,161, June 26, 1966; Monsanto Technical Bulletin I-270; and unpublished information.

298. D. Dyroff, unpublished data.
299. F. Liebau and H. Williams, *Angew. Chem.*, **76**, 303 (1964); *Angew. Chem. Intern. Ed. Engl.*, **3**, 315 (1964).
300. R. Mehrotra and V. Gupta, *J. Polymer Sci.*, **54**, 613 (1961); *Kolloid Z.*, **184**, 30 (1962); R. Mehrotra and P. Vyas, *J. Polymer Sci. A*, **3**, 2535 (1965).
301. E. Thilo, A. Winkler, and H. Hofsäss, *J. Prakt. Chem.*, **7**, 46 (1958); R. Mehrotra and C. Oza, *Indian J. Chem.*, **4**, 356 (1966).
302. K. Jost, *Acta Cryst.*, **14**, 779 (1961).
303. E. Thilo and R. Rätz, *Z. Anorg. Chem.*, **260**, 255 (1949).
304. D. Harrison, *J. Electrochem. Soc.*, **107**, 217 (1960).
305. E. Thilo, *Angew. Chem.*, **67**, 141 (1955).
306. P. Silber and S. Jualmes, *Compt. Rend.*, **254**, 4034 (1962).
307. I. Schulz, *Z. Anorg. Allgem. Chem.*, **287**, 106 (1956).
308. S. Ohashi and J. Van Wazer, *J. Am. Chem. Soc.*, **80**, 1010 (1958).
309. D. Reynolds, W. Hill, and K. Jacob, *J. Assoc. Offic. Agr. Chemists*, **27**, 559 (1944).
310. J. Brown and F. Hummel, *J. Electrochem. Soc.*, **111**, 660 (1964).
311. D. Corbridge, *Acta Cryst.*, **8**, 520 (1955).
312. I. Mardirosova and G. Bukhalova, *Russ. J. Inorg. Chem. (Engl. Transl.)*, **11**, 1275 (1966).
313. R. Mehrotra and V. Gupta, *J. Indian Chem. Soc.*, **39**, 97 (1962).
314. K. Jost, *Acta Cryst.*, **16**, 623 (1963).
315. B. Raistrick and J. Raitt, *Chem. Eng.*, **70**, 62 (1963).
316. W. Hilmer and K. Dornberger-Schiff, *Acta Cryst.*, **9**, 87 (1956).
317. F. Liebau, *Z. Physik. Chem. (Leipzig)*, **206**, 73 (1956).
318. P. Kierkegaard, *Arkiv Kemi*, **18**, 521 (1962).
319. V. Chiola and C. Vanderpool, *J. Electrochem. Soc.*, **112**, 456 (1965).
320. K. Dornberger-Schiff, F. Liebau, and E. Thilo, *Acta Cryst.*, **8**, 752 (1955).
321. R. Liddell, *J. Am. Chem. Soc.*, **71**, 207 (1949).
322. A. Boullé, *Compt. Rend.*, **206**, 915 (1938).
323. K. Jost, *Acta Cryst.*, **14**, 844 (1961).
324. E. Thilo, G. Schulz, and E. Wichmann, *Z. Anorg. Allgem. Chem.*, **272**, 182 (1953).
325. K. Jost, *Acta Cryst.*, **16**, 640 (1963).
326. K. Jost, *Acta Cryst.*, **15**, 951 (1962); *Soviet Physics Crystallography (Engl. Transl.)*, **6**, 670 (1962).
327. E. Griffith and I. Kodner, U.S. Pat. No. 3,312,523, April 4, 1967.
328. E. Griffith, *J. Am. Chem. Soc.*, **78**, 3867 (1956).
329. K. Jost, *Acta Cryst.*, **17**, 1539 (1964).
330. A. Zvorykin, *Russ. J. Inorg. Chem. (Engl. Transl.)*, **8**, 138 (1963).
331. D. Corbridge, *Acta Cryst.*, **9**, 308 (1956).
332. J. Sarver and F. Hummel, *J. Electrochem. Soc.*, **106**, 500 (1959).
333. F. Katnak and F. Hummel, *J. Electrochem. Soc.*, **105**, 125 (1958).
334. A. Westman and P. Gartaganis, *J. Am. Ceram. Soc.*, **40**, 293 (1957).
335. A. Westman, M. Smith, and P. Gartaganis, *Can. J. Chem.*, **37**, 1764 (1959).
336. M. Murphy, M. Smith, and A. Westman, *J. Am. Ceram. Soc.*, **44**, 97 (1961).
337. A. Westman and M. Murphy, *J. Am. Ceram. Soc.*, **44**, 475 (1961).
338. M. Murphy and A. Westman, *J. Am. Ceram. Soc.*, **45**, 401 (1962).

339. T. Meadowcraft and F. Richardson, *Trans. Faraday Soc.*, **59**, 1564 (1963); **61**, 54 (1965).
340. V. Vargin and T. Tsekhomskaya, *Zh. Prikl. Khim.*, **33**, 2633 (1960).
341. L. Blair and M. Beals, *J. Am. Ceram. Soc.*, **34**, 110 (1951).
342. S. Ohashi and T. Matsumura, *Bull. Chem. Soc. (Japan)*, **35**, 501 (1962).
343. J. Rothermel, K. Sun, and A. Silverman, *J. Am. Ceram. Soc.*, **32**, 153 (1949).
344. F. Carli, *Atti II Congresso Nazl. Chim. Pura Appl.*, **1926**, 1146; *Chem. Abstr.*, **22**, 2119 (1928).
345. C. Davis and C. Monk, *J. Chem. Soc.*, **1949**, 413.
346. L. Jones, *Anal. Chem.*, **14**, 536 (1942).
347. F. Cramer and H. Hettler, *Chem. Ber.*, **91**, 1181 (1958).
348. M. Pouchot, I. Tordjmann, and A. Durif, *Bull. Soc. Franc. Mineral. Crist.*, **89**, 405 (1966); *Chem. Abstr.*, **66**, 23114s (1967).
349. R. Andrieu, R. Diament, A. Durif, M. Pouchot, and D. Tran qui, *Compt. Rend. Ser. AB*, **262B**, 718 (1966).
350. G. Morey, *J. Am. Chem. Soc.*, **74**, 5783 (1952).
351. E. Eanes and H. Ondik, *Acta Cryst.*, **15**, 1280 (1962).
352. E. Griffith and J. Van Wazer, *J. Am. Chem. Soc.*, **77**, 4222 (1955).
353. H. Grunze and E. Thilo, *Z. Anorg. Allgem. Chem.*, **281**, 284 (1955).
354. H. Ondik and J. Gryder, *J. Inorg. Nucl. Chem.*, **14**, 240 (1966).
355. E. Thilo and U. Hauschild, *Z. Anorg. Allgem. Chem.*, **261**, 323 (1950).
356. H. Ondik, *Acta Cryst.*, **18**, 226 (1965).
357. E. Thilo and M. Willis, *Chem. Ber.*, **9**, 1213 (1953).
358. R. Gross, J. Gryder, and G. Donnay, Paper presented at 128th Am. Chem. Soc. Meeting, Minneapolis, Minnesota, September, 1955.
359. H. Ondik, S. Block, and C. MacGillavry, *Acta Cryst.*, **14**, 455 (1961).
360. H. Ondik, *Acta Cryst.*, **17**, 1139 (1964).
361. K. Andress, K. Fischer, and W. Gehring, *Z. Anorg. Chem.*, **260**, 331 (1949).
362. V. Jarchow, *Acta Cryst.*, **17**, 1253 (1964).
363. F. d'Yvoire, *Compt. Rend.*, **251**, 2182 (1960).
364. L. Pauling and J. Sherman, *Z. Krist.*, **96**, 481 (1937).
365. I. Born, *Angew. Chem.*, **67**, 409 (1955).
366. F. Warschauer, *Z. Anorg. Allgem. Chem.*, **36**, 137 (1903).
367. A. Durif, C. Martin, I. Tordjmann, and D. Tranqui, *Bull. Soc. Franc. Mineral. Crist.*, **89**, 439 (1966); *Chem. Abstr.*, **66**, 109201e (1967).
368. D. Barney and J. Gryder, *J. Am. Chem. Soc.*, **77**, 3195 (1955).
369. K. Jost, *Acta Cryst.*, **19**, 555 (1965).
370. V. Illarionov, R. Ozerov, and E. Kil'disheva, *Russ. J. Inorg. Chem. (Engl. Transl.)*, **5**, 1352 (1960).
371. A. Shtin, *Izv. Sibirsk. Otd. Akad. Nauk SSSR*, **1958**, 29; *Chem. Abstr.*, **53**, 12906i (1959).
372. G. Budova and N. Voskresenskaya, *Russ. J. Inorg. Chem. (Engl. Transl.)*, **6**, 704 (1961).
373. D. Kurbatov and N. Demenev, *Zh. Prikl. Khim.*, **29**, 1747 (1956); *Chem. Abstr.*, **51**, 7210e (1957).
374. A. Shtin, *Izv. Sibirsk. Otd. Akad. Nauk SSSR*, **1961**, 68; *Chem. Abstr.*, **55**, 20743d (1961).
375. R. Jameson and J. Salmon, *J. Chem. Soc.*, **1954**, 28.

376. S. Sigov and G. Sadykova, *Uzbeksk. Khim. Zh.*, **1959**, 18; *Chem. Abstr.*, **54**, 6283f (1960).
377. A. Durif, J. Grenier, M. Pouchot, and D. Tranqui, *Bull. Soc. Franc. Mineral. Crist.*, **89**, 273 (1966); *Chem. Abstr.*, **65**, 14545d (1966).
378. E. Staritzky and L. Asprey, *Anal. Chem.*, **29**, 984 (1957).
379. R. Repinsky, C. McCarty, E. Zemyan, and K. Drenck, *Phys. Rev.*, **86**, 793 (1952).
380. R. Douglass, *Acta Cryst.*, **15**, 505 (1962).
381. R. Ripan and G. Marcu, *Comun. Acad. Rep. Populare Romine*, **7**, 323 (1957); *Chem. Abstr.*, **52**, 814f (1958).
382. R. Ropp, *J. Electrochem. Soc.*, **111**, 538 (1964).

AUTHOR INDEX

Numbers in parentheses are reference numbers and show that an author's work is referred to although his name is not mentioned in the text. Numbers in *italics* indicate the pages on which the full references appear.

A

Aaron, A., 177(157), 179(157), *216*
Abel, E. W., 28(171), *42*
Adams, R. M., 47, 48(9), 50, 54(4), 55(4, 31,36), 56(36), 57(4,36,45), 58–60(4), 71(4), 85(4), *97–99*
Adamson, A. W., 5(52), 14, *40, 41*
Adler, R. G., 92, *102*
Aftandalian, V. D., 51(18), 53, 87(137), *98, 101*
Ahrland, S., 29(176), 32(176), *42*
Aia, M., 175(130,137), 177(137), 183(182), 189(182), 192(130), 198(182), 200(137,182), 209(182), 210(137,182), *215, 216*
Akalan, S., 192(259), *218*
Aleksanyan, V., 177(153), *216*
Alexander, J., 163(26), *212*
Al-Janabi, M. Y., 5(58), 6(58), *40*
Allen, A. D., 11(110), *41*
Almy, G. M., 46(142), *101*
Altwicker, E. R., 83(118), *101*
American Society for Testing and Materials, 170(63), *213*
Andersen, W. C., 37(201), *43*
Anderson, J. S., 38(207), *43*
Anderson, S. E., 3(34), *39*
Anderson, W. A., 148(207), *156*
Andress, K., 205(361), 207(361), *221*
Andrew, E. R., 148(199), *156*
Andrieu, R., 203(349), 204(349), 209(349), 210(349), *221*
Ang, H. G., 122(126), *154*
Angelici, R. J., 28(170,172), 29(173), *42*
Apte, B., 194(290), *219*
Arai, Y., 171(102), 172(102), *214*
Argyle, I., 186(218), 200(218), 211(218), *217*
Arquette, G., 179(162), 193(162), *216*
Arseniev, A., 169(293), *219*
Ashby, E. C., 54(33a), *98*
Askenasy, P., 192(260), *218*
Asprey, L., 210(378), *222*
Audrieth, L., 164(35), 174(123), 185(217), 191(123), 192(216,275), 195(275), 199(123), 202(35), 206(35), *213, 215, 217, 219*
Avdnevskaya, K., 184(206), *217*

B

Baddley, W. H., 2(21,22), 31(184), 32(21, 22), 36(22), *39, 43*
Bader, G., 10(99), *41*
Baer, H., 176(141), *215*
Bailar, J. C., Jr., 6(64), 10(82), 28(168), *40, 42*
Baird, M. C., 23(155), *42*
Baker, D. J., 9, *40*
Balareff, D., 171(127), 192(127), *215*
Ball, D. L., 5(56), *40*
Barca, R., 19(136), *41*
Barney, D., 206(368), *221*
Baskin, Y., 160(23), 163(23), 211(23), *212*
Basolo, F., 1, 2(4–7,10,11,21,22), 3(4,5), 4(6,7,36,39), 7(65–67), 8(74), 10(82), 18(130), 19(134), 20(5), 21(6,7), 25(36, 164), 26(164), 28(170), 29(178), 30(180, 181), 31(36,164,182,184), 32(10,11,21, 22), 33(11), 34(11), 35(195), 36, 38(208), *38–43*
Basset, H., 175(140), 176(140,146), 192(146), *215*
Bath, S. S., 11(105), *41*
Batha, H. D., 51(10a), *97*
Baudler, M., 105(21), 108, 111(51), 112

223

(60), 118(96), 119(96), 124, 125(96), 128 (96), *152–156*, 170(87,88), *214*
Baxter, R., 178(165), 179(165), 181(165), 185(165), 199(165), *216*
Baylis, A. B., 94, 95(159), 96, *102*
Beall, H. A., 65, 74(107), *100*
Beals, M., 202(341), *221*
Beck, P., 119(118), 128(118), *154*
Beck, W., 24, 37(162), *42*
Beckers, H. G., 126(139), *155*
Bedwell, W., 175(140), 176(140), *215*
Beeli, 112(56), *153*
Beg, M. A. A., 128(141), *155*
Belavskaia, E. M., 57(49), *99*
Bell, R., 164(35), 193(281), 202(35), 206(35), *213*, *219*
Belluco, U., 35(195), *43*
Belyaev, I., 173(111), *214*
Benjamin, L. E., 65(106), 74(106), 81(106), *100*
Bennett, F. W., 115(87), 122(87), *153*
Berger, E. R., 8(69,71), *40*
Bergman, A., 173(112,113), 178(113), *215*
Bergmann, J. G., 2(4,5), 3(4,5), 20(5), *38*
Berkan, Z., 112(66), *153*
Bertini, I., 2(15,23), 3(23), 34(15,23), *39*
Beryl, S., 177(160), 179(160), 180(160), *216*
Besson, A., 124(136), *155*
Beukenkamp, J., 170(81,82), 194(286), *214*, *219*
Beutner, K., 24(160), *42*
Beyer, A., 126(140), *155*
Bianchi, D., 104(178), *156*
Biberacher, G., 170(76), *214*
Bifano, C., 6(63), *40*
Binder, H., 104(11,15,16), 105(11,19,190), 107(11), 119(15,16), 126, 133, 137(15,16), 145(11), 147(16), 149(11), 150(11), *152*, *156*
Birk, J. P., 2(28), 5(28), *39*
Bjorklund, C., 186(220), *217*
Blair, L., 202(341), *221*
Blakley, L. M., 19(138), 28(138), *42*
Blaser, B., 105(22), 106(24,27,28,35), 108(35), 113, 117, *152*
Blay, N. J., 71(91), 72(91b), 84(120,123), 85(123), *100*, *101*
Bleyer, B., 182(163), 188(163), *216*

Block, S., 206(359), 207(359), *221*
Bloomfield, P. R., 122(129), 123(129), *154*
Bobtelsky, M., 169(58), 175(135), 177(135), 185(212), 188(176), 192(212), 195(176,212), *213*, *215–217*
Boer, F. P., 61(55), 62(55), 77, *99*
Bogolyubov, G. M., *156*
Boichinova, E., 186(225), *218*
Boie, I., 105(18), *152*
Boldebuck, E. M., 52, 53(17), 56, 57(17), 59(17), *98*
Bond, A. C., 48, 50(7), *97*
Bonneman-Bemia, P., 191(255), 193–195 (255), *218*
Boone, J. L., 66(76), 74(76b), *99*
Borer, K., 92(155), *102*
Borgardt, M., 108(187,188), *156*
Borlera, M., 159(6), 186(6,229,230), 187(6), 211(6), *212*, *218*
Born, I., 205(365), 206(365), *221*
Boryta, D., 164(32), *213*
Boulle, A., 163(22), 174(122), 183(190,196,199), 185(22), 188(242), 191(265), 192, 193(199,283), 194(199), 198(199), 199(322), 201(199), 205(199), 206(199), *212*, *215*, *217–220*
Boulouch, R., 112(55), *153*
Bovard, R. M., 53(19), *98*
Bowkley, H., 64(70), *99*
Boyd, F., 191(257), 192(257), *218*
Bragg, J. K., 61(153), 92(154), *102*
Brault, A. T., 36(196), *43*
Briscoe, H. V. A., 38(207), *43*
Broja, G., 55, *98*
Brown, A. E., 47(50), *99*
Brown, E., 175(131,133,134), 176(131,133), 178(133,134), 180(133,134), 181(133,134), 196(131), *215*
Brown, H. C., 47, 48(8), 50, 51(8,12,15), 54, 55(29e,37), 56, *97–99*
Brown, J., 183(192), 198(310), 209(192, 310), *217*, *220*
Brown, W., 175(131), 176(131), 187(235), 196(131), *215*, *218*
Buchwald, H., 107(38), 114, 118(101), 119(101,119), 120(123), 122(123), 129(145), 130(145), 137(76), *152–155*
Buckingham, D. A., 19(133), 20(143), 28, 38(210,211), *41–43*

Buckler, S. A., 118(100), 119(100), 122 (100), 123(100), 138(100), *154*
Budanova, V., 165(40), 183(40), *213*
Budenz, R., 119(111), *154*
Budova, G., 211(372), *221*
Bürger, H., 147(174), *155*
Bukhalova, G., 198(312), 199(312), 209 (312), *220*
Bunton, C., 171(96), *214*
Burdese, A., 159(6), 186(6,229,230), 187 (6), 211(6), *212, 218*
Burg, A. B., 47, 55(29c), 60(143), 66(76), 68, 72(100), 73, 74, 77, 80, *97–102*, 104 (9,10), 105(17), 106(10,33), 107(33,40), 114, 115(9,10,77,78), 116, 119(9,10,112–115), 122(9,10,127,128,134), 123(9,10, 112,128), 124(9,10,112,128), 132(77, 149), 133(150), 140(9,94,113), 144(77), *152–155*
Burmeister, J. L., 1, 2(10,11,21,26,29), 5 (54,55,58), 6(54,58), 32(10,11,21,29, 187), 33(11,189), 34(11), 35(194), 36 (29), *38–40, 43*
Busch, D. H., 13, *41*
Busch, N., 170(83), *214*
Bush, J. D., 52, *98*
Butler, I. S., 28(171), *42*
Buxton, R., 167(50), 207(50), 208(50), *213*
Buyers, A., 185(216), 192(216), *217*

C

Cady, G., 166(43), *213*
Cady, J., 192(272), 193(272), *219*
Callaghen, J., 84(124), *101*
Callis, C. F., 148(175,207), *155, 156*, 158 (1), 191(262), *212, 219*
Calvo, C., 166(201), 167(201), 176(143), 183(201), 184(201), 185(210), 187(201, 210,239), *215, 217, 218*
Campbell, D., 188(177), *216*
Candlin, J. P., 5(53,62), *40*
Caprioli, G., 35(193), *43*
Carlgren, O., 10(81), *40*
Carli, F., 202(344), *221*
Carlin, R. L., 5(47), *39*
Carpenter, R. A., 52, *98*
Carrison, L., 167(45), 188(45), *213*
Carroll, R. L., *156*

Carter, R. P., *156*
Catone, D. L., 10(93), 11(93), *40*
Cattalini, L., 35(195), *43*
Cave, W., 182(178), 188(178), 196(178), 203(178), 207(178), *216*
Ceausescu, D., 183(191), *217*
Chaimovich, H., 171(96), *214*
Chalk, A. J., 10(94), *40*
Chamberlain, D. L., Jr., 54(28), *98*
Chamberlain, M. M., 6(64), *40*
Chambers, J., 185(210), 187(210), *217*
Chang, L., 186(225), *218*
Chapman, A., 170(66), *213*
Chatt, J., 3(33), 9(75,76), 10(84–86,102), 29(176), 32(176), *39–42*
Chen, W., 191(257), 192(257), *218*
Chernyaev, I. I., 22, 26(165), *39, 42*
Chiola, V., 199(319), *220*
Chiras, S. J., 55(38), *98*
Christen, P. J., 130(146), *155*
Chulski, T., 112(62), *153*
Clapper, T. W., 92, *102*
Clark, H. C., 37(203), *43*, 128(141), *155*
Clark, T., 160(7), 203(7), *212*
Clarke, R. P., 61(153), 92(154), *102*
Cleve, P. T., 10(81), *40*
Coates, R., 171(100), 172(100), 189(100), 190(100), *214*
Cohen, M. S., 84(125), 86(130), *101*
Cohen, S. T., *156*
Coleman, J. S., 36(200), *43*
Collman, J. P., 3(35), 10(97,98,103,104), 11, 12, 37(104), *39–41*
Cook, C. D., 22, *42*
Corbridge, D., 158(2), 170(62,64,65), 174 (116), 177(116), 178(62), 187(116), 190 (62), 191(62,251,252), 196(64), 198 (311), 199(311), 200(331), 203(2,64), 207(64), *212, 213, 215, 218, 220*
Cowley, A. H., 113(73), 119(116,121), 122 (121), 123(116,121), 147(116), 150(116), *151, 153, 154, 156*
Cramer, F., 203(347), *221*
Cramer, R., 10(96), 11(96), *40*
Cromer, D., 187(236), *218*
Cross, R. J., 10(87), *40*
Crutchfield, M., 143(197), *156*, 170(67), 191(262), *213, 219*

D

Daly, J. J., 107(41), 108(41), 116(41), 118 (41,107), *152, 154*
Datars, W., 185(210), 187(210), *217*
Davies, D., 191(251), *218*
Davies, N. R., 29(176), 32(176), *42*
Davis, C., 202(345), *221*
Davis, R. E., 47, *99*
DeAcetis, W., 83(116), *101*
Delepine, M., 8, *40*
Demenev, N., 212(373), *221*
Dewald, W., 178(166), 191(266), *216, 219*
Diament, R., 203(349), 204(349), 209 (349), 210(349), *221*
Dickerson, R. E., 80(110), 82(117), *101*
DiGiorgio, P. A., *100*
Dillard, C. R., 64, *102*
DiLuzio, J. W., 9(78), 10(78,91), *40*
Ditter, J. F., 61(82), 63(82), 74(82), 76 (82), 77, 80, 92(82), *100*
Dixon, K. R., 37(203), *43*
Dobbers, J., 108(188), *156*
Dobrott, R. D., 61(62), 62(62), 90, *99, 101*
Dobson, J., 61(56,57,144), 62(56,57,144), 67, 68, 74, 77, 78(151), 79, 81, 88(151), *99, 100, 102*
Döll, G., 125(152,153), *155*
Dörken, C., 113, *154*
Doi, T., 168(52), 170(52), *213*
Dombrovskii, N., 164(34), 174(117), 189 (248), 191(248), *213, 215, 218*
Domine-Berges, M., 192(258), *218*
Donnay, G., 206(358), 207(358), *221*
Donohue, J., 107(191), *156*
Dornberger-Schiff, K., 199(316,320), *220*
Douglass, R., 210(380), 211(380), *222*
Dragulescu, C., 180(170), 182(185), 183 (191), 187(170), *216, 217*
Drenck, K., 210(379), *222*
Dreyer, H., 179(169), 185(169), 209(169), *216*
Dubost, M., 176(144), 193(283), *215, 219*
Duke, B. J., 56(41), *99*
Dungan, C. H., 143(197), *156*
Dunks, G. B., 72(98), 73, *100*
Dunstan, I., 71(91), 84(120,123), 85(123), *100, 101*

Dupin, A., 183(190), *217*
Du Pont de Nemours, E. I., & Co., 49 (27), *98*
Dupuis, T., 177(150), *215*
Durgin, C., 174(124), *215*
Durif, A., 203(348,349), 204(349), 205 (367), 206(367), 209(349,377), 210(349, 377), *221, 222*
Dutt, N. K., 5(51), *40*
Duval, C., 177(150), 179(167), 180(167), *215, 216*
Duval, R., 179(167), 180(167), *216*
Dwyer, F. P., 13, 21, 36, *41–43*
Dymon, J., 190(253), 191(253), *218*
Dymova, T. N., 49(25), *98*
Dyroff, D., 157, 164(36), 190(36), 191 (36), 201(298), *213, 220*

E

Eanes, E., 204(351), *221*
Ebel, J., 170(73,83), *214*
Edwards, J., 160(13), 167(13), 189(13), *212*
Edwards, J. O., 4(42), 5(47), *39*
Edwards, L. J., 54(28,32b), 56(43), 61 (59), 63(59), 70, 71(150), 72(93), 73, 74, 87, 89(136), 92(136), 95–97, *98–102*
Egan, E., 171(91), 192(269), 202(269), *214, 219*
Eimer, L., 14(120), *41*
Eliazyan, L., 177(161), 183(161,194), *216, 217*
Elleman, D. D., 147(202), *156*
Elliott, J. R., 52, 53(17), 56, 57(17), 59 (17), *98*
Ellis, P., 14(121), 19(136), *41*
Eméleus, H. J., 115(87), 122(87), *153*
England, J., 191(257), 192(257), *218*
Enrione, R. E., 61(55), 62(55), 77, 93, 94 (157), *99, 102*
Enslin, F., 179(169), 185(169), 209(169), *216*
Epstein, M., 118(98,100), 119(98,100), 122(98,100), 123(98,100), 128(98), 138 (98,100), 147(98), 149(98), *154*
Erlenmeyer, H., 118(106), 119(106), *154*
Ermolenko, N., 192(274), 193(284), *219*
Espenson, J. H., 2(28), 5(28,60), *39*
Essen, L. N., 27(167), *42*

Etherington, T. L., 70(84), *100*
Etienne, J., 163(22), 183(196), 185(22), *212*, *217*
Evers, E. C., 104(3), 115(80), *151*, *153*

F

Falius, H., 105(20), 117, *152*
Farona, M. F., 2(14), 34(14,191,192), *38*, *43*
Farr, T., 171(99), 189(247), 190(247), 196(294), *214*, *218*, *219*
Faust, G., 192(267), 198(267), 208(267), *219*
Faust, J. P., 83(113), *101*
Fay, R. C., 37(205), *43*
Fedneva, E. M., 53(20), *98*
Fehlhammer, W. P., 24, 37(162), *42*
Fehlner, T. P., 66, 93, 94, *100*, *102*
Feldmann, W., 170(78), *214*
Feshchenko, N. G., 139(157,158), *155*
Figgis, B., 71(91), 72(91c), *100*
Fild, M., 118(108), 122(108), 147(108), *154*
Filinov, F., 165(40), 183(40), 187(231), *213*, *218*
Finch, A., 104(4), 140(4,166), *152*, *155*
Finholt, A. E., 48, 50(7), 54, 97, *98*
Fischer, E. O., 23, *42*
Fischer, J., 177(148), *215*
Fischer, K., 205(361), 207(361), *221*
Fleitmann, T., 164(38), *213*
Fleming, J., 171(99), 189(247), 190(247), *214*, *218*
Flodin, N. W., 60(143), *102*
Fluck, E., 103, 104(1,7,11,15,16), 105(11, 19,190), 106(34), 107(7,11), 108(45), 119(15,16), 122(135), 126, 133, 137(15, 16), 140(45), 142(45), 143(7,196), 144 45,168), 145(11,45,168), 146(45,168), 147(16,168,171,172,174), 148(45,193, 203), 149(11,45,168), 150(7,11), *151*, *152*, *155*, *156*
Folting, K., 61(66), 62(66), 89, *99*, *101*
Ford, M. D., 70(88,150), 71(85b,150), 73 (103), 74(103), *100*, *102*
Forstner, J. A., 61(61), 62(61), *99*
Forsythe, M. W., 73(103), 74(103), *100*
Foster, R., 174(124), *215*
Foster, W. E., 54(33a), *98*

Fox, E., 192(272), 193(272), *219*
Frary, F. C., 111(53), *153*
Fraser, R. T. M., 2(27), 5(27,49,50,61), 6 (27,61), 15, *39–41*
Frazier, A., 171(101), 172(101), 175(131, 133,134), 176(131,133), 178(133,134), 180(133,134), 181(133,134), 189(101), 190(101), 196(131), 201(101), *214*, *215*
Frazier, S. E., 137(195), *156*
Friedman, L. B., 61(62), 62(62), 72(101), 90, *99*, *101*
Fritzsche, F., 139(161), *155*
Fröhlich, H. O., 148(176), *156*
Frommer, S., 175(152), *215*
Fu, Y. C., 46(142), *102*
Fujioka, G., 166(43), *213*

G

Gaines, D. F., 61(54,56), 62(54,56), 65 (73,74), 72, 75, 76(74), 79, *99*, *100*
Gallaghen, J., 84(122), *101*
Gallagher, M. J., *155*
Garner, C. S., 37(204), *43*
Garner, W., 160(14), 161(14), 164(14), *212*
Garrett, J. M., 37(202), *43*
Garrett, A. B., 71(90), 83(118), *100*, *101*
Garrick, F. J., 18, *41*
Gartaganis, P., 189(245), 202(245,334, 335), *218*, *220*
Garvan, F. L., 36(197), *43*
Gaver, R. W., 19(137), *41*
Geanangel, R., 61(67), 63(67), 74, 75, *99*
Gehring, W., 205(361), 207(361), *221*
Gellman, A. D., 27(167), *42*
Genth, F. A., 2(1), 4(1), *38*
Gerhart, F. J., 72, *100*
Germann, F. E. E., 112(59), *153*
Getsinger, J., 175(132), *215*
Gibbons, S. G., 76(64), *99*
Gibbs, W., 2(1), 4(1), *38*
Giesbrecht, E., 179(171), 185(208,216), 188(243), 192(216,273,275), 194(208, 273,287,289), 195(243,273,275,278,279, 289), *216–219*
Gilbert, J. R., 56(41), *99*
Gilbreath, J. R., 48(8), 50, 51(8), *97*
Gillard, R. D., 9, 18, *40*, *41*
Gingrich, N. S., 109(48), *152*

AUTHOR INDEX

Giran, H., 171(90), 196(90), *214*
Glemser, O., 118(108), 122(108), 147(108), *154*
Glockling, F., 10(87), *40*
Goehring, M., 168(56), *213*
Goerrig, D., 55, *98*
Goetze, U., 147(174), *155*
Goddard, J. B., 32(185), *43*
Gombler, W., 140(164), *155*
Gonick, E., 14(124), *41*
Good, C. D., 51(10a,10b), *97*
Goodgame, D. M. L., 2(9), 3(31), *38, 39*
Goodspeed, N., 83(113), *101*
Gorbunov, A. I., 59(51), *99*
Gorton, E. M., 14(119), *41*
Goryacheva, V., 173(112,113), 178(113), *215*
Goubeau, J., 104(178), *156*
Graham, J. R., 29(173), *42*
Graham, T., 160(8), *212*
Grande, J., 170(82), *214*
Grant, L. R., 114, 115(78), *153*
Grassi, R., 19(135), *41*
Gray, H. B., 4(43), 18(130), *39, 41*
Grekov, S., 165(42), 169(59), 177(59), 179(59), 181(59), 184(59,204), 185(42), 188(42), *213, 217*
Grenier, J., 209(377), 210(377), *222*
Griffith, E., 162(19), 166(44), 167(50), 170(68), 199(327), 200(44,328), 204(328, 352), 206(328), 207(50,328), 208(50), *212, 213, 220*
Griffiths, J. E., 122-124(128), *154*
Grimes, R. N., 72(94), 86(133), 87, *100, 101*
Grimmer, A., 196(295), 198(295), *219*
Groenweghe, L. C. D., 144(167), 145(167), *155*
Gross, R., 206(358), 207(358), *221*
Grossman, J., 14(125), *41*
Grossmann, C., 128(142), *155*
Grünewald, R., 104(8), 137(8), *152*
Grünewald, W., 118(102), 119(102), 122(131,132), 123(131,132), 149(102), *154*
Grunze, H., 170(75), 172-174(103), 192, 199(103), 204(353), 205(353), 206(353), *214, 221*
Grunze, I., 163(31), 166(31), 170(31), 175(31), 176(31), 183(31), 188(31), 197-201(31), 203-207(31), 209-211(31), *213*
Gryder, J., 204(354), 206(358,368), 207(358), *221*
Guérin, H., 189(179), 211(179), *216*
Gupta, J., 194(288,290), *219*
Gupta, V., 201(300), *220*
Guzairov, R., 165(42), 185(42), 188(42), *213*
Gyarfas, E. C., 13, *41*
Gysling, H. J., 2(26,29), 32(29), 33(189), 36(29), *39, 43*

H

Haider, S., 177(154), 196(154), *216*
Haim, A., 2(19,20), 5(19,57,59), 6(57), 7(57), 16, 19, *39-42*
Hajek, B., 186(221), *217*
Hall, L. H., 61-63(65), 87, 88, 90, *99*
Halpern, B., 38(209), *43*
Halpern, J., 2(8), 4(40), 5(8,53,62), 6(8), 19(138), 28(138), *38-40, 42*
Hammaker, G. S., 2(6,7), 4(6,7), 21(6,7), *38*
Hammamoto, K., 196(296), *219*
Hammar, H., 48(9), 50(9b), *97*
Hammerström, K., 118(96), 119(96), 125(96), 128(96), *154*
Handler, W., 175(152), *215*
Harker, D., 82, *101, 102*
Harkort, H., 192(268), *219*
Harris, R., 172(107), 173(107), *214*
Harris, R. H., 37(201), *43*
Harris, S. W., 71(90), 83(118), *100, 101*
Harrison, B. C., 64, 67, 68(72), *99, 100*
Harrison, D., 187(232), 188(241), 197(304), 211(232,304), *218, 220*
Harrod, J. F., 10(94), *40*
Harsfall, R. B., Jr., 46(142), *101*
Hart, F. A., 3(33), *39*, 144(169), *155*
Hart, R. R., 109, *152*
Hartlapp, G., 163(28), *212*
Hartman, F. A., 29(174), *42*
Hartsuk, J. A., 72(101), *100*
Harwell, K. E., 72(96), *100*
Harwood, H. J., 129, 131(148), *155, 156*
Haselbeck, H., 186(228), *218*
Hasserodt, U., 139(161), *155*
Haszeldine, R. N., 115(87), 122(87), *153*

Hatfield, J., 189(247), 190(247), *218*
Hatfield, W. E., 37(205), *43*
Hauschild, U., 204(355), *221*
Hawley, W., 172(107), 173(107), *214*
Hawthorne, M. F., 61(60,68), 62(60), 84, 88, *99*, *101*
Heck, R. F., 10(88), *40*
Hefferan, G. T., 70(150), 71(150), *102*
Hefferson, G. T., 73(103), 74(103), *100*
Heimbach, P., 12(115), *41*
Henderson, S., 177(149), *215*
Henderson, W. A., 118(98,100), 119(98,100,109), 122(98,100), 123(98,100,109), 128(98), 138, 147(98), 149(98), *154*
Hendricks, S., 176(145), 192(272), 193(272), 198(145), *215*, *219*
Henkel, H., 10(83), *40*
Henneberg, W., 164(38), *213*
Henry, P. M., 2(5), 3(5), 20(5), *38*
Herrick, C. S., 70(84), *100*
Hertzberg, G., 46(142), *102*
Herzog, A., 160(13), 167(13), 189(13), *212*
Hettler, H., 203(347), *221*
Hewertson, W., 120(122), 121(122), *154*
Hicks, V., 173(108), 174(108), 190(108), *214*
Hieber, W., 10(99,101), 24(160), *41*, *42*
Hileman, J. C., 29(175), *42*
Hill, G. R., 46(142), *102*
Hill, K., 104(2), 111(2,50), *151*, *153*
Hill, O., 164(35), 202(35), 206(35), *213*
Hill, W., 176(145), 192(267,272), 193(272), 198(145,267,309), 208(267), *215*, *219*, *220*
Hillman, M. J., 83, *101*
Hilmer, W., 199(316), *220*
Hisar, R., 161(15), *212*
Hitchman, M. A., 2(9), 3(31), *38*, *39*
Hites, R. D., 67, *100*
Hoekstra, R., 47(2), *97*, *98*
Hoffman, C., 175(137), 177(137), 183(181), 189(181), 200(137), 210(137), *215* *216*
Hoffman, M., 119(117), 122(117), 123(117), 144(206), *154*, *156*, 182(187), 197(187), *217*
Hoffmann, H., 104(8), 119(118), 122(130), 128(118), 137(8), *152*, *154*
Hofsäss, H., 202(301), *220*

Hollenberg, I., 118(108), 122(108), 147(108), *154*
Holmes, W., 163(30), *213*
Holtzman, R., 48(89), 52(89), 59(89), 69(89), 71(89), 84(89), 85, *100*
Holyer, R. H., 30(179), *43*
Hooge, F. N., 130(146), *155*
Hopmann, R., 47(50), *99*
Horner, L., 119(118), 128(118), *154*
Hosking, J. W., 12(112), *41*
Hotop, W., 192(268), *219*
Hough, W. V., 54(28), 70(88,150), 71(150), 72(93), 73(103), 74(103), 97, *98*, *100*, *102*
House, D. A., 37(204), *43*
Houten, S. van, 104(13), *152*
Howarth, O. W., 2(13), *38*
Howell, P. A., 80(110), *101*
Hubbard, C. D., 30(179), *43*
Hubbard, F., 174(119), *215*
Huber, H., 178(166), 191(263), *216*, *219*
Huff, W. J., 113(67), *153*
Hughes, R. L., 48(89), 52(89), 59(89), 69(89), 71(89), 84(89), 85, *100*
Huheey, J. E., *151*
Huhti, A., 189(245), 202(245), *218*
Hummel, F., 176(142), 183(192), 186(218, 222), 187(232), 188(241), 192(270), 198(310), 200(218,332,333), 209(192,310), 210(332), 211(218,232,332,333), *215*, *217–220*
Hunt, H. R., Jr., 14(119), *41*
Hunt, J. P., 20, *42*
Hunt, R. M., 64, *99*
Hurd, D. T., 49, *98*
Hutchinson, J., 175(140), 176(140), *215*
Hutter, J., 184(203), *217*

I

Ibers, J. A., 11(106–108), *41*
Ichikawa, A., 189(250), *218*
Iliceto, A., 35(193), *43*
Illarionov, V., 211(370), *221*
Indelli, A., 193(280), 195(292), *219*
Ippolitova, E., 188(174), *216*
Irani, R., 162(17), 177(147), 186(227), *212*, *215*, *218*
Irgolic, K., 148(200), *156*

Issleib, K., 104(7), 106(34), 107(7), 114, 115(81,82,85,86,91,92), 118(97,99), 119(97,117), 120(82), 121, 122(82,91,117, 130,135), 123(117), 125(86,151–154), 126(86), 129, 131(91,92), 139(159,160), 140(79,154,160), 143(7), 144(168,206), 145(168), 146(168), 147(168,171,172), 148(203), 149(168), 150(7), *152–156*

J

Jablonski, W., 175–177(136), 198(136), *215*
Jackson, C. B., 53(19), 55(31), *98*
Jacob, K., 176(145), 198(145,309), *215*, *220*
Jacquinot, M., 185(217), *217*
Jameson, R., 212(375), *221*
Jander, G., 170(72), *214*
Jarchow, V., 206(362), 207(362), *221*
Jauhal, G. S., 22, *42*
Jeffes, J., 163(30), *213*
Jenkins, H., 175(129), *215*
Jenkins, I. D., *155*
Jennings, M. A., 34, *43*
Jørgensen, S. M., 2(2), 4, *38*
Johns, J. W. C., 46(142), *102*
Johnson, A. B., 46(142), *102*
Johnson, R. C., 7(67), 8(68–71), *40*
Jolly, W. L., 55, *98*
Jones, L., 205(346), *221*
Jones, L. H., 36(199), *43*
Jones, R. C., 148(175), *155*
Jordon, R. B., 20, *42*
Joshi, K. K., 116(94), 140(94), *154*
Jost, K., 197(302), 199(314,323,325,326), 200(326,329), 208(369), *220, 221*
Jualmes, S., 197(306), *220*
Judd, G. F., 64(146), 83(114), *101, 102*
Julean, J., 180(170), 187(170), *216*
Jung, H., 107(39), *152*
Jung, W., 113(68), *153*
Juvinell, G. L., 147(202), *156*

K

Kabachnik, M. I., 128, 130(143), *155*
Kaczmarczyk, A., 61(58), 62(58), 87, *99*
Kaelble, E., 191(262), *219*
Kang, J. W., 10(98), 11, *40, 41*
Karateev, D., 168(55), 210(55), *213*

Karbe, K., 170(72), *214*
Karl-Kroupa, E., 170(74), 177(151), 197(151), 200(151), 207, *214, 215*
Kasparek, F., 162(18), 203–207(18), *212*
Kasper, J. S., 82, *101, 102*
Katnak, F., 200(333), 211(333), *220*
Kato, A., 189(250), *218*
Katz, J. J., 48(8), 50, 51(8), *97*
Keilin, B., 77, *100*
Keller, P., 61(57,144), 62(57,144), 74(144), 77, 81, *99, 102*
Kerschbaum, F., 163(25), *212*
Kertes, S., 175(135), 177(135), 185(212), 188(176), 192(212), 195(176,212), *215–217*
Kettle, S. F. A., 30(179), *43*
Keulen, E., 106(31), *152*
Kibby, C. L., 47(50), *99*
Kiehl, S., 160(9), 173(9), 199(9), *212*
Kierkegaard, P., 185(214), 187(238), 199(318), *217, 218, 220*
Kil'disheva, E., 211(370), *221*
Kilner, M., 29(174), *42*
Kilpatrick, M., 188(177), *216*
Kime, N., 11(108), *41*
Kimlick, J. E., 12(114), *41*
King, A., 190(253), 191(253), *218*
King, E. L., 5(56), *40*
King, R. B., 10(100), 23, 24, *41, 43*
Kirchner, H., 187(235), *218*
Kirk, N., 70(84), 71(85a), *100*
Kirsanov, A. V., 139(157,158), *155*
Kiser, R. W., 141(184), *156*
Kislova, A., 173(112), *215*
Klein, M. J., 64, 67, 68(72), 69, *99, 100*
Klement, R., 168(57), 169(60), 170(80), 175(138), 177(156,159), 178(159), 179(138,156), 180(138,156,159), 186(228), 191(264), 196–198(57), 199(156), 200(57), 205(57), 206(57), 208(156), *213–216, 218, 219*
Klygin, A., 187(234), *218*
Klyucharov, Ya., 180(173a), 197(173a), 209(173a), *216*
Kobayashi, E., 172(105), 174(125), 200(105), 210(105), *214, 215*
Kobayashi, M., 193(285), 195(277), *219*
Kobayashi, T., 182(186), *217*
Kodner, I., 199(327), *220*

Köhler, H., 116, 119(95), *154*
Köpf, H., 113(70), *153*
Koerner v. Gustorf, E. A., 12(113), *41*
Koester, R., 58, *99*
Kohn, M., 182(184), *216*
Kolitowska, J. H., 112(66), *153*
Kolloff, R., 170(84), *214*
Kolpakova, A., 187(231), *218*
Koma, K., 161(16), 200(16), *212*
Kor, G., 185(213), *217*
Korinth, J., 162(20), 184(20,202), *212*, *217*
Korte, F., 139(161), *155*
Koski, W. S., 46(142), 61–63(65), 66, 87, 88, 90, 93, 94, *99*, *100*, *102*
Kotlensky, W., 61(52), 62(52), 74(105), 79, *99*, *100*
Kowalska, E., 160(11), *212*
Kowalski, W., 160(11), *212*
Kraft, G., 177(148), *215*
Kratzer, R., 80, *101*
Krauss, H.-L., 107(39), *152*
Krech, K., 114, 115(92), 125(154), 131(92), 140(79,154), *153*, *155*
Kreidler, E., 192(270), *219*
Kreshkov, A., 168(55), 210(55), *213*
Kruck, T., 23, *42*
Kruse, W., 5(48), *39*
Kubota, M., 11(111), 12(112), *41*
Kuchen, W., 107(38), 114, 118(101,102), 119(101,102,119), 120(123), 122(123, 131,132), 123(131,132), 126(139), 129 (145), 130(145), 137(76), 149(102), *152–155*
Kudcharker, M., 148(200), *156*
Kuebler, N. A., 109, *152*
Kulakova, V. N., 119(110), *154*
Kulasingam, G. C., 2(24), 3(24), *39*
Kundra, S., 194(288), *219*
Kurbatov, D., 212(373), *221*
Kurnakov, N. S., 26, *42*
Kurzen, F., 54(28), *98*
Kuznetsov, V., 181(173), 184(173), *216*

L

LaFouchardiere, A. de, 187(233), *218*
Landesman, 72(99), *100*
Lane, A. P., 140(162), *155*
Lang, K., 55, *98*
Langenfeld, B., 115(83), 116, 120(124), 122(83,124), 129, 130(83), 131(83), *153*, *154*
Langford, C. H., 4(43), 18(130), *39*, *41*
Langguth, R., 177(151), 186(219), 197 (151,219), 200(151,219), 201(219), 207 (219), *215*, *217*
LaPlaca, S. J., 11(106,107), *41*
Laudenklos, H., 54(28), *98*
Lawless, E. W., 48(89), 52(89), 59(89), 69(89), 71(89), 84(89), 85, *100*
Lazar-Jucu, D., 183(191), *217*
Lecompte, J., 188(242), *218*
LeDonche, L., 187(233), *218*
Lee, C. S., 14, *41*
Lehr, J., 171(101), 172(101), 175(131,133, 134), 176(131,133), 178(133,134), 180(133,134), 181(133,134), 189(101), 190(101), 192(269), 196(131), 201(101), 202(269), *214*, *215*, *219*
Leininger, E., 112(62), *153*
Leitsin, V., 165(42), 169(59), 177(59), 179 (59), 181(59), 184(59,204), 185(42), 188 (42), *213*, *217*
Lelong, B., 184(205), 186(205), 188(242), 194(205), 195(205), 202(205), *217*, *218*
Lessman, O., 179(169), 185(169), 209 (169), *216*
Letcher, J., 143(197), *156*, 170(67), *213*
Leung, Y. C., 104(5), *152*
Levchenko, E. S., 112(61), *153*
Lewis, J., 32(186), *43*
Lewis, J. S., 61(58), 62(58), 87, *99*
Lewis, L. L., 56(43), *99*
Lewis, R., 86(132), 87, *101*
Lichtenwalter, M., 72(96), *100*
Liddell, R., 199(321), 204(321), *220*
Liebau, F., 199(317,320), 200(299), 201 (299), 205(299), 206(299), 210(299), *220*
Linck, R. G., 6(63), *40*
Linde, L. M. van der, 130(146), *155*
Lindenbaum, S., 170(81), *214*
Lindner, E., 142(181), *156*
Linevsky, T., 64(70), *99*
Lipscomb, W. N., 47(3), 61, 62(55,62,66), 65, 72(94,101), 74(107), 77, 80(110), 82 117), 84(121), 86(132), 87, 89, 90, 97, 97, *99–102*, 107(192), *156*
Littlewood, A. B., 92(155), *102*

Lobdell, A., 163(27), *212*
Loeb, N., 170(85), *214*
Lohr, L. L., Jr., 84(121), *101*
Lowe, E., 170(64), 196(64), 203(64), 207(64), *213*
Lowe, J. V., 84(122), *101*
Lucht, C. M., 82, *101*, *102*
Ludlum, K. H., 76(63), *99*
Lukaszewicz, K., 185(211), *217*
Lumry, R. W., 4(44), *39*
Lundgren, D., 170(85), *214*
Lutsenko, I. F., 140(163), 141(163), *155*
Lutz, C. A., 60, 61(53), 62(53), 66, 67(77), 68, 71(85c), 75, 76(53), *99*, *100*
Lynden-Bell, R. M., 147(201), *156*
Lyons, J., 171(93), *214*

M

Maasböl, A., 23, *42*
McCarty, C., 210(379), *222*
McCarty, L. V., 61(153), 92(154), *100*, *102*
McCleverty, J. A., 10(89), *40*
McCollough, F., 171(98), 172(98), *214*
McCoubrey, J., 163(30), *213*
McCullogh, C., 173(114), 192(114), *215*
McCullough, J., 170(68), *213*
McCullough, R. L., 36(199), *43*
McCune, H., 179(162), 187(240), 193(162), 195(240), *216*, *218*
McElroy, A. D., 54(28), 57(45), *98*
MacGillavry, C., 206(359), 207(359), *221*
McGilvery, J., 160(10), 162(10), 189(10, 249), 190(10), 191(10), *212*, *218*
McGinnety, J., 11(108), *41*
McIntosh, A., 175–177(136), 198(136), *215*
Mack, J. L., 84(122), *101*
McKeag, A., 175(128,129), *215*
McKinstry, H., 188(241), *218*
MacLean, C., 185(213), *217*
Macquire, R. G., 69, *100*
McWhinnie, W. R., 2(24), 3(24), *39*
Mahler, W., 104(9,10), 105(17), 106(10), 115(9,10), 119(9,10,112,114), 122 (9,10, 127), 123(9,10,112,127), 124(9,10,112), 140(9), 145(204), 148(204), *152*, *154*, *156*
Mai, J., 112(57), *153*

Maier, L., 104(6), 107(36,37,41), 108(37, 41), 115(84,89), 116(37,41), 118(41,104, 105,107), 119(37,104), 122(104), 123 (104), 128(37), 129(84,170), 130, 131 (147), 132, 133(104), 138(6), 140(37), 143, 144(84,147,167,170), 145(167), 146 (147), 148(205), *151–156*
Makart, H., 186(224), *218*
Makhlauf, J., 61(59), 63(59), 87, *99*
Malowan, J., 171(92), *214*
Manatt, S. L., 147(202), *156*
Mangold, D. J., 83, *101*
Mann, F. G., 144(169), *155*
Mann, H., 175(132), *215*
Marcu, G., 210(381), *222*
Mardirosova, I., 198(312), 199(312), 209(312), *220*
Mark, V., 143(197), *156*, 170(67), *213*
Markina, V. Yu., *99*
Markowitz, M., 164(32), 172(106,107), 173(106,107), 198(106), *213*, *214*
Martin, C., 205(367), 206(367), *221*
Martin, R., 189(179), 211(179), *216*
Marzilli, L. G., 38(210,211), *43*
Mash, D., 177(149), *215*
Mason, W. R., III, 8(68,71), *40*
Matsumura, T., 202(342), *221*
Matsuo, T., 183(198), *217*
Mawby, R. J., 30(180), *43*
May, F. H., 48(9), 50(9b), *97*
Mead, E. J., 54(28), *98*
Meadowcraft, T., 202(339), *221*
Medalia, A. I., 14(120), *41*
Meek, D. W., 3(32), 34(32), *39*
Mehrotra, R., 199(313), 201(300), 202(301), *220*
Mehta, S., 185(207), *217*
Meinel, L., 104(14), 141(14), 149(14), *152*
Melardi, E., 192(273), 194(273), 195(273, 278,279), *219*
Mengel, M., 170(87), *214*
Menzi, K., 179(168), *216*
Merlin, J., 164(33), 189(33), 203–205(33), *213*
Merz, K., 187(235), *218*
Mesmer, R., 177(147), 186(227), *215*, *218*
Metcalf, J., 174(126), 190(261), 191(256), *215*, *218*, *219*

Metropolitan Vickers Electrical Co., 54 (21), *98*
Meyers, H., 5–7(45), *39*
Mezey, E. J., 71(90), *100*
Michaelis, A., 112(58), 116, 119(95), *153, 154*
Middlesworth, E., 176(142), *215*
Mikheeva, V. I., 49(25), 53(20), *98, 99*
Miller, H. C., 51(18), 53, 87(137), *98, 101*
Miller, J. M., 122(126), *154*
Miller, N. E., 61(61), 62(61), 90, *99*
Milobedzki, T., 112(66), *153*
Mitcherling, B., 118(97), 119(97), *154*
Miura, M., 167(48), 168(48,52), 170(52), *213*
Moedritzer, K., 144(167), 145(167), *155*
Monk, C., 189(246), 202(345), *218, 221*
Monsanto Co., 171(95), *214*
Mooney, R., 175(130,137), 177(137), 182 (189), 183(181), 189(181), 192(130), 200 (137), 210(137), 215–217
Moore, E. B., Jr., 82(117), *101*
Moore, J. W., 19(139), *42*
Morey, G., 173(109,110), 191(110,254, 257), 192, 199(109), 203(350), 204(110), *214, 218, 221*
Mori, M., 37(206), *43*
Moriguchi, Y., 167(48), 168(48), *213*
Morin, C., 174(118), 176(118), 188(242), 189(118), 191(118,265), 192(118,258), 193(283), *215, 218, 219*
Morrey, J. R., 46(142), *102*
Morris, D., 30(180), *43*
Morris, D. E., 24(159), 29, *42*
Morris, M. L., 7(66), *40*
Moser, H. C., 141(182–184), *156*
Müller, B., 182(163), 188(163), *216*
Müller, D.-W., 125(151), *155*
Müller-Schiedmayer, H., 106(29), 107(29), 108(29), 114(29), 119(29), 122(29), 123 (29), 128(29), 140(29), *152*
Muetterties, E. L., 23, 31(183), *42, 43*, 47 (5), 48(5), 49(26), 51(18), 53, 61(61), 62 (61), 84(5), 85(5), 87(137), 90, *97–99, 101*
Murmann, R. K., 20(145), 22(145,147), *42*
Murphy, M., 202(336–338), *220*
Murray, K. J., 55(37), *98*

Murray, L. J., 55(37), *98*
Muschi, J., 10(101), *41*
Myers, H. W., 59, *99*

N

Nador, B., 187(237), *218*
Nagai, S., 171(102), 172(102), *214*
Nakamura, S., 2(8), 5(8,53), 6(8), *38, 40*
Name, R. G. van, 113(67), *153*
Neff, J. A., 83(115,119), *101*
Nelson, A., 171(97), *214*
Nelson, G., 171(93), *214*
Nessler, F., 192(260), *218*
Neumann, H. M., 14(119), *41*
Newbury, R., 14(125), *41*
Nicole, M. J., 56(44), *99*
Nicpon, P., 3(32), 34(32), *39*
Niebergall, H., 115(83), 116, 120(124), 122(83,124), 129, 130(83), 131(83), *153, 154*
Nielsen, M. L., 147(173), *155*
Nielsen, R. P., 137(195), *156*
Nielson, M., 182(178), 188(178), 196(178), 203(178), 207(178), *216*
Nikol'skaya, N., 187(234), *218*
Nixon, J. F., 119(113), 140(113), *154*
Noack, M., 23, *42*
Nöth, H., 104(14), 108(43), 115(43,88), 121(88,125), 122(43), 131(88), 141(14), 149(14), *152–155*
Norbert, A., 174(121), *215*
Nordman, C. E., 61, *102*
Norman, A. D., 66, 68, 96, *100, 102*
Norman, J. H., 83, *101*
Norton, F. J., 61(153), 79, 92(154), *100, 102*
Novobilský, V., 148(193), *156*
Nowotny, H., 183–188(193), *217*
Nyholm, R. S., 10, 32(186), *40, 43*

O

Ogata, H., 14(125), *41*
O'Grady, E., 191(256), *218*
Ohashi, S., 170(79,89), 192(271), 193 271), 198(271,308), 202(342), *214, 219–221*
Oliver, J., 60(143), 69(148), 71(148), *102*
Olsen, I. I., 19(133), 20(143), 28(169), *41, 42*

Onak, T. P., 72, 73, 87(133), *100*, *101*
Ondik, H., 204(351,354,356), 206(359, 360), 207(359,360), *221*
Osipov, V., 176(139), *215*
Osterheld, R., 172(106), 173(106), 174 (123), 177(151), 186(219), 191(123), 197 (151,219), 198(106), 199(123), 200(151, 219), 201(219), 207(219), *214*, *215*, *217*
Osterroht, C., 122(133), 123(133), *154*
Otani, S., 168(52), 170(52), *213*
Oyama, K., 161(16), 200(16), *212*
Oza, C., 202(301), *220*
Ozerov, R., 211(370), *221*

P

Palchak, R. J. F., *101*
Palenik, G. J., 107(191), *156*
Palmer, R. A., 19(138), 28(138), 38(208), *42*, *43*
Panteleeva, E., 187(231), *218*
Park, C. D., 70(150), 71(150), *102*
Parry, R. W., 45, 89(136), 92(136), 95, 96 (136), *101*, 148(198), *156*
Parshall, G. W., 10(96), 11(96), *40*, 87 (137), *101*
Partridge, E., 173(108), 174(108), 190 (108), *214*
Parvin, K., 122(129), *154*
Pass, F., 118(103), 119(103), 122(103), 123(103), 128(103), *154*
Patel, N., 185(207), *217*
Pauling, L., 205(364), *221*
Payne, D. S., 140(162), *155*
Pearl, C. E., 84(125), 86(130), *101*
Pearson, R. G., 2(4,5), 3(4,5), 4(36,39), 7 (66), 8(74), 18(130), 19(134,139), 20(5), 25(36,164), 26(164), 29(177), 31(36, 164), 32(177), 35(195), *38–43*
Pearson, R. K., 48(9), 50, 56(43), 96, 97, *99*, *102*
Pease, R. N., 61(53), 92(154), *102*
Pecile, C., 33, *43*
Pekarek, V., 188(244), *218*
Penfold, B. R., 106(30), *152*
Penneman, R. A., 36(199,200), *43*
Peppard, D. F., 28(168), *42*
Perrier, M., 179(171), 194(287), *216*, *219*
Peters, D. E., 2(27), 5(27), 6(27), *39*
Peterson, H., Jr., 36(200), *43*

Peterson, L. K., 106(33), 107(33), 119 (115), *152*, *154*
Petrov, A. A., *156*
Petushkova, S., 181(173,183), 184(173, 183), 188(172), 189(183), *216*
Petz, E., 177(156), 179(156), 180(156), 199(156), 208(156), *216*
Pfluger, C. E., 37(205), *43*
Pfrengle, O., 161(14a), 170(77), *212*, *214*
Phillips, C. S. G., 73, 74, 92(155), *100*, *102*
Phillips, D. A., 61(53), 62(53), 75, 76(53), *99*
Pinnell, R. P., 113(73), 119(116,121), 122 (121), 123(116,121), 147(116), 150(116), *151*, *153*, *154*
Pinsker, J., 113(69), 139(69), *153*
Piper, T. S., 37(205), *43*
Pitochelli, A. B., 61(60), 62(60), 88, *99*, *101*
Pitsch, M., 112(58), *153*
Pöe, A. J., 2(10), 28(170), 32(10), 36(196), *38*, *42*, *43*
Pollart, K. A., 129, 131(148), *155*, *156*
Popov, A., 186(226), *218*
Popp, R., 168(57), 196–198(57), 200(57), 205(57), 206(57), *213*
Porthault, M., 164(33), 189(33), 203–205 (33), *213*
Postnikova, G. B., 140(163), 141(163), *155*
Pouchot, M., 203(348,349), 204(349), 209 (349,377), 210(349,377), *221*, *222*
Pressley, G. A., Jr., 94, 95(159), 96, *102*
Priebe, E., 115(85), *153*
Prijs, B., 118(106), 119(106), *154*
Probst, J., 112(63), *153*
Prodan, E., 194(276,291), 195(276,291), *219*
Prodan, L., 192(274), 193(284), 194(291), 195(291), *219*
Pustinger, J., 147(173), *155*, 182(178), 188 (178), 196(178), 203(178), 207(178), *216*
Putnam, R. F., 59, *99*

Q

Quagliano, J. V., 25(163), 31(163), *42*
Quimby, O., 168(53), 170(61), 187(240), 189–191(53), 195(240), *213*, *218*

R

Rätz, R., 174(120), 195(120), 197(303), 215, 220
Raistrick, B., 163(29), 199(29,315), 212, 220
Raitt, J., 163(29), 199(29,315), 212, 220
Rakuzin, M., 196(293), 219
Ranby, P., 175(129), 177(149), 215
Randolph, C. L., 56(40), 98
Rao, B. C. S., 54(28), 98
Rapp, L. R., 47(2), 97, 98
Rauhut, M. M., 113(71,72), 119(71), 147(71), 153
Ray, P. R., 5(51), 40
Raymond, K. N., 36, 43
Read, I. A., 56(41), 99
Reesor, J. W. B., 115(90), 118(90), 119(90), 128(90), 153
Reid, I. K., 21(146), 42
Reinhardt, H., 104(178), 156
Reiset, J., 10(79), 40
Remy, P., 183(199), 193(199), 194(199), 198(199), 201(199), 205(199), 206(199), 217
Repinsky, R., 210(379), 222
Rettew, R. R., 8(70), 40
Reynolds, D., 176(145), 192(267), 198(145,267,309), 208(267), 215, 219, 220
Reynolds, W. L., 4(44), 39
Rhodes, R. E., 10(92), 11(92), 40
Richards, R. E., 2(13), 38
Richardson, F., 202(339), 221
Rigden, J. S., 46(142), 102
Rieman, W., III, 167(47), 170(81), 213, 214
Ripan, R., 210(381), 222
Ritter, D. M., 60, 61(53), 62(53), 66, 67(77), 68, 69(148), 71(85c,148), 75, 76(53), 99, 100, 102
Robertson, B., 166(201), 167(201), 183(201), 184(201), 187(201), 217
Robin, M. B., 109, 152
Roedel, C. F., 52, 53(17), 56, 57(17), 59(17), 98
Roper, W. R., 10(97,103,104), 37(104), 40, 41
Ropp, R., 175(137), 177(137), 182(188,189), 183(181,182), 189(181,182), 198(182), 200(137,182), 209(182), 210(137,182,382), 215–217, 222
Rose, H., 111(49), 153
Rosenheim, A., 113(69), 139(69), 153, 175(152), 215
Rossel, T., 170(86), 214
Rothbart, H., 167(47), 213
Rothermel, J., 202(343), 221
Roy, R., 176(142), 215
Royen, P., 104(2), 111(2,50), 151, 153, 162(20), 184(20,202), 212, 217
Rudolph, K., 119(111), 140(164), 154, 155
Rudolf, R. W., 148(198), 156
Rund, J. V., 8, 40
Rutenberg, A. C., 20(144), 42
Ryschkewitsch, G. E., 37(202), 43, 71(90), 100

S

Sabatini, A., 2(15,23), 3(23), 34(15,23), 39
Sadykova, G., 212(376), 222
Saffir, P., 15, 41
Saglietto, G., 195(292), 219
Saito, Y., 37(206), 43
Sallièr Dupin, A. de, 163(22), 185(22), 212
Salmon, J., 212(375), 222
Sambeth, J., 168(56), 213
Sanderson, R. T., 47, 55(29c), 97, 98
Sandhu, J. S., 72(100), 73, 100
Sandoval, A. A., 141(182–184), 156
Sansoni, B., 170(80), 214
Sargeson, A. M., 19(133), 20(141–143), 21(146), 28(169), 38(209–211), 41–43
Sarver, J., 183(197), 186(197), 200(332), 210(332), 211(332), 217, 220
Satrapa, H., 20(143), 42
Sauer, R., 171(94), 214
Schaeffer, R., 61(52,54,56,57,144), 62(52,54,56,57,144), 65(73,74), 66–68, 74, 75, 76(63,74), 77, 78(151), 79, 80(110), 81, 85, 88(151), 93, 94(157), 96, 99–102
Schechter, W. H., 52, 55(31), 98
Schegaleva, T. A., 57(49), 99
Schindlbauer, H., 118(103), 119(103,120), 122(103), 123(103), 128(103,120), 154
Schlabacher, W., 55, 98
Schleicher, A., 10(83), 40

Schlesinger, H. I., 47, 48, 50, 51(8,12,14, 15), 54, 55(29c,29e), 56, 60(143), 77, 97–99, *102*
Schless, H., 142(181), *156*
Schmich, M., 51(11), *97*
Schmidt, H., 178(166), *216*
Schmidt, L., 108(186), 111(51), *153*, *156*
Schmidt, M., 113(70), *153*
Schmidt, U., 105(18), 122(133), 123(133), *152*, *154*
Schmidtke, H.-H., 2(16–18), 19(134), *39*, *41*
Schmitt, T., 55, *98*
Schmulbach, C., 162(17), *212*
Schmutzler, R., 115(93), 131(93), 146(209), *154*, *156*
Schott, H., 12(115), *41*
Schubert, A. E., 70(84), *100*
Schubert, F., 55, *98*
Schubert, L., 25(163), 31(163), *42*
Schuelke, U., 167(46), 207(46), 208(46), *213*
Schultz, R. D., 56(40), *98*
Schultze, D., 163(21), *212*
Schulz, G., 199(324), *220*
Schulz, I., 185(215), 197(307), 200(307), *217*, *220*
Schumann, H., 113(70), *153*
Schunn, R. A., 31(183), *43*
Schwartz, C., 167(45), 188(45), *213*
Schwarz, F., 193(282), *219*
Schwarzenbach, G., 14, *41*
Schwarzenberg, A., 164(39), 175(39), *213*
Sclar, C., 167(45), 188(45), *213*
Scott, A., 160(10), 162(10), 189–191(10), *212*
Sears, P., 161(104), 172(104), 200(104), 201(104), *214*
Seel, F., 119(111), 140(164), *154*, *155*
Seeley, G. R., 60(143), 69(148), 71(148), *102*
Seichter, F. S., 118(98), 119(98), 122(98), 123(98), 128(98), 138(98), 147(98), 149(98), *154*
Seidel, W., 115(82,91), 118(99), 120(82), 121, 122(82,91), 131(91), 136(189), *153*, *156*
Semsel, A. M., 113(71,72), 119(71), 147(71), *153*

Senoff, C. V., 11(110), *41*
Serebrennikov, V., 177(161), 183(161,195), 184–187(195), *216*, *217*
Setkina, O., 186(225), *218*
Seyffert, H., 24, *42*
Shapiro, I., 51, 56(39), 72(99), 77, 97–100
Sharova, A., 188(175), *216*
Shaw, B. L., 9(75,76), 10(84–86), *40*
Sheinkman, I. E., 112(61), *153*
Sheka, A., 177(158), 179(158), 180(158), 185(158), 187(158), 188(158), *216*
Shen, C. Y., 157, 158(1), 160(12,297), 163(24), 164(12,37), 167(49), 171(93), 190(261), 191(256), 192(12), 200(297), 201(297), *212–214*, *218*, *219*
Shepeleva, E. S., 128(143), 130(143), *155*
Sherman, J., 205(364), *221*
Shevchenko, G., 177–181(155), 186(155), *216*
Shima, M., 196(296), *219*
Shoolery, J. N., 148(175,207), *155*, *156*
Shore, S. G., 61(67), 63(67), 74, 75, *99*
Shriver, D. F., 3(34), *39*
Shriver, S. A., 3(34), *39*
Shtin, A., 188(175), 211(371), 212(374), *216*, *221*
Siegal, B., 84(122,124), *101*
Siegel, M., 175(132), *215*
Sigida, N., 173(111), *214*
Sigov, S., 212(376), *222*
Silber, P., 174(121), 197(306), *215*, *220*
Silverman, A., 202(343), *221*
Simpson, P. G., 61(66), 62(66), 80(110), 89, *99*, *101*
Sinke, E. J., 94(159), 95(159), *102*
Sinyavskaya, E., 177(158), 179(158), 180(158), 185(158), 187(158), 188(158), *216*
Sisler, H. H., 71(90), *100*, 137(195), *156*
Skoblo, L., 180(173a), 197(173a), 209(173a), *216*
Skolnik, S., 51(11), *97*
Sloan, T. E., 2(25), *39*
Smajkiewicz, R., 185(211), *217*
Smedal, H. S., 24, *42*
Smith, C. J., Jr., 54(22), *98*
Smith, E., 170(69), *213*
Smith, G., 173(108), 174(108), 190(108), *214*
Smith, G. B. L., 51(11), *97*

Smith, I. C., 48(89), 52(89), 59(89), 69 (89), 71(89), 84(89), 85, *100*
Smith, J., 171(101), 172(101), 175(131, 133,134), 176(131,133), 178(133,134), 180(133,134), 181(133,134), 189(101), 190(101), 196(131), 201(101), *214, 215*
Smith, M., 202(335,336), *220*
Smith, P. W., 32(186), *43*
Snover, J. A., 55(37), *98*
Soborovskii, L. Z., 119(110), *154*
Solomon, I. J., 64, 67, 68(72), 69, *99, 100*
Solov'eva, G. S., 59(51), *99*
Spandau, H., 126(140), *155*
Spencer, C. J., 107(192), *156*
Speter, M., 112(64), *153*
Spielman, J. R., 61(82), 63(82), 68, 74 (82), 76(82), 77, 80, 92(82), *100*
Spies, L., 10(83), *40*
Spitsyn, V., 188(174), *216*
Spoor, N. L., 38(207), *43*
Ständeke, H., 108(187,188), *156*
Stafford, F. E., 94, 95(159), 96, *102*
Stafford, S. L., 10(100), *41*
Stafiej, S. F., 65(106), 74(106), 81(106), *100*
Stahlheber, E., 160(297), 200(297), 201 (297), *219*
Stang, A. F., 72(93), 97, *100*
Staritzky, E., 187(236), 210(378), *218, 222*
Steward, E., 175(128), *215*
Stewart, H., 164(32), *213*
Stewart, R. D., 92, *102*
Stock, A. E., 46, 54(28), 97(1), 66, 75, 82, 97, *98*, 111(52,54), *153*
Stone, B., D., 2(4), 3(4), *38*
Stone, F. G. A., 10(100), *41*
Stone, P., 192(269), 202(269), *219*
Strabel, H., 108(187,188), *156*
Stranks, D. R., 2(30), 4(38), 21, *39*
Strauss, U., 170(69,70), *213*
Strecker, W., 128(142), *155*
Street, E. H., 104(3), 115(80), *151, 153*
Strobel, J., 147(173), *155*
Stuhlmann, F., 170(88), *214*
Subbaraman, P., 194(288,290), *219*
Sun, J., 3(35), *39*
Sun, J.-Y., 11(111), *41*
Sun, K., 202(343), *221*
Sutherland, D., 5(55), *40*

Sutin, N., 2(19,20), 5(19), *39*
Swanson, C., 171(98), 172(98), *214*

T

Takacs, E. A., 65(106), 74(106), 81(106), *100*
Tananaev, I., 177(155), 178(155,165), 179 (155,165), 180(155), 181(155,165,173, 183), 184(173,183,206), 185(165,209), 186(155), 188(172), 189(183), 199(165), *216, 217*
Tarayan, V., 177(161), 183(161,194), *216, 217*
Tarr, B. R., 10(82), *40*
Taube, H., 4(37), 5(45,46,48), 6(45,46), 7(45), 15, 16, 19, 20(141,142,144), 145, 22(145), *39, 41, 42*
Taylor, J. R., 53(19), *98*
Taylor, R. C., 148(198), *156*
Tebbe, F., 65(73), 66, *99*
Tekster, E., 187(231), *218*
Terem, N., 192(259), *218*
Tewari, P. H., 19(137), *41*
Thilo, E., 158(3,4), 163(21,31), 166(31), 167(46), 170(4,31,75,78), 171(94), 172 (103), 173(103), 174(103,120), 175(31), 176(31), 183(31), 188(31), 192, 195 (120), 196(295), 197(4,31,303,305), 198 (4,31,295), 199(31,103,305,320,324), 200 (4,31), 201(3,31), 202(4,301), 203, 204 (4,31,353,355,357), 205(31,353), 206(31, 353), 207(31,46), 208(46), 209–211(31), *212–215, 219–221*
Thirlwell, L., 170(66), *213*
Thomas, C. D., 109(48), *152*
Thomas, F., 170(62), 178(62), 190(62), 191 (62), *213*
Thomas, G., 115(86), 125(86), 126(86), *153*
Thorsteinson, E. M., 29(178), 30(180), *42*
Tien, T., 186(222), *218*
Tierney, P. J., 56(42), *98, 99*
Timms, P. L., 73, 74, *100*
Todd, A., 168(54), *213*
Todd, J. E., 66, *100*
Tomlinson, R., 163(30), *213*
Tordjmann, I., 203(348), 205(367), 206 (367), *221*
Tramer, A., 2(12), *38*

Tranqui, D., 203(349), 204(349), 205 (367), 206(367), 209(349,377), 210(349, 377), 221, 222
Traxler, R. N., 112(59), 153
Treadwell, W. D., 112(56), 153
Treitler, T., 170(70), 213
Tribunescu, P., 182(185), 217
Trimm, D. L., 5(62), 40
Trömel, G., 192(268), 219
Trotz, S. I., 83(116), 101
Truszkowski, A., 160(11), 212
Tsao, M.-S., 19(136), 41
Tsekhomskaya, T., 202(340), 221
Turco, A., 33, 35(195), 43
Turnbull, K. R., 38(209), 43
Tzschach, A., 115(81), 129, 139(159), 153, 155

U

Ueda, S., 161(16), 200(16), 212
Ueno, Y., 170(89), 214
Utsumi, S., 196(296), 219

V

Van Alten, L., 60(143), 102
Vanderpool, C., 199(319), 220
Vandersall, H., 161(104), 172(104), 200(104), 201(104), 214
Van Ghemen, M., 106(29), 107(29), 108(29), 114(29), 119(29), 122(29), 123(29), 128(29), 140(29), 152
Van Houten, S., 106(32), 152
Van Wazer, J. R., 105(23), 108, 143(197), 148(175,207), 152, 155, 156, 158, 159(5), 162(17), 166(5), 169, 170(5,67,68, 71,74,79), 174(5), 177, 178(5), 179(5), 181(5), 182(5), 188, 192(271), 193(271), 198(271,308), 202, 203, 204(352), 207, 212–214, 219–221
Vargin, V., 202(340), 221
Vasil'ev, G., 183–187(195), 217
Vasil'ev, V., 183(200), 185(200), 217
Vasil'eva, V., 178(165), 179(165), 181(165), 185(165,209), 199(165), 216, 217
Vaska, L., 9(78), 10(78,90–93), 11(92,93, 105), 40, 41
Vastine, F., 11(111), 41
Veleker, T., 175(137), 177(137), 200(137), 210(137), 215

Venanzi, L. M., 2(13), 38
Verdier, J., 174(122), 215
Vesely, V., 188(244), 218
Vetter, H.-J., 108(43), 115(43,88), 121(88, 125), 122(43), 131(88), 152-155
Vicentini, G., 188(243), 194(287,289), 195(243,289), 218, 219
Vilceanu, N., 180(170), 187(170), 216
Viltange, M., 178(164), 180(164), 185(164), 186(164), 194(164), 204(164), 206(164), 216
Vlcek, A. A., 7(65), 40
Voellenkle, H., 183–188(193), 217
Vogel, F., 112(65), 153
Vos, A., 104(12), 106(31), 152
Voskresenskaya, N., 177(160), 179(160), 180(160), 211(372), 216, 221
Vrieze, K., 10, 40
Vyas, P., 201(300), 220

W

Wakefield, 171(91), 214
Walker, A. O., 60(143), 102
Wallace, G., 160(9), 173(9), 199(9), 212
Walter, M. K., 45
Wandel, E. J., 83(119), 101
Wang, F. E., 86(132), 87, 101
Wannagat, U., 24, 42
Warschauer, F., 205(366), 206(366), 221
Wartik, T., 64, 99
Waser, J., 104(5), 152
Watanabe, T., 37(206), 43
Watson, H. R., 10(102), 41
Watson, L., 174(126), 215
Watson, W. H., 113(74), 119(74), 120(122), 121(122), 153, 154
Watters, J., 177(157), 179(157), 216
Wawersik, H., 30(181), 31(182), 43
Wegmann, J., 179(168), 216
Weidenbaum, K. J., 2(22), 32(22), 36(22), 39
Weil, T., 118(106), 119(106), 154
Weilmuenster, E., 71(90), 83(118), 100, 101
Weiss, H. G., 51(11), 56(39), 97, 98
Weiss, J., 108(42), 152
Wendlandt, W., 186(226), 218
Wenschuh, W., 139(160), 140(160), 155

AUTHOR INDEX

Werner, A., 2(3), 4, 10(80), 13, 22, *38*, *40–42*
West, J., 173(115), *215*
Westman, A., 202(334–338), *220*
Weymouth, H., 167(47), *213*
Wheatley, P. J., 80(110), *101*, 130, *156*
Wiberg, E., 106(29), 107(29), 108(29), 114, 119(29), 122(29), 123(29), 128, 140 (29), *151*, *152*
Wiberley, S. E., 76(63), *99*
Wichmann, E., 199(324), *220*
Wiebenga, E. H., 104(12,13), 106(32), *152*
Wieker, W., 168(51), 170(51), 196(295), 198(295), *213*, *219*
Wilcox, H. K., 19(137), *41*
Wilke, G., 12(115), *41*
Wilkins, R. G., 4(41), 14(121), 30(179), *39*, *41*, *43*
Wilkinson, G., 10(89), 23(155), *40*, *42*
Williams, H., 200(299), 201(299), 205 (299), 206(299), 210(299), *220*
Williams, J., 71(91), 72(91b), *100*
Williams, M. J. G., 14(121), *41*
Williams, R. E., 61(82), 63(82), 72, 74 (82), 76, 77, 80, 86(129), 92(82), *100*, *101*
Williams, R. L., 71(91), 72(91b,91c), 84 (120,123), 85(123), *100*, *101*
Willis, M., 204(357), *221*
Wilmarth, W. K., 5–7(57), 19(135–137), *40*, *41*
Wineman, P., 170(69), *213*

Winkler, A., 202(301), *220*
Wittmann, A., 183–188(193), *217*
Wojcicki, A., 2(14,25), 29(174), 34, *38*, *39*, *42*, *43*
Woodward, G., 171(100), 172(100), 189 (100), 190(100), *214*
Worms, K.-H., 106(24,27,28,35), 108(35), 113, *152*
Wright, D. A., 106(30), *152*
Wright, G. F., 115(90), 118(90), 119(90), 128(90), *153*
Wunz, P. R., 86(131), *101*
Wynn, V. T., 148(199), *156*

Y

Yatsimirskii, K., 183(200), 185(200), *217*
Yoza, N., 170(89), *214*
d'Yvoire, F., 161(41), 165(41), 168(41), 180(41), 182(41), 184(41), 189(180), 192 (41), 193(41,180), 194(41), 197(41,180), 198(41), 199(41,180), 201(41), 205(41, 363), 211(180), *213*, *216*, *221*

Z

Zavrazhnova, D., 187(234), *218*
Zemyan, E., 210(379), *222*
Ziegler, K., 58, *99*
Zingaro, R. A., 148(200), *156*
Zinov'ev, Y. M., 119(110), *154*
Zvankova, Z. V., 3
Zvorykin, A., 200(330), *220*
Zweifel, G., 55(37), *98*

SUBJECT INDEX

A

Alkali metal pyrophosphates, 171–173
Alkali metal tripolyphosphates, 189–192
  methods of preparation of, table, 190
Alkaline earth dihydrogen orthophosphates, 175
Alkaline earth pyrophosphates, 173–175
  methods of preparation of, table, 174
Amines, optically active, resolution of, 37–38
Ammonium pyrophosphates, 171
Asymmetric induction, in optically labile complex ions, 13

B

Biphosphines, physical properties of, table, 115
Boron hydrides, higher, early preparation of, 46–47
  polyhedral rearrangements and exchange reactions of, 96–97
  preparation of, 60–91
  known, table, 62–63
  unidentified, 88

C

Calcium pyrophosphates, as plant nutrients, 175
Carboalkoxocarbonyl complexes, preparation of, 23
Carbon monoxide, coordinated, direct conversion to cyanate ion, 24
  direct conversion to cyanide, 23–24
Catalysis, in the resolution of optically active complex ions, 13–14
Cesium pyrophosphates, 173
Cobalt aquo complexes, addition of carbon dioxide to, forming carbonato complexes, 22
Cobalt complexes, reaction with nitrite ion, retention of configuration in, 22
Cobalt(III) complexes, hexacoordinate, rearrangement during reactions of, 27–28
Complex ions, tervalent, optically active, resolution of, 14
Condensed phosphates, *see* Phosphates, condensed
Coordination complexes, with a carbene, 23
  substitution reactions of, *see* Substitution reactions
Coordination compounds, oxidation–reduction reactions of, *see* Oxidation–reduction reactions
  synthesis of, application of reaction mechanisms to, 1–38
Counterion stabilization, examples of, table, 37
Counterions, large, as stabilizer of metal complexes, 36–37
Cyclophosphines, physical data of, 119
  preparation of, tables, 118, 123, 128

D

d^6 complexes, substitution reactivity of, 7
d^8 complexes, solid-state isomerization, stability of, 34
Decaborane-9 free radical, 87
Decaborane-14, direct synthesis of, 82
  structure of, 82
Decaborane-16, 86–87
  structure of, 86
Decaboranes, 82–87
  alkylated, synthesis of, 83–85
  halogenated, synthesis of, 85–86
Diborane, isotopically substituted, 60
  preparation of, 48–60
  based on borohydride decomposition, 54–57
  miscellaneous methods of, 58–60
  using metal hydrides, 48–54
  purification of, 60
  pyrolysis of, 64–65, 92

Diboranes, alkylated, 60
Durants salt, 15

E

Electron transfer, via bridging ligands, 6
Exchange reactions, 7

H

Heptaboranes, 76
Hexaborane-10, isomeric, proposed mode of formation of, 75
  preparation of, 73–75
  structure of, 73
Hexaborane-12, 75–76
  proposed structure for, 76

I

Icosaborane-16, 90
  structure of, 91
Icosaborane-26, 90–91
Isomerism, *cis–trans*, in square-planar complexes, 26–27
  linkage, 4, 21
    of metal complexes, table, 2–3
    of square-planar complexes, 32
  optical, 21
Isooctadecaborane-22, 89–90
  structure of, 90

K

Kurnakov's test, in structural assignment of geometric isomers, 26

L

Ligands, ambidentate, 4, 6
  bridging, 3, 6, 30
  chelating, 3
  nonbridging, 6
Lithium pyrophosphates, 172

M

Metal carbonyl complexes, reaction of, 29–31
  catalysis in, 29
Metal complexes, counterion stabilization of, 36–37
  isolation of, by counterion technique, 36–37
  noncarbonyl, reaction with chelating ligands, rate-determining step in, 30
Metal nitrosyl complexes, reactions of, 29–31
  catalysis in, 29
Metal phosphides, reactions with phosphorus halides or halogens, 120–126
Metaphosphates, higher, 207–208
  methods of preparation of, table, 208
Metaphosphoric acids, 202
Metaphosphoric acids and their salts, 202–208
Mixed metal pyrophosphates, 175–182
  methods of preparation of, table, 176
Molecular nitrogen, in coordination complexes, 11
Molecular nitrogen complexes, 24

N

Nitrite, coordinated, evidence for bonding through nitrogen in, 22
Nonaborane-15, 79–81
  structure of, 81
Nonaboranes, 79–81

O

Octaborane-12, 77–78
  structure of, 77
Octaborane-14, structure of, 79
Octaborane-18, 79
n-Octadecaborane-22, 88–89
  structure of, 89
Octadecaboranes, 88–90
Organophosphorus halides, reactions with metal hydrides, 127–128
  reactions with tertiary phosphines, 136–138
Oxidation–reduction reactions, 4–17
  catalyzed by metal complexes, 12
  of coordination complexes, bridged activated complexes in, 6, 7, 14
  catalysis by alcohols, 8
  catalysis by coordination complexes, 8
  inner sphere, examples of, table, 5
  inner sphere mechanism for, 4–13
  of d^6 complexes, 10
  of d^8 complexes, 10

SUBJECT INDEX

metal-ion promoted, 12
one-electron oxidants in, 14–17
outer sphere mechanism for, 13–17
two-electron oxidants in, 15–17
Oxidative addition reactions, of d^8 complexes, 9, 10

P

Pentaborane-9, alkylated, synthesis of, 71–72
  purification of, 71
  structure of, 69
  synthesis of, 69–73
Pentaborane-11, alkyl-substituted, synthesis of, 68–69
  isotopically labeled, synthesis of, 68
  structure of, 67
  synthesis of, 67–69
Pentaboranes, 67–73
  deuterated or halogenated, 72
Pentadecaborane, 88
Phase equilibrium approach, in the characterization of condensed phosphates, 169
Phosphate glasses, 202
Phosphates, branched, 208
  condensed, 158–212
    characterization of individual products of, 169–170
      by the phase equilibrium approach, 169
    definition of, 159
    methods of characterization of, 168–170
    preparation by crystallization of equilibrium products, 166–167
    preparation from solution in a solvent, 164–166
    reactions involving the breakage of the P—O—P linkage, 164
    separation of mixtures of, 167–168
    of uncertain anion type, 208–212
  containing no hydrogen, dehydration of, 162–163
  definition of, 158
Phosphides, oxidation of, 139
Phosphine oxides, primary, condensation of, 138

primary or secondary, reactions with aminophosphines, 132–133
  reactions with organohalophosphines, 113–119
  reactions with organomercury compounds, 141
  reactions with phosphoryl halides, 142
Phosphines and biphosphines, contributions in ppm to the chemical shift from ligands of, table, 151
Phosphonic acids, reactions with potassium, 126–127
Phosphoric acids, condensed, 158–212
  condensed, methods of preparation of, 160–168
    preparation by formation of nonaqueous by-product, 163
    preparation by low-temperature dehydration, 162
    preparation by thermal dehydration, 160
    factors affecting reaction rate and purity, 162
Phosphorus halides, reactions with metals, 120–126
Phosphorus, phosphonous or phosphinous esters, reactions with organophosphorus(III) halides, 133–136
  red and white, 111–113
Phosphorus-31 nuclear magnetic resonance data, 143–151
Phosphorus–phosphorus bonds, chemical shifts and coupling constants for compounds containing, table, 144–150
  compounds containing, table, 104–108
  formation of, 109–110, 113–143
    by electric discharge, 140–141
    by rearrangement reactions, 141
    by thermal decomposition of phosphines, 139
Phosphoryl halides, reactions and phosphines, 142
Platinum(II) complexes, reaction with nucleophilic reagents, electrophilic catalysis in, 35
Platinum(II) square-planar complexes, the *trans* effect in, 26

Polyphosphates, tetra and higher, 196–202
  methods of preparation of, table, 197–200
  of single chain length, 196, 201–202
Potassium diphenylhypophosphonate, preparation of, 127
Potassium diphenyltetrathiohypophosphonate, preparation of, 127
Principle of microscopic reversibility, in the reaction of cobalt complexes, 22
Pyrophosphates, containing no alkali metal, 182–189
  methods of preparation of, 182–188
  mixed metal, 175–182
  methods of preparation of, table, 176
Pyrophosphoric acid, crystalline forms of, 171
  preparation of, 171
  rate of crystallization of, 171
Pyrophosphoric acid and its salts, 171–173

R

Reducing agents, as catalysts in oxidation–reduction reactions, 8
Reductants, one-electron, 7
  two-electron, 7
Rhodium(III) complexes, synthesis of, use of lower valent rhodium species in, 9
Rubidium pyrophosphates, 173

S

Sodium orthophosphates, preparation by thermal dehydration, variation of electrical resistance during, 161
Square-planar complexes, electronic configuration of, 31
  reactions of, 31–36
Steric interactions, in the stability of octahedral systems, 34
Substitution reactions, of coordination complexes, kinetic versus thermodynamic control in, 26
  of metal complexes, 17–36
    catalysis in, 18
    discussion of S_N1CB mechanism in, 18–19
  S_N1 reaction in, 19
  S_N1, S_N2, and S_E, definition of, 17
  without metal–ligand cleavage, 20–25
  of octahedral complexes, hydrolysis and anation of, 18–20
S_N1CB, of square-planar complexes, 32
S_N2, of square-planar complexes, 31

T

Tetraalkylbiphosphines, preparation of, 121
Tetraborane, alkyl and halo substituted, preparation of 66–67
  isotopically labeled, preparation of, 66
  preparation of, by high pressure conversion of diborane, 64
  synthesis of, 61–67
Tetraethylbiphosphate, preparation of, 129
Tetrametaphosphates, 205, 207
  methods of preparation of, table, 205–206
Tetraorganobiphosphine disulfides, physical properties of, table, 131
  preparation of, table, 130, 132
Tetraorganobiphosphines, preparation of, table, 122
Tetraphenylbisphosphine, preparation of, 114
Thiocyanate, coordinated, evidence for bonding through nitrogen in, 22
Thiophosphonic acids, reactions with potassium, 126–127
Thiophosphonic anhydrides, reactions with tertiary phosphines, 136–138
Thiophosphonic halides, reactions with organomagnesium halides, 128–132
Thiophosphoryl halides, reactions with organomagnesium halides, 128–132
Thiosulfatopentaamminecobalt(III), chromium(II) reduction of, 6
trans effect, 25–29
  definition of, 25
  in platinum(II) complexes, order of ligand directing ability in, 26
Trimetaphosphates, 203, 205
  methods of preparation of, table, 203–204
Tripolyphosphates, other than alkali metal, 192–196

methods of preparation of, table, 193–195
Tripolyphosphoric acid, 189
Tripolyphosphoric acid and derivatives, 189–196

U

Undecaboranes, 87–88

Urea, as flux and dehydration agent, in the preparation of alkali metal polyphosphates, 161

V

Vaska's compound, 9

CUMULATIVE INDEX, VOLUMES 1-5

	VOL.	PAGE
Actinides, Compounds of (Cunningham)	3	79
Aldimine Compounds, Metal Derivatives of (Martin)	1	59
Anhydrous Metal Nitrates (Addison and Logan)	1	141
Boron Hydrides (Parry and Walter)	5	45
Boron–Nitrogen Compounds (Geanangel and Shore)	3	123
Chromium Subgroup, Halide and Oxyhalide Complexes of Elements of (Fowles)	1	121
Compounds Containing P—P Bonds (Fluck)	5	103
Condensed Phosphoric Acids and Condensed Phosphates (Shen and Dyroff)	5	157
Coordination Compounds, Optically Active (Kirschner)	1	29
Coordination Compounds, Synthesis of (Burmeister and Basolo)	5	1
Coordination Polymers (Bailar)	1	1
Cyclopentadienyl and Arene Metal Carbonyls (Pruett)	2	187
Divalent Radicals of Group IV (Timms)	4	59
Fluorides, Binary (Muetterties and Tullock)	2	237
Fluorine Compounds of the Platinum Metals (Bartlett)	2	301
Fluorocarbon Complexes, of Transition Metals (Bruce and Stone)	4	177
Halide and Oxyhalide Complexes of Elements of the Titanium, Vanadium, and Chromium Subgroups (Fowles)	1	121
Halogen and Halogenoid Derivatives of the Silanes (MacDiarmid)	1	165
Hydrides, of Groups IV and V (Jolly and Norman)	4	1
Hypohalites and Compounds Containing the —OX Group (Williamson)	1	239
Ketimine Compounds, Metal Derivatives of (Martin)	1	59
Metal Alkoxides (Bradley)	2	169
Metal Carbonyls (Hileman)	1	77
Metal Carbonyls, Cyclopentadienyl and Arene (Pruett)	2	187
Metal Halides, in Low Oxidation States (Corbett)	1	1
Nitrates, Anhydrous Metal (Addison and Logan)	1	141
Optically Active Coordination Compounds (Kirschner)	1	29
Organogermanium Compounds (Hooton)	4	85
Orthophosphates, Orthophosphoric Acids and (Shen and Callis)	2	139
Orthophosphoric Acids and Orthophosphates (Shen and Callis)	2	139
Oxide Bronzes (Banks and Wold)	4	237
P—P Bonds, Compounds Containing (Fluck)	5	103
Phosphazene Compounds (Shaw, Keat, and Hewlett)	2	1
Platinum Metals, Fluorine Compounds of the (Bartlett)	2	301
Polymers, Coordination (Bailar)	1	1
Saline Hydrides (Messer)	1	203
Silicon–Nitrogen Compounds (Aylett)	2	93
Sulfur–Nitrogen–Fluorine Compounds (Glemser)	1	227
Sulfur Oxyfluorides, and Related Compounds (Ruff)	3	35
Sulfur- and Phosphorus-Bridged Complexes of Transition Metals (Hayter)	2	211

	VOL.	PAGE
Titanium Subgroup, Halide and Oxyhalide Complexes of Elements of (Fowles)	1	121
Transition Metals, Fluorocarbon Complexes of (Bruce and Stone)	4	177
Transition Metals, Sulfur- and Phosphorus-Bridged Complexes of (Hayter)	2	211
Vanadium Subgroup, Halide and Oxyhalide Complexes of Elements of (Fowles)	1	121
Xenon, Compounds of (Appelman and Malm)	2	341

4-27-71/q